CHEMISTRY RESEARCH AND APPLICATIONS

NATURAL HETEROCYCLES

EXTRACTION AND BIOLOGICAL ACTIVITY

CHEMISTRY RESEARCH AND APPLICATIONS

Additional books in this series can be found on Nova's website
under the Series tab.

Additional e-books in this series can be found on Nova's website
under the e-book tab.

CHEMISTRY RESEARCH AND APPLICATIONS

NATURAL HETEROCYCLES

EXTRACTION AND BIOLOGICAL ACTIVITY

K. L. AMETA

AND

SAKSIT CHANTHAI

EDITORS

nova publishers

New York

Library of Congress Cataloging-in-Publication Data

Natural heterocycles : extraction and biological activity / [edited by] K. L. Ameta and Saksit Chanthai (Department of Chemistry, Faculty of Arts, Science and Commerce, Mody Institute of Technology and Science (Deemed University) Lakshmangarh, Sikar, Rajasthan, India and others).
pages cm. -- (Chemistry research and applications)
Includes bibliographical references and index.
ISBN 978-1-63463-424-3 (hardcover)
1. Heterocyclic compounds. 2. Biochemistry. I. Ameta, K. L., editor. II. Chanthai, Saksit, editor.
QD400.N38 2015
572'.4--dc23

2014041538

Published by Nova Science Publishers, Inc. † New York

CONTENTS

PREFACE

Simple organic molecules are the basic building blocks of an extremely large portion of what is found in nature and in human beings. Heterocyclic compounds played a vital role in biological processes and are wide spread as natural products. They are widely found in nature particularly in nucleic acids, plant alkaloids, anthocyanins and flavones as well as in haem and chlorophyll. Additionally some vitamins, proteins, hormones contain aromatic heterocyclic system. Heterocycles have enormous potential as the most promising molecules as lead structures for the design of new drugs.

Heterocyclic chemistry is one of the most complex and intriguing branch of organic chemistry and heterocyclic compounds constitute the largest and most varied family of organic compounds. Many broader aspects of heterocyclic chemistry are recognized as disciplines of general significance that impinge on almost all aspects of modern organic chemistry, medicinal chemistry and biochemistry. Heterocyclic compounds offer a high degree of structural diversity and have proven to be broadly and economically useful as therapeutic agents. Heterocyclic compounds are organic compounds containing at least one atom of carbon and at least one element other than carbon, such as sulfur, oxygen or nitrogen within a ring structure. Since in heterocycles non-carbons usually are considered to replace carbon atoms, they are called heteroatoms e.g. different from carbon and hydrogen. It is believed that heteroatom atom pairs are primarily responsible for the biological activity of the molecules.

Thus, the purpose of the present book is to provide a succinct summary of methods for the extraction of various different sized bioactive natural heterocycles including their biological importance as anticancer, antifungal, antibacterial, anti-HIV, anti-inflammatory, antioxidant and many more.

The Editors

ABOUT THE EDITORS

Dr. K. L. Ameta received his doctorate degree in Organic Chemistry from M. L. Sukhadia University, Udaipur, India in the year 2002. Presently, he is working as Assistant Professor of Chemistry, Faculty of Arts, Science and Commerce, Mody Institute of Technology and Science (Deemed University) Lakshmangarh, Rajasthan, India. He has vast experience of teaching both graduate and post graduate level students. His research area involves synthesis, characterization and biological evaluation of different sized bioactive heterocyclic systems. In addition, to this he has keen interests in natural heterocycles, heterogeneous catalyzed organic synthesis and photocatalysis. He has published a number of research articles in various journals of national and international repute.

Dr. Saksit Chanthai received his doctorate degree in Chemistry from Sophia University, Tokyo, Japan under the Hitachi Scholarship Foundation in the year 1997. Presently, he holds routinely both teaching load and research works as Associate Professor of Chemistry, Department of Chemistry, Faculty of Science, Khon Kaen University, Khon Kaen, Thailand. His teaching background and research experiences involve an Analytical Chemistry for both undergraduate and graduate students. In particularly, his research fields are subject to instrumental methods of analysis: trends in sample preparation and analytical method developments and validation for foods/food products and drinks analysis. He has participated numbers of national and international chemistry conferences including their published research articles.

CONTRIBUTORS

Adriana Neske	Instituto de Química Orgánica, Facultad de Bioquímica, Química y Farmacia. Universidad Nacional de Tucumán, Ayacucho 471, Tucumán (4000), Argentina.
Alicia Bardón	Instituto de Química Orgánica, Facultad de Bioquímica, Química y Farmacia. Universidad Nacional de Tucumán, Ayacucho 471, Tucumán (4000), Argentina.
Amit Dodiya	Division of Medicinal Chemistry, Department of Chemistry, Maharaja Krishnakumarsinhji Bhavnagar University, Bhavnagar 364 002, Gujarat, India.
Anil K Sharma	Department of Chemistry, Jagannath University, Rampura, Chaksu, Jaipur-303901 (Rajasthan), India
Archana D. Chapolikar	Department of Chemistry, Govt. College of Arts & Science, Aurangabad-431001(MS), India
Arpit Kumar Pathak	Department of Chemistry, University College of Science, M L S University, Udaipur-313001(RAJ.)
Chabutai S. Patil	Deartmenp. of Chemistry, Deogiri College, Aurangabad-431005(MS), India.
Chetna Ameta	Department of Chemistry, University College of Science, M L S University, Udaipur-313001(RAJ.)
Dharma Kishore	Department of Chemistry, Banasthali University, Banasthali 304022, India.
Digambar D. Gaikwad	Department of Chemistry, Govt. College of Arts & Science, Aurangabad-431001(MS), India
Eduardo Alberto Parellada	Instituto de Química Orgánica, Facultad de Bioquímica, Química y Farmacia. Universidad Nacional de Tucumán, Ayacucho 471, Tucumán (4000), Argentina.
Elena Cartagena	Instituto de Química Orgánica, Facultad de Bioquímica, Química y Farmacia. Universidad Nacional de Tucumán, Ayacucho 471, Tucumán (4000), Argentina.
Farediah Ahmad	Department of Chemistry, Faculty of Science, Universiti Teknologi Malaysia, Skudai, 81310 Johor Baru, Johor, Malaysia

Jatesadaporn Yaisoongnern	Department of Chemistry, Faculty of Science, Khon Kaen University, Khon Kaen 40002, Thailand
K. L. Ameta	Department of Chemistry, FASC, Mody Institute of Technology & Science, Lakshmangarh-332311, Rajasthan, India
Khong Hen Yen	School of Chemistry and Environment Studies, Faculty of Applied Sciences, Universiti Teknologi Mara (UiTM) Sarawak, 94300 Kota Samarahan, Sarawak, Malaysia
Kuldeep K Sharma	Department of Chemistry, Jagannath University, Rampura, Chaksu, Jaipur-303901 (Rajasthan), India
Lilian Di Toto Blessing	Instituto de Química Orgánica, Facultad de Bioquímica, Química y Farmacia. Universidad Nacional de Tucumán, Ayacucho 471, Tucumán (4000), Argentina.
Maya V. Katariya	Department of Chemistry, Deogiri College, Aurangabad-431005(MS), India.
N C Desai	Division of Medicinal Chemistry, Department of Chemistry, Maharaja Krishnakumarsinhji Bhavnagar University, Bhavnagar 364 002, Gujarat, India.
Prapha Arnnok	Department of Chemistry, Faculty of Science, Khon Kaen University, Khon Kaen 40002, Thailand
Priyanka Jain	Department of Chemistry, Banasthali University, Banasthali 304022, India.
Rajendra P. Pawar	Department of Chemistry, Deogiri College, Aurangabad-431005(MS), India
Rajendra P. Pawar	Department of Chemistry, Deogiri College, Aurangabad (MS) India 431 005
Saksit Chanthai	Department of Chemistry, Faculty of Science, Khon Kaen University, Khon Kaen 40002, Thailand
Saroj R. Bembalkar	Department of Chemistry, Deogiri College, Aurangabad-431005(MS), India.
Seema Kothari	Department of Chemistry, Pacific College of Engineering, PAHER University, Udaipur-313003, India.
Shipra Bhardwaj	Government P.G. College, Kota-324001, Rajasthan, India
Shiv Narayan Paliwal	Department of Chemistry, Career Point Technical Campus,Career Point University, Rajsamand-313329, Rajasthan, India.
Sonia Díaz	Instituto de Química Física, Facultad de Bioquímica, Química y Farmacia. Universidad Nacional de Tucumán, Ayacucho 471, Tucumán (4000), Argentina.
Sudesh Kumar	Department of Chemistry, Banasthali University, Banasthali 304022, India.
Sunil U. Tekale	Department of Chemistry, Deogiri College, Aurangabad (MS) India 431 005
Wan Mohd Nuzul Hakimi Wan Salleh	Department of Chemistry, Faculty of Science, Universiti Teknologi Malaysia, Skudai, 81310 Johor Baru, Johor, Malaysia.

In: Natural Heterocycles
Editors: K. L. Ameta and Saksit Chanthai
ISBN: 978-1-63463-424-3

Chapter 1

Annonaceous Acetogenins As a Model of Bioactive Heterocyclic Derivative Structures. Extraction and Biological Evaluation

Adriana Neske[1,•], Elena Cartagena[1], Sonia Díaz[2], Alicia Bardón[1], Eduardo Alberto Parellada[1] and Lilian Di Toto Blessing[1]

[1]Instituto de Química Orgánica, Facultad de Bioquímica, Química y Farmacia, Universidad Nacional de Tucumán, Argentina
[2]Instituto de Química Física, Facultad de Bioquímica, Química y Farmacia, Universidad Nacional de Tucumán, Tucumán, Argentina

Abstract

Most annonaceous acetogenins are characterized by an unbranched C32 or C34 fatty acid ending in a γ-lactone, with a long alkyl chain that usually presents hydroxyls groups, tetrahydrofurans and tetrahydropyrans. While many studies have addressed the properties and synthesis of natural acetogenins due to their attractive biological activities, a number of analogues have also been described. This review covers the isolation, and biological evaluation of acetogenins. These compounds represent a class of bioactive compounds whose cytotoxicity seems to be related to the conformation adopted in the membrane revealed by the intermolecular interactions between the headgroups of membrane (specifically phosphate groups) and the hydroxylated-THF of ACG. In addition, the hydrocarbon chain of ACGs tend to reside within the hydrophobic region between membrane monolayers composed by the acyl chains of phospholipids. The analysis of cells and supernatants obtained from cultures of polycyclic aromatic hydrocarbon degrading bacteria by chromatographic techniques and mass spectrometry allowed us to determine that the ACGs are located in the bacterial pellet, as was expected given their

• E-mail: aneske@fbqf.unt.edu.ar.

theoretical log P values. The localization of the ACGs, without structural modifications, in the bacterial pellet extracts, corroborates that they act only as stressors, by inducing a stress response related with an increase in AI production which in turn results in more biofilm formation.

Keywords: Annonaceous acetogenins, lipid membranes, naphthalene degradation

1. Introduction

1.1. Edible fruit sources of heterocyclic structures with pharmacological and environmental interest. Annonaceous acetogenins: A new class of polyketides, the annonaceous acetogenins, has been isolated from annonaceous plants growing in tropical and subtropical regions. Since the isolation of the first acetogenin, uvaricin, more than 400 members of the family have been found and characterized (Figure 1) (Gu et al., 1995, 1997; Zeng et al., 1996; Zafra-Polo et al., 1996; Cavé et al., 1997; Alali et al., 1999; Johnson et al., 1999). Until 2005, the number of ACGs was 417, 40 of them being linear, 16 epoxy, 206 mono-THF, 146 bis-THF, 1 tri-THF, and 8 other ACGs belonging to the THP group (Bermejo et al., 2005). The common skeleton is most often characterized by an unbranched C32 or C34 fatty acid ending in a γ-lactone. Several oxygenated functions, such as hydroxyls, ketones, epoxides, tetrahydrofurans (THF) and tetrahydropyrans (THP), may be present, as well as double and triple bonds (Figure 1). Thus several types of ACGs have been characterised, based on the nature of the functional groups which are present. These including mono-THF, adjacent bis-THF, nonadjacent bis-THF, tri-THF, adjacent tetrahydrofuran-tetrahydropyran (THF-THP), nonadjacent THF-THP, mono-THP, and acetogenins containing only γ-lactones. Biogenetically, it has been suggested that the THF or THP cores are generated by polyepoxidation of an unconjugated polyene followed by domino cyclizations.

2. Extraction, Isolation and Purification of Natural ACGs

The classic extraction of ACGs from plants is carried out by successive solvent extractions with increasingly polar solvents, or by liquid/liquid partition from an initial alcoholic extract. The separation and identification of ACGs is then performed by chromatography on silica gel or by preparative RP-HPLC with refractive index (RI) and ultraviolet (UV) detectors, or by mass spectrometry (MS), with or without separating the individual components by liquid chromatography (LC-MS) (Zeng et al., 1996; Cavé et al., 1997; Alali et al., 1999). Positional isomeric and epimeric ACGs have been successfully separated by preparative HPLC. Countercurrent chromatography (CCC or CPC) has also been used for isolating ACGs. This method is highly efficient since large amounts of mixtures and crude extracts, when purified by this method, have yielded pure compounds. Again, positional isomers, epimers and homologous ACGs have been successfully separated by CPC (Bermejo et al., 2005). ACG with α,β-unsaturated γ-lactones corresponding to the and those with β-hydroxylated γ-lactones, can be detected by TLC revealed with Kedde's reagent (methanol solution of 3,5-dinitro benzoic acid in alkaline medium).

This technique can be used during the process of separation and purification for determining the presence or absence of ACG on either column fractions or HPLC peaks.

This procedure involves the application of a solution of ACG on a silica gel TLC plate using a thin capillary glass tube to generate a spot of no more than 2 mm in diameter. The plate is then placed in a developing chamber containing a suitable mobile phase. Because different compounds ascend the TLC plate at different rates, separation is achieved. UV light can be used to observe these substances since they absorb this radiation. Kedde´s reagent will produce a characteristic pink - magenta color in the presence of ACG soon after heating the plate for a few seconds at 60°C. As an example we present a diagram of the procedures necessary for the obtaining of a methanolic fraction from which squamocin and laherradurin, among other ACG, can be obtained (Scheme 1).

3. Structural Elucidation of ACGs

The structural elucidation of ACGs has already been summarized in previous reviews (Zeng et al., 1996; Cavé et al., 1997; Alali et al., 1999). It consists on the analysis of mass spectra to determine the exact molecular formula, followed by Electronic Ionization Mass Spectrometry (EI-MS) and Fast Atom Bombardment (FAB-MS), which allow determining the position of the functional groups on the alkyl chain.

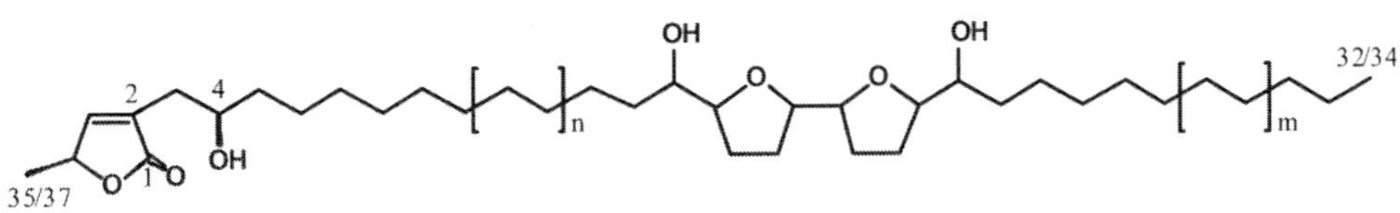

Figure 1. General structure of ACGs.

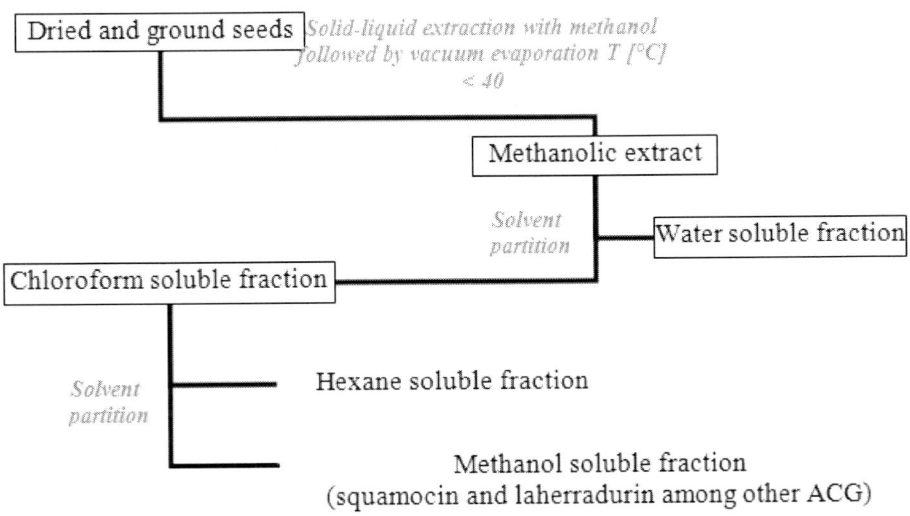

Scheme 1. Preliminary procedures for the obtaining of the methanolic fraction for the separation and purification of ACG (mono and bis THF) present in the seeds of *Annona cherimolia*.

Elucidation of the relative stereochemistry of the stereogenic centers by detailed analysis of the 1H- and 13C-NMR spectra is now straightforward, thanks to the possible comparisons of the spectra between those of the natural products and synthetic models. To determine the absolute configurations, the advanced Mosher ester methodology has allowed several authors to determine unambiguously the absolute configuration of the carbinols present in the vicinity of THF rings, and thus to assign the whole configuration of the THF units. For isolated carbinols, the use of 2-NMA (2-napthylmethoxy acetic acid) esters is of great interest, since the influence of the aromatic rings can extend to five carbon–carbon bonds (Duret et al., 1997).

Concerning the absolute configuration of the terminal γ-lactone, several methods have been used and described in the literature, but they all required the degradation of the natural ACGs in order to obtain lactic acid derivatives, which can be analyzed by chromatography. Recently, it has been proposed to analyze ACGs by 1H-NMR spectroscopy, in the presence of a chiral solvating agent (CSA), and deduce the absolute configuration of the terminal γ-lactone by analyzing the chemical shift differences of the carbinolic proton of the lactone in the presence of the (R)- and (S)-CSA (Latypov et al., 2002). Circular dichroism has also been used for the determination of the absolute configuration of the terminal γ-lactone (Gawronski and Wu, 1999).

ACGs are poor chromophores, so they can be tentatively identified as any peak in a chromatogram that has only one λmax, at or about 210 nm. May be used an HPLC with a C-18 (octadecyl silica, or ODS) semipreparative scale column and RI detection with a methanol gradient as mobile phase to analyze ACGs from different species (Bermejo et al., 2005).

Once an ACG has been purified, Fourier-transform infrared (FTIR), NMR and MS are important methods and complementary techniques of analyzing it. FTIR can tell whether functional groups are present in the compound being analyzed. It is especially sensitive for –OH groups. FTIR produces relatively broad peaks that are characteristic of functional groups, such as the OH at about 3400 cm-1, CH2 at about 2922 cm-1, a THF ring at about 2850 cm-1 and a butenolide at about 1744 cm-1 (Bermejo et al., 2005).

ACGs can be identified by characteristic peaks in the NMR spectrum. An unsaturated methylated γ-lactone will produce 13C-NMR peaks at about 173.9, 134.3, 148.0, 77.4 and 19.2 ppm due to carbons 1 (O–C=O), 2 (C=CH), 33 (C=CH) and 34 (CH), respectively, and 1H-NMR peaks at 7.0 (H-33, doublet), 5.00 (H-34, quartet and doublet C=CH) and 1.41 ppm (H-35, doublet, 7.0 Hz). The –CH2– that is attached to the lactone (C 3), produces a 1H-NMR peak at 2.25 (triplet, J = 7.0 Hz) and a 13C-NMR peak at 25.0 ppm (Melot et al., 2009). A saturated γ-lactone will produce 13C-NMR peaks at about 34.4, 44.2, 79.0 and 178 ppm, due to the CH CH2 CH and O–C=O of carbons 2, 3, 4 and 1, respectively (Gallardo and Aragon, 1998). The THF ring will produce 1H-NMR peaks at or near 1.65 and 1.98 ppm and 13C-NMR peaks at about 28.8 and 29.7 ppm. Those ACGs that have an oxirane will produce 1H-NMR peaks at or near 2.90 and 2.93 ppm and 13C-NMR peaks at about 56.8 and 57.3 ppm. Some of them are connected to a –CH2-CH2-CH=CH, with 1H-NMR peaks at about 1.61, 2.22, 5.39 and 5.41 and 13C-NMR peaks at about 27.9, 24.3, 128.1 and 130.8 ppm. It should be noted that the exact chemical shifts depend on the deuterated solvent in which the sample is dissolved, its concentration and moisture content (Melot et al., 2009).

Often, MS can provide the molecular weight and the molecular formula. NMR can tell the number of hydrogens and carbons that are present, and can also be used to determine the conformation of ACGs (Latypov et al., 2002). Since ACGs have –OH groups, they were identified by their MS characteristic fragmentation patterns (Gu et al., 1997).

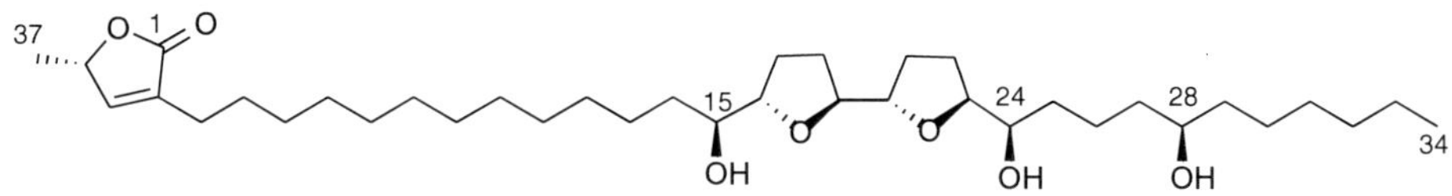

Figure 2. ACG squamocin.

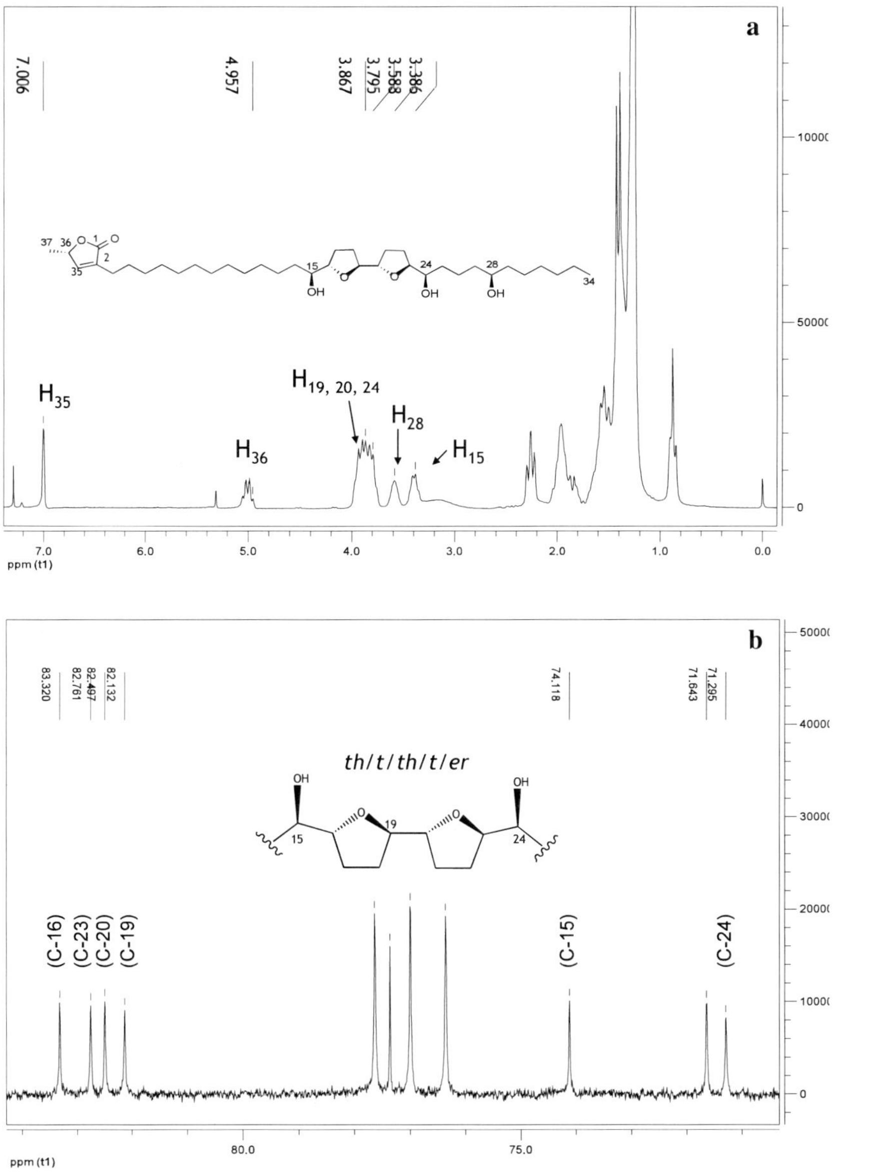

Figure 3. NMR spectra of squamocin. (a) ^{1}H-NMR 200 MHz ($CDCl_3$). Main diagnostic signals are highlighted. (b) ^{13}C-NMR 50 MHz ($CDCl_3$). Signals that allow us to determine the relative configuration of oxygenated domain of squamocin are highlighted.

They lose waters from molecular ions and sodium adducts (FAB-MS) of molecular ions. This is abbreviated by the following for the loss of the first three waters: [M + H – H2O]+, [M + H – 2H2O]+, [M + H – 3H2O]+, [M + Na – H2O]+, [M + Na – 2H2O]+ and [M + Na – 3H2O]+.

Particularly, the ACG squamocin (Figure 2) was extracted from the seeds of Annona cherimolia, a tropical tree of the Annonaceae family.

After, exhaustively purified by reverse phase high performance liquid chromatography (RP-HPLC) with RI detector, the characterization of squamocin was assessed by spectroscopic techniques ^{1}H-NMR, ^{13}C-NMR (Figure 3 a,b), and MS (Figure 4 a,b) in comparison with previously reported data.

4. Chemical Derivation and Synthesis of ACGs

An overview of recent progress on the total synthesis of acetogenins from Annonaceae during the past 12 years is provided. These include mono-tetrahydrofurans, adjacent bis-tetrahydrofurans, nonadjacent bis-tetrahydrofurans, tri-tetrahydrofurans, adjacent tetrahydrofuran-tetrahydropyrans, nonadjacent tetrahydrofuran-tetrahydropyrans, mono-tetrahydropyrans, and acetogenins containing only γ-lactone.

a

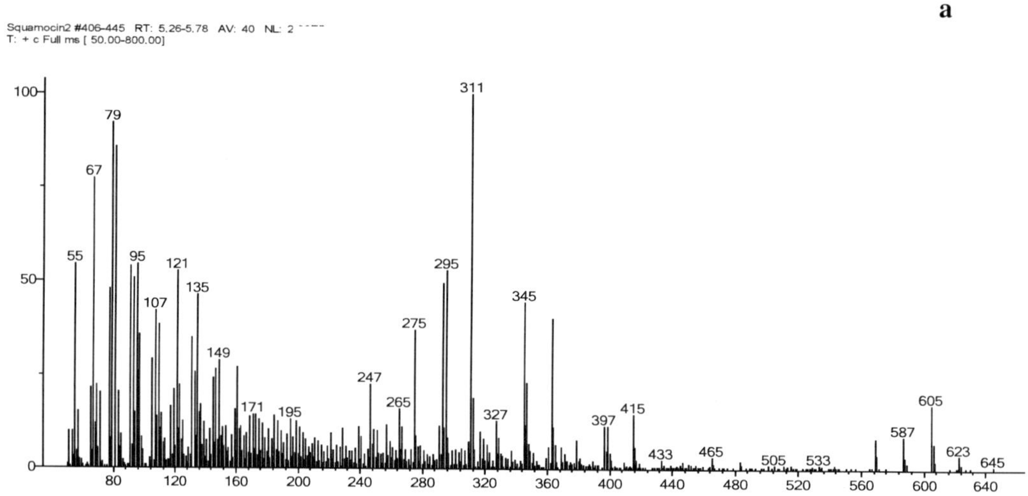

b

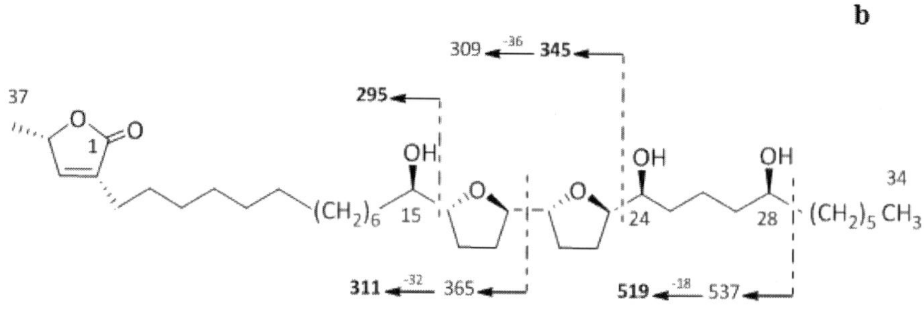

Figure 4. (a) Squamocin MS. (b) Squamocin MS fragmentation pattern.

ACGs have been considerable interesting for synthetic chemists and many total syntheses have been reported in the literature. These syntheses have been reviewed previously by several groups (Bermejo et al., 2005; Huang et al., 2003; Jiang et al., 2004; Konno et al., 2004; Fujita et al., 2005; Liu et al., 2008), including a substantial recent review by Li et al. (2008). Due to the significant number of total syntheses of the non-adjacent compounds, we felt that a detailed and up to date review dedicated to this acetogenin sub-classes would provide a valuable resource for researchers working with acetogenins.

It is apparent from the examples discussed above that the groups of ACGs containing non-adjacent bis-cyclic ether cores have attracted substantial interest from synthetic chemists. This interest has derived, in part, due to the potent biological activity of these acetogenins. However, it can be seen from the diversity of chemistries employed that these non-adjacent systems have also stimulated and inspired the development of synthetic methodology and are likely to continue to do so for some time to come.

5. Cell Membrane Target of ACGs and Their Relationship to Bioactivity

ACGs display a wide range of biological activities like a potent cytotoxicity by specific inhibition of complex I of the mitochondrial respiratory chain (Londershausen et al., 1991). Such cytotoxicity is evident in their insecticide, acaricide, deworming, herbicide (Zafra-Polo et al., 1996; Zeng et al., 1996) and tumor cell growth inhibition activities (Oberlies et al., 1995, 1997) and implies a membrane level interaction. Furthermore, acetogenin cytotoxicity would be strongly related to the conformation adopted in the membrane (Shimada et al., 1998 a,b).

Several strategies for the total synthesis of natural acetogenins and their analogues have been reported, motivated by their unique structural features and attractive biological activities (Li et al., 2008; Hu et al., 2006; Casiraghi et al., 1998; Marshall et al., 1997; Figadère et al., 1995, 1996; Hoppe 1995).

6. ACGs as Bacterial Biofilm Promoters

Biofilms are sessile microbial communities whose cells are anchored to a substrate or to an interface of other bacteria that produce and exhibit an altered phenotype with respect to their growth rate and gene transcription. These bacteria are embedded in a matrix of extracellular polymeric substances that they themselves produce (Rodney et al., 2002). Biofilm associated microorganisms play a crucial role in terrestrial and aquatic nutrient cycling and environmental contaminant biodegradation (Li et al., 2009).

But, why should we try to increase biofilm formation in bacteria that degrade recalcitrant and toxic compounds? Biofilm-mediated bioremediation can be more efficient and safer than that produced by planktonic organisms. This is because biofilm cells have a better chance for adaptation and survival, especially during periods of stress, because they are protected within the matrix (Decho, 2000).

The processes of quorum sensing and biofilm formation have been extensively related in the literature. A variety of bacteria employs quorum sensing (QS) to coordinate communal behavior among which biofilm formation plays a key role. QS consists in the regulation and coordinated expression of genes in response to cell density (Gonzalez and Keshavan, 2006). This phenomenon is caused by the production, release and detection of small signal molecules called type 1 autoinducers (AI-1). In Gram negative bacteria they are N-acylhomoserine lactones (AHLs) (Bassler and Losick, 2006). AHLs play an important role in the regulation of biofilm formation in some bacteria (Waters and Bassler, 2005).

There are, in the best of our knowledge, two examples of strains of the genus Pseudomonas which have proved able to increase their biofilm formation by the addition of the ACG squamocin or laherradurin (2.5 μg ml-1). These strains are P. aeruginosa PA100 (Cartagena et al., 2007) and P. plecoglossicida J26 (Parellada et al., 2011, 2012).

Squamocin, laherradurin and AHLs (Figure 5), structurally share the γ-lactone moiety. Despite their structural similarities, squamocin is not an autoinducer agonist. Instead they are able to generate stress by association with cell materials resulting in an increased production of AI-1 which in turn amplifies the effects of quorum sensing producing higher biofilm formation rates. (Parellada et al., 2011, 2013).

6.1. Location and molecular integrity of ACG after the bacterial biofilm stimulation: As reported by Parellada et al. (2011, 2012) squamocin and laherradurin (Figure 5), can indirectly affect QS mechanism by inducing a stress related increase in autoinducers (AHLs) production (56, 30 %, respectively), which results in more biofilm formation. These results shown that the ACG laherradurin and squamocin, differentially, stimulate the production of AHL (Scheme 2).

But the question arises regarding their location and the integrity of their molecular structures after that biofilm formation has been stimulated.

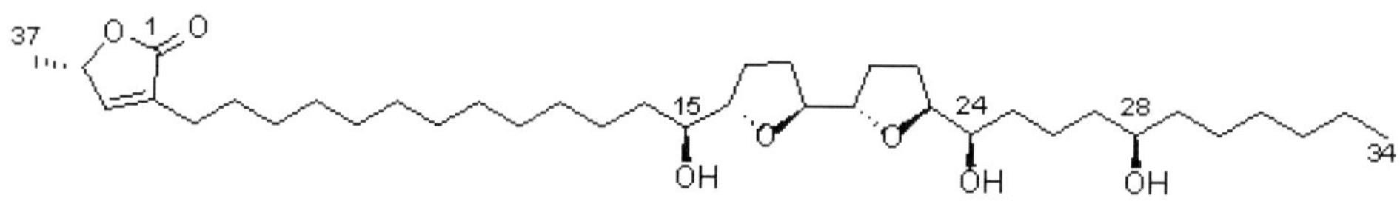

a

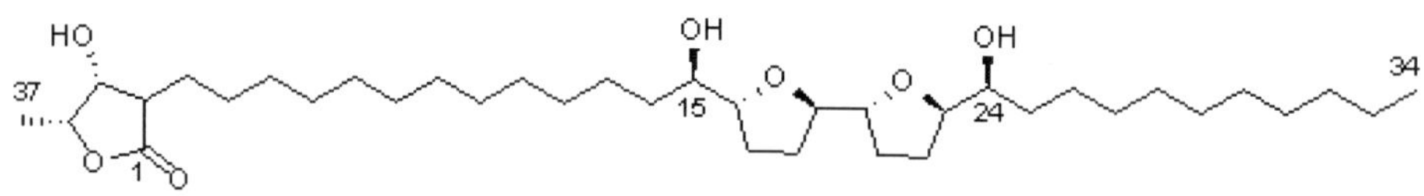

b

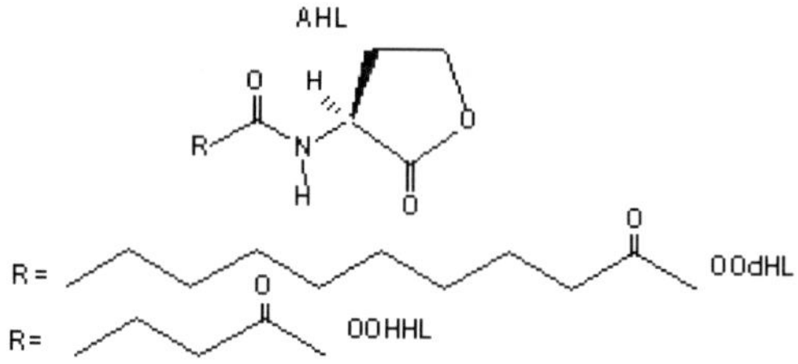

c

Figure 5. Squamocin (a), laherradurin (b), and AHLs (c).

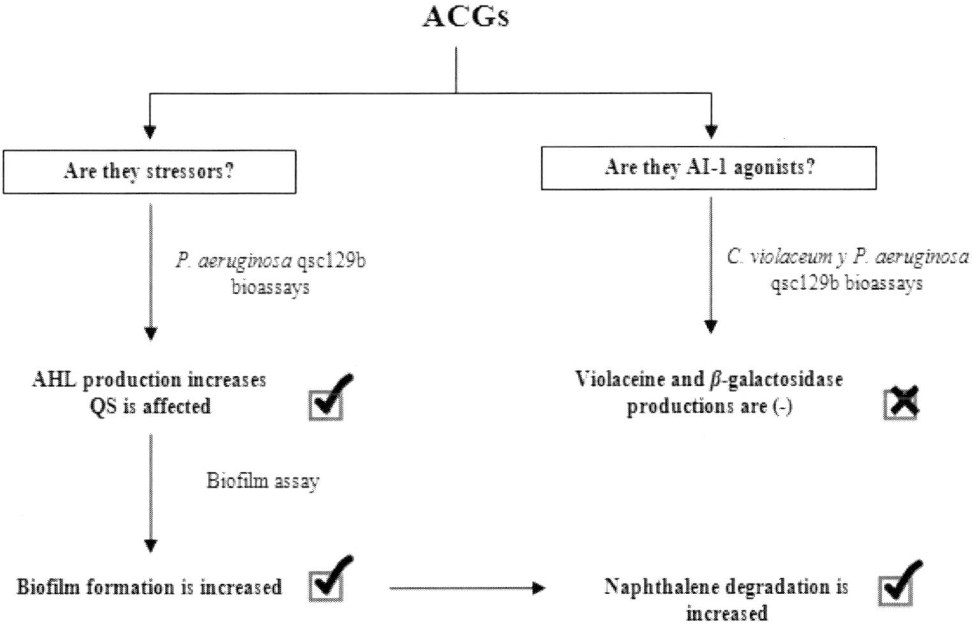

Scheme 2. Possible mode of action of the ACGs in the stimulation of biofilm formation using QS reporter strains.

Thin Layer Chromatography, Liquid Chromatography and Mass Spectrometry assays could determine whether the ACGs produce biofilm stimulation while remain diluted in the liquid media or by close association with lipophilic cell materials.

Following this approach Parellada et al. (2013) analyzed the CH2Cl2 extracts from both cells and supernatants obtained from cultures of P. plecoglossicida J26 grown in presence of 2.5 μg ml-1 of laherradurin or squamocin. They argued that the absence of ACGs in the supernatants of bacterial cell cultures grown in their presence would indicate that these lipophilic molecules are being retained within bacterial membranes or other lipophilic cell materials.

As expected, TLC with Kedde´s reagent did not reveal the presence of any of the ACGs in extracts of supernatants. The RP-HPLC analysis showed no peak with retention time similar to any of the ACG standards.

On the other hand, positive reactions with Kedde's reagent revealed the presence of the ACGs in the pellet extracts. The RP-HPLC analysis showed peaks with retention times corresponding to the ACGs that were then subjected to EI-MS analysis. The presence of laherradurin and squamocin in the pellet extracts was confirmed by comparison of their molecular ions and EI-MS fragmentation patterns with those corresponding to the ACG standards at the same conditions (Scheme 3).

To support this findings, the presence of ACGs (hydrophobic compounds, MW > 500 Dalton) in the bacterial pellet extracts was deduced by theoretical calculations of their octanol/water partition coefficients (log P of laherradurin = 8.539; log P of squamocin = 9.818). These log P values suggest that these ACGs might reside in the lipid membranes of cells with a great ability to interact with the liposomal membrane interface. Furthermore, ACG cytotoxicity would be strongly related to the conformation adopted in the membrane (Shimada et al., 1998 a,b).

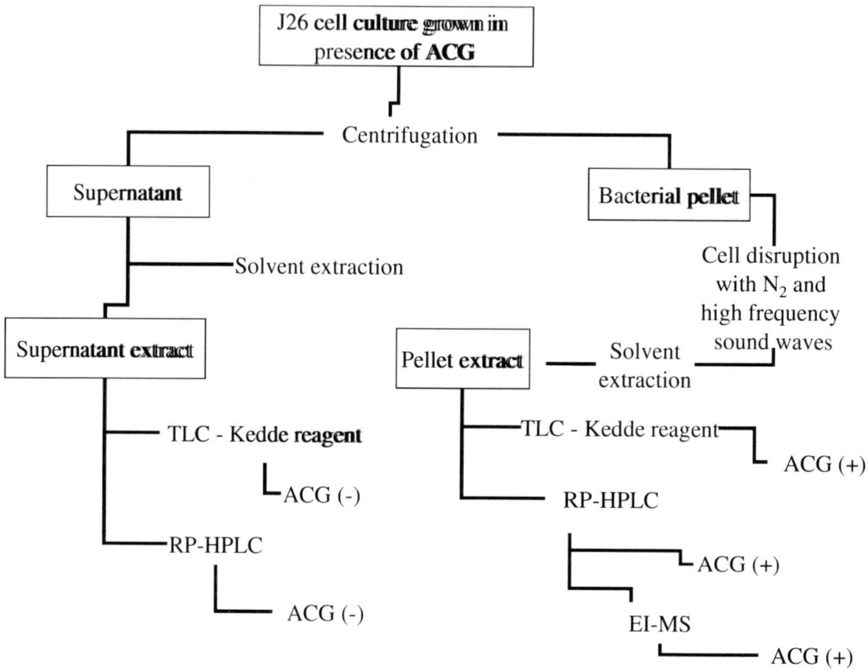

Scheme 3. Chromatographic and spectroscopic methods that confirm the presence of ACGs in bacterial pellets, obtain from cell cultures grown in their presence. (+) Involves the formation of a pink-magenta compound by reaction with Kedde´s reagent, coincidence or similarity in retention times during a chromatographic run or excellent correlation between fragmentation patterns.

In this regard, Bombasaro et al. (2011), and Di Toto Blessing et al. (2012), indicated that specific sites of interaction of the ACGs in the lipid bilayers are in the lipid headgroups (carbonyl ester linking the glycerol backbone and phosphate) of POPC through hidrogen bonds with THF/flanking hydroxyls of ACG.

The localization of the ACGs, without structural modifications, in the bacterial pellet extracts, corroborates that they act only as stressors (Parellada et al., 2013). Further studies are required for the elucidation of the specific sites of interaction of ACGs with prokaryotic cell materials.

7. Optimization of PAH Biodegradation by Natural ACGs

Polycyclic aromatic hydrocarbons (PAHs) are a diverse class of hydrophobic organic compounds consisting of two or more fused aromatic rings. Due to their carcinogenic, mutagenic and toxic properties, some PAHs are considered priority pollutants for remediation by the Environmental Protection Agency United States (EPA) and the European Environment Agency (Keith and Telliard, 1979; Smith et al., 1989). Among PAHs, naphthalene has been commonly taken as a model compound for biodegradation studies (Ferrero et al., 2002).

Many species of various bacterial genera are promising for the degradation of PAHs. Among them, P. plecoglossicida J26. Squamocin and laherradurin stimulate P. plecoglossicida J26 biofilm formation leading to increased consumption of naphthalene.

Several independent assays have shown that 2.5 μg ml-1 of squamocin or laherradurin can increase biofilm formation by 67 or 48%, respectively.

In these assays these ACGs were able to nearly triple naphthalene consumption compared to the control test (Parellada et al., 2012).

8. Natural ACGs and Their Interactions with Model Membranes of POPC

Despite a number of studies on their structure–activity relationship (SAR) in exerting biological effects (Friedrich et al., 1994; Tormo et al., 2003; Konno et al., 2004; Bermejo et al., 2005; Duval et al., 2006), little is known about the mechanism of action at molecular level of ACGs, namely how these compounds interact with the membrane bilayers. Based on the results of 1H-NMR spectroscopic and differential scanning calorimetry studies of ACGs in liposomal membranes, Shimada et al. (1998 a,b) have proposed the model of an active conformation of these inhibitors in the membrane environment. A THF ring(s) with their flanking hydroxyl groups resides near the glycerol backbone of phosphatidylcholine and acts as a hydrophilic anchor at the membrane surface. On the other hand, the lactone ring interacts directly with the protein receptor site of complex I by lateral diffusion in the mitochondrial membrane interior.

Molecular dynamics (MD) calculations of diacetyl-guanacone in a fully hydrated l-palmitoyl-2-oleoyl-syn-glycero-3-phosphatidylcholine (POPC) bilayer have been reported by Barrachina et al. (2007). Also, it was reported the conformational behavior of a bis-THF ACG, bullacin B, by using a combined decane/water system simulating the biological membrane (Bombasaro et al., 2008). MD simulation reveals the presence of hydrogen bonds, which are critical to adopt a particular conformation, and indicates the important role of THF rings and the flanking acetyl groups for the geometry of this molecule.

Membrane environment determines and limits the conformations and the location of lipophilic compounds acting at a membrane receptor in the lipid bilayer. Therefore, to determine the interactions of these ACGs in those membranes seems to be necessary to understand their role as potent cytotoxic compounds. Previous reports have indicated that the effects caused by certain compounds on the hydration patterns surrounding phosphate and carbonyl groups of phospholipids can be studied by FTIR spectroscopy (Díaz et al., 1999, 2003; Hübner and Blume, 1998).

Due to the complex composition found in cell membranes, the use of experimental model systems (liposomes), of predetermined lipid composition, has proved to be a suitable way to explore the variety of physicochemical, electrical and mechanical properties of lipid membranes. The biological effects on the cells are strongly determined by the interaction of these substances with cellular membranes.

The elucidation of the structural characteristics of the ACG molecules involved in these processes is carried out by Fourier Transform Infrared spectroscopy (FTIR). This technique is a powerful tool to understand the interactions between phospholipid membranes and biologically important molecules such as ACG´s, as well as to explore the effect on the hydration patterns surrounding on the different polar groups of the lipid membrane (Díaz et al., 1999; Hübner and Blume, 1998).

Dehydration of a biological membrane usually results in such massive disorders that its structure and functions are irreversibly damaged. Phospholipid polar head groups normally are hydrated to some extent, and are separated from each other by these water molecules. Thus, when phospholipids are dehydrated, the packing density of the head groups would be expected to increase, thereby increasing opportunities for van der Waals interactions among the hydrocarbon chains. The properties of hydration of the lipid bilayer were measured by following the frequency of the vibration modes of the carbonyl and phosphate groups in the interphasial region.

The peak frequencies of the component bands reported for the two populations of carbonyls: the non-bonded (free) (centered at 1742.5 cm−1) and bonded (hydrated) (centered at 1724 cm−1) populations in the gel state, and the non-bonded (1737 cm−1) and bonded (1722 cm−1) populations in the fluid state (Díaz et al., 1999; Hübner and Blume, 1998; Mantsch and McElhaney et al., 1991). The shifts of these two populations were studied as a function of ACG/lipid ratio of POPC in the gel and liquid crystalline state. According to FTIR analysis these populations do not correspond to the C=O groups at the sn1 and sn2 position. That is, there is a dynamical equilibrium between C=O in the sn1 position fluctuating between the non-bonded and bonded populations, and the same happens with sn2 carbonyl groups. H-bond compounds appear to affect these two populations (Hübner and Blume, 1998; Cacela and Hincha, 2006).

In FTIR measurements in fully hydrated phosphatidylcholine in the liquid-crystalline state, the characteristic phosphate group vibrational band assigned to the PO2− antisymmetric stretching mode (vas PO2−) is centered at 1229.5 cm−1, while the band assigned to the PO2− symmetric stretching mode (vs PO2−) is centered at 1085.0 cm−1. It is widely accepted that the frequency of the vibration (vas PO2−) is very sensitive to lipid hydration, mainly because of direct H binding to charged phosphate oxygens (Pohle and Selle, 1996; Arrondo et al., 1984; Jendrasiak and Smith, 2001; Luzardo et al., 2000). C-H stretching vibrations give rise to bands in the region 2800-3100 cm-1. CH2 bands, antisymmetric stretch is centered at 2920 cm-1 and symmetric stretching is centered at 2850 cm-1. The vibrational bands of the terminal methyl groups CH3, is centered at 2956 cm-1, antisymmetric stretching and 2870 cm-1 symmetric stretching. These bands overlap with stretch methylenes. Deformation vibrations of methylene groups produce bands around 1470 cm-1.

In the IR spectra at different molar ratios of ACG/POPC, the frecuency shifts of the PO2- and C=O stretching vibration bands were analyzed in comparison with the corresponding bands of POPC without addition of the ACGs. Comparison between the POPC/ACG´s systems allow us to better understanding of the nature of the interactions that take place in the hydrophilic region of the membrane, and helping us to predict the degree of insertion of ACG´s in the hydrophobic region according to the structural characteristics of the bilayers. The frequencies of the CH2 and CH3 symmetric and antisymmetric stretching modes of vibration of the lipid acyl chains were not affected by mono or bis-THF at the temperatures and conditions assayed.

The results revealed intermolecular interactions between the headgroups of POPC (specifically phosphate groups) and the hydroxylated-THF of ACGs mono and bis THF. These interactions indicated that the phosphate groups of the lipid membrane form stronger hydrogen bonds with hydroxylated-THF rings of ACGs than with water molecules which suggest a strong dehydration in its environment, suggesting that mono THF ACGs would replace the hydration water with subsequent formation of hydrogen bonds in the same environment in all molar ratios assayed. Particularly, in the bis THF ACG squamocin, the results suggest that, in addition to water loss, there is an H-bond formation with phosphates (Di Toto Blessing et al., 2012).

Solute H bonding is possible with the PO2- and C=O groups of diacyl lipids and has been investigated in detail for a membrane-bound disaccharide (Díaz et al., 1999; Popova and Hincha, 2003).

It is rather natural that the hydrogen-bonding around the phosphate group is enriched in the presence of the hydroxylated mono and bis THF ring analogues because they are hydrogen-bond donors and/or acceptors.

We identify ACGs mono and bis THF specific areas of interaction with PO2- groups in the lipid interphase, but there are no evidence of interactions of the investigated solutes with the C=O group. This indicates that ACGs and analogues do not penetrate this deeply into the membrane (Di Toto Blessing et al., 2012).

Variations in absorption of IR frequencies produced by competition from reduced models with water molecules at the interface were determined by quantum-mechanical calculations of the main functional groups of the ACGs (γ-lactone, α, β-unsaturated) and artificial lipid membranes (phosphate and carbonyl). The methodology for optimizing patterns and subsequent frequency analysis using ab initio calculations employs the basis set 6-31G*. All calculations were performed using the Gaussian 03 program and performed either in a vacuum as in the presence of oriented water molecules with lone electron pairs. Frequency variations in the reduced stretch models show the same trend in the presence of an explicit solvent such as water, than that observed in the experimentally determined by FTIR spectroscopy (Table 1) (Di Toto Blessing et al., 2012).

To understand the molecular details of the interaction of mono THF ACGs and bis THF ACGs with the lipid membrane we performed MD simulations in which these ACGs interact with POPC in the fully hydrated state of the lipid.

Table 1. Stretching frequencies of the carbonyl groups and phosphate

Submodel	In vacuum (cm^{-1})	With explicit solvent (cm^{-1})
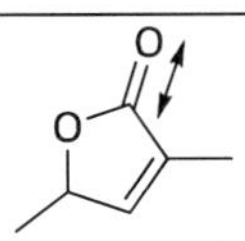	2021.09	1937.41
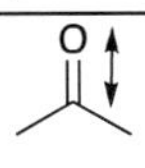	1994.18	1955.85
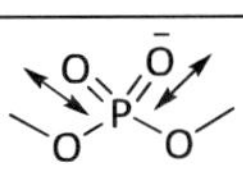	1180.46 1396.28	1188.87 1324.74
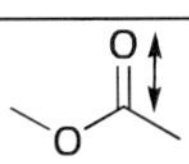	2010.08	1958.67

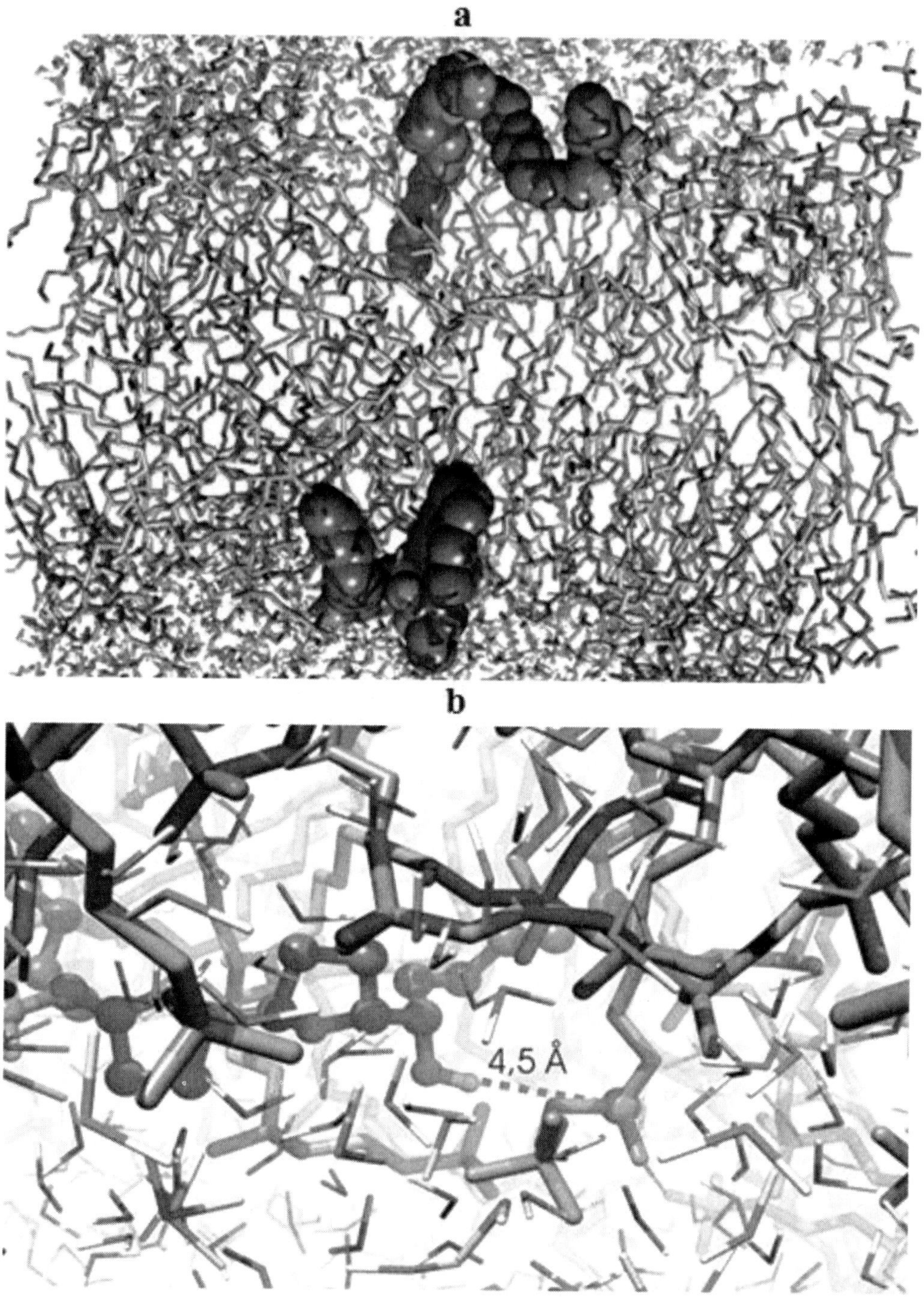

Figure 6. (a) Snapshot of rolliniastatin-1/POPC system at 15 ns of the MD simulation. ACG is represented in spacefill, while the lipid molecules are in sticks. (b) Snapshot displaying the hydrogen bond interaction between the OH of ACG and the phosphate group of POPC bilayer.

Thus, we focused our attention on the intermolecular interactions between the lipid headgroups (glycerol and phosphate) of POPC and the THF/flanking hydroxyls of ACG mono and bis THF.

This study reveals that the OH groups of ACGs might preferably interact with the phosphate groups of membranes in the liquid crystalline state. In agreement, as was revealed by FTIR analysis, these interactions might involve the displacement of surrounding water molecules and the formation of hydrogen bonds with the phosphate groups.

To get a better view of the interaction ACGs bis THF/POPC, figure 6 was plotted. This figure shows that OH groups of ACG are located close to the phosphate groups, with hydrogen atoms pointing toward the phosphate oxygens of the POPC molecule.

It is noteworthy that the knowledge based on both approaches, experimental and theoretical, will increase the comprehension of the action mechanism of ACGs in the lipid membranes.

CONCLUSION

ACGs are a relatively new class of bioactive naturally occurring products. The difficulty of isolating these compounds and elucidating their structures makes them a challenging target for total synthesis. Their wide spectrum of biological properties is probably the most intriguing and exciting domain, and the future will show whether it is possible to disclose the structure-activity relationship, probably on the basis of synthetic derivatives. Furthermore, it will be useful to look for simplifications of the structure without loss of activity. As a result of these investigations, it will not be surprising if ACGs or related compounds with structural modifications might, in the near future, play a significant role in cancer therapy *via* an original mechanism of action. Hence we believe it is worthwhile to observe further developments in the field of ACGs.

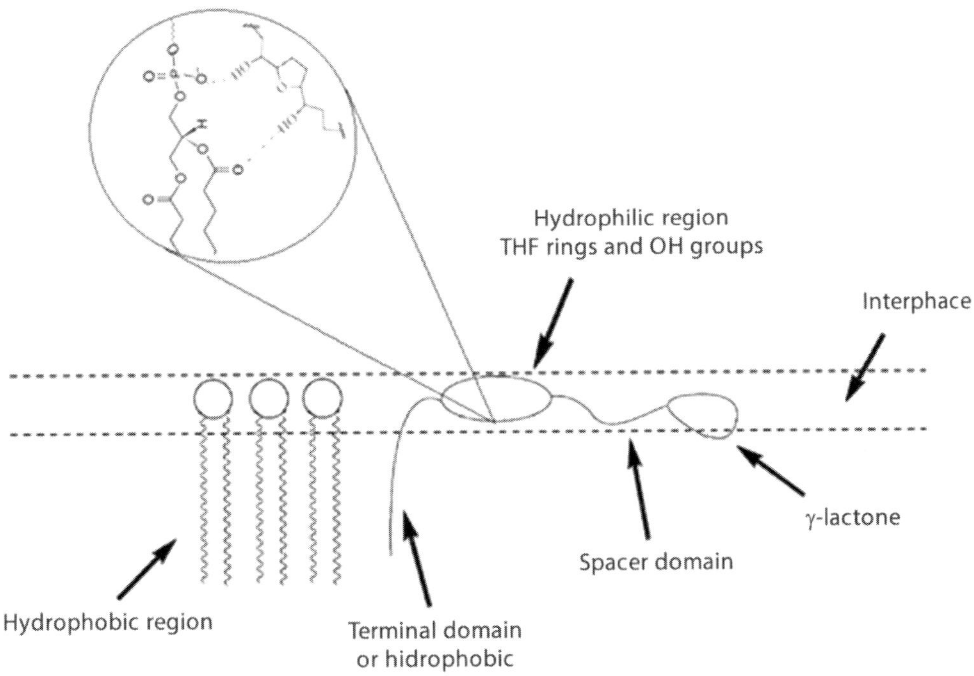

Scheme 6. ACGs mechanism of action in the artificial lipid membranes of POPC.

On the other hand, many of these compounds have demonstrated the ability to strongly stimulate biofilm formation of PAHs degrading bacteria, such as *Pseudomonas plecoglossicida* J26. The evidence gathered in this sense supports the use of ACGs to improve the efficiency of naphthalene degradation by these bacteria.

The FTIR results revealed intermolecular interactions between the headgroups of POPC (specifically phosphate groups) and the hydroxylated-THF of ACGs mono and bis THF. These interactions indicated that the phosphate groups of the lipid membrane form stronger hydrogen bonds with hydroxylated-THF rings of ACGs than with water molecules which suggest a strong dehydration in its environment. To better understand these experimental results, MD simulations of these ACGs embedded in a fully hydrated POPC lipid bilayer were carried out. Our simulations emphasize the role of the THF rings, through their flanking hydroxyl groups, to form the hydrogen bond interactions acting as hydrophilic anchor in the lipid membrane. Our results are in complete agreement with previous reported simulations (Barrachina et al., 2007; Bombasaro et al., 2008) and give an additional support for the stepwise binding mechanism proposed by Shimada et al. (Shimada et al., 1998a,b) On the other hand, the results reported here could be regarded as interesting complementary data for FTIR measurements. It is noteworthy that the knowledge based on both approaches, experimental and theoretical, will increase the comprehension of the ACGs mechanism of action in the artificial lipid membranes of POPC (Scheme 6).

References

Alali, F. Q.; Liu, X. X.; McLaughlin, J. L. 1999. Annonaceous Acetogenins: Recent Progress. *J. Nat. Prod.*, 62, 504-540.

Arrondo, J. L. R; Goni, F. M.; Macarulla, J. M. 1984. Infrared spectroscopy of phosphatidylcholines in aqueous suspension: a study of the phosphate group vibrations. *Biochim. Biophys. Acta.*, 794, 165-168.

Barrachina, I.; Royo, I.; Baldoni, H.; Chahboune, N.; Suvire, F.; DePedro, N.; Zafra Polo, M. C.; Bermejo, A.; Cabedo, N.; Saez, J.; Tormo, J. R.; Enriz, R. D.; Cortes, D. 2007. New antitumoral acetogenin 'Guanacone type' derivatives: Isolation and bioactivity. Molecular dynamics simulation of diacetyl-guanacone. *Bioorg. Med. Chem.,* 15, 4369-4381.

Bassler, B. L. and Losick, R. 2006. Bacterially speaking. *Rev. Cell,* 125, 237-246.

Bermejo, A.; Figadère, B.; Zafra-Polo, M. C.; Barrachina, I.; Estornell, E.; Cortes, D. 2005. Acetogenins from Annonaceae: recent progress in isolation, synthesis and mechanisms of action. *Nat. Prod. Rep.*, 22, 269-303.

Bombasaro, J. A.; Masman, M. F.; Santágata, L. N.; Freile, M. L.; Rodríguez, A. M.; Enriz, R. D. 2008. A Comprehensive Conformational Analysis of Bullacin B, a Potent Inhibitor of Complex I. Molecular Dynamics Simulations and Ab Initio Calculations. *J. Phys. Chem. A.,* 112, 7426-7438.

Bombasaro, J. A.; Di Toto Blessing, L.; Díaz S.; Neske, A.; Suvire, F; Enriz, R. D; Rodriguez, A. M. 2011. Theoretical and experimental study of the interactions of annonaceous acetogenins with artificial lipid bilayers. *J. Mol. Struct.*, 1003, 87-91.

Cacela, C. and Hincha, D. K. 2006. Low Amounts of Sucrose Are Sufficient to Depress the Phase Transition Temperature of Dry Phosphatidylcholine, but Not for Lyoprotection of Liposomes. *Biophys. J.*, 2831-2842.

Cartagena, E.; Álvarez Colom, O.; Neske, A.; Valdez, J. C.; Bardón A. 2007. Effects of plant lactones on the production of biofilm of *Pseudomonas aeruginosa*. *Chem. Pharm. Bull.*, 55, 22-25.

Casiraghi, G.; Zanardi, F.; Battistini, L.; Rassu, G. 1998. Current advances in the chemical synthesis of annonaceous acetogenins and relatives. *Chemtracts,* 11, 803-827.

Cavé, A., Figadère, B.; Laurens, A.; Cortes, D. 1997. Acetogenins from Annonaceae, in Progess in the Chemistry of Organic Natural Products, eds. W. Herz, G. W. Kirby, R. E. Moore, W. Seglich and Ch. Tamm, Springer-Verlag,Wien. 70, pp. 81-288.

Decho, A. W. 2000. Microbial biofilms in intertidal systems: An overview. Continent. *Shelf Res.,* 20, 1257-1273.

Díaz, S. B.; Amalfa F.; Biondi de López A. C.; Disalvo, E. A. 1999. Effect of water polarized at the carbonyl groups of phosphatidylcholines on the dipole potential of lipid bilayers. *Langmuir,* 15, 5179-5182.

Díaz, S. B.; Biondi de López, A. C.; Disalvo, E. A. 2003. Dehydration of carbonyls and phosphates of phosphatidylcholines determines the lytic action of lysoderivatives. *Chem. Phys. Lipids.,* 122, 153-157.

Di Toto Blessing, L.; Álvarez Colom, O.; Popich, S.; Neske, A.; Bardón, A. 2010. Antifeedant and Toxic Effects of Acetogenins from Annona montana on *Spodoptera frugiperda. J. Pest Sci.,* 83, 307-310.

Di Toto Blessing, L.; Ramos, J.; Díaz, S.; Ben Altabef, A.; Bardón, A.; Brovetto, M.; Seoane, G.; Neske, A. 2012. Insecticidal Properties of Annonaceous Acetogenins and Their Analogues. Interaction with Lipid Membranes. *Nat. Prod. Commun.*, 7, 1215-1218.

Duret, P.; Waechter, A. I.; Margraff, R.; Foucault, A.; Hocquemiller, R.; Cavé, A. 1997. High-speed contercurrent chromatography: a promising method for the separation of the Annonaceous acetogenins. *J. Liq. Chrom. Rel. Technol.*, 20, 627-635.

Duval, R. A.; Lewin, G.; Peris, E.; Chahboune, N.; Garofano, A.; Dröse, S.; Cortes, D.; Brandt, U.; Hocquemiller, R. 2006. Heterocyclic analogues of squamocin as inhibitors of mitochondrial complex I. On the role of the terminal lactone of annonaceous acetogeninas. *Biochemistry,* 45, 2721-2728.

Ferrero, M.; Llobet-Brossa, E.; Lalucat, J.; Roselló-Mora, R.; Bosch, R. 2002. Coexistence of two distinct copies of naphthalene degradation genes in *Pseudomonas* strains isolated from the West Mediterranean region. *Applied Environmental Microbiology,* 68, 957-962.

Figadère, B. 1995. Syntheses of acetogenins of annonaceae: A new class of bioactive polyketides. *Acc. Chem. Res.,* 28, 359-365.

Figadère, B. and Cavé, A. 1996. Total stereoselective synthesis of acetogenins of annonaceae: A new class of bioactive polyketides. In Studies in Natural Products Chemistry; Atta-ur-Rahman, Ed.; Elsevier: Amsterdam, The Netherlands, Volume 18, pp. 193-227.

Friedrich, T.; Van Heek, P.; Leif, H.; Ohnishi, T.; Forche, E.; Kunze, B.; Jansen, R.; Trowitzsch-Kienast, W.; Hofle, G.; Reichenbach, H.; Weiss, H. 1994. Two binding sites of inhibitors in NADH:ubiquinone oxidoreductase (complex I) - Relationship of one site with the ubiquinone-binding site of bacterial glucose:ubiquinone oxidoreductase. *Eur. J. Biochem.*, 219, 691.

Fujita, D.; Ichimaru, N.; Abe, M.; Murai, M.; Hamada, T.; Nishioka, T.; Miyoshi, H. 2005. Synthesis of non-THF analogs of acetogenin toward simplified mimics. *Tetrahedron Lett.,* 46, 5775-5779.

Gallardo, T. and Aragon, R. 1998. Acetogenins from *Annona glabra* seeds. *Phytochem.*, 47, 811-816.

Gawronski, J. and Wu, Y. C. 1999. Anote on the determination of absolute configuration of acetogenins by circular dichroism. *Pol. J. Chem.*, 73, 241-243.

Gonzalez, J. E. and Keshavan N. D. 2006. Messing with bacterial quorum sensing. *Microbiol. Mol. Biol. Rev.,* 70, 859-875.

Gu, Z.-M.; Zhao, G.-X.; Oberlies, N. H.; Zeng, L.; McLaughlin, J. L. 1995. Annonaceous acetogenins: Potent mitochondrial inhibitors with diverse applications. In Recent Advances in Phytochemistry; Arnason, J. T., Mata, R., Romeo, J. T., Eds.; Plenum Press: New York, NY, USA, Vol. 29, pp. 249-310.

Gu, Z. M.; Zhou, D.; Wu, J.; Shi, G.; Zeng, L., McLaughlin, J. L. 1997. Screening for Annonaceous acetogenins in bioactive plant extracts by liquid chromatography/mass spectrometry. *J. Nat. Prod.*, 60, 242-8.

Hoppe, R. and Scharf, H. D. 1995. Annonaceous acetogenins - synthetic approaches towards a novel class of natural products. *Synthesis,* 1447-1464.

Hu, T. S.; Wu, Y. L.; Yao, Z. J. 2006. Recent progress on the chemical synthesis of annonaceous acetogenins and their structurally modified mimics. In Medicinal Chemistry of Bioactive Natural Products; Liang, X.-T., Fang, W.S., Eds.; John Wiley & Sons, Inc.: West Sussex, UK, pp. 399-441.

Huang, G.-R.; Jiang, S.; Wu, Y.-L.; Jin, Y.; Yao, Z.-J.; Wu, J.-R. 2003. Induction of cell death of gastric cancer cells by a modified compound of the annonaceous acetogenin family. *Chem. Bio. Chem.*, 4, 1216-1221.

Hübner, W. and Blume, A. 1998. Interactions at the lipid-water interface. *Chem. Phys. Lipids,* 96, 99-123.

Jendrasiak, G. L. and Smith, R. L. 2001. The effect of the choline head group on phospholipid hydration. *Chem. Phys. Lipids,* 113, 55-66.

Jiang, S.; Li, Y.; Chen, X.-G.; Hu, T.-S.; Wu, Y.-L.; Yao, Z.-J. 2004. Parallel fragment assembly strategy towards multiple-ether mimicry of anticancer annonaceous acetogenins. *Angew. Chem. Int. Ed.*, 43, 329-334.

Johnson, H. A., Oberlies, N. H.; Alali, F. Q.; McLaughlin, J. L. 1999. Thwarting resistance: Annonaceous acetogenins as new pesticidal and antitumor agents. In Biologically Active Natural Products: Pharmaceuticals; Cutler, S. J., Cutler, H. G., Eds.; CRC Press: Washington, DC, USA, pp. 173-183.

Keith, L. H. and Telliard, W. A. 1979. Priority pollutants: a perspective view. *Environmental Science and Technology,* 13, 416-423.

Konno, H.; Hiura, N.; Makabe, H.; Abe, M.; Miyoshi, H. 2004. Synthesis and mitochondrial complex I inhibition of dihydroxy-cohinbin A, non-THF annonaceous acetogenin analogue. *Bioorg. Med. Chem. Lett.,* 14, 629-632.

Latypov, S.; Franck, X.; Jullian, J. C.; Hocquemiller, R.; Figadère, B. 2002. NMR determination of absolute configuration of butenolides of Annonaceous type. *Chem. Eur. J.*, 8, 5662-5666.

Li, N.; Shi, Z.; Tang, Y.; Chen, J.; Li, X. 2008. Recent progress on the total synthesis of acetogenins from annonaceae. *Beilstein J. Org. Chem.,* 4, N°. 48.

Li, M. Y.; Zhang, J.; Lu, P.; Xu, J. L.; Li, S. P. 2009. Evaluation of biological characteristics of bacteria contributing to biofilm formation. *Pedosphere,* 19, 554-561.

Liu, H.-X.; Shao, F.; Li, G.-Q.; Xun, G.-L.; Yao, Z.-J. 2008. Tuning the acyclic ether moiety of anticancer agent AA005 with conformationary constrained fragments. *Chem. Eur. J.,* 14, 8632-8639.

Londershausen, M.; Leicht, W.; Lieb, F.; Moeschler, H.; Weiss, H. 1991. Molecular mode of action of annonins. *Pest. Sc.,* 33, 427-438.

Luzardo, M. C.; Amalfa, F.; Núñez, A.; Díaz, S. B.; Biondi de Lopez, A. C.; Disalvo, E. A. 2000. Effect of trehalose and sucrose on the hydration and dipole potential of lipid bilayers. *Biophys. J.*, 78, 2452-2458.

Mantsch, H. H. and McElhaney, R. N. 1991. Phospholipid phase transitions in model and biological membranes as studied by infrared spectroscopy. *Chem. Phys. Lipids,* 57, 213-226.

Marshall, J. A.; Hinkle, K. W.; Hagedorn, C. E. 1997. Recent developments in the synthesis of annonaceous acetogenins. *Isr. J. Chem.*, 37, 97-107.

Melot, A.; Fall, D.; Gleye, C.; Champy, P. 2009. Apolar Annonaceous Acetogenins from the Fruit Pulp of *Annona muricata. Molecules,* 14, 4387-4395.

Oberlies, N. H.; Jones, J. L.; Corbett, T. H.; Fotopoulos, S. S.; McLaughlin J. L. 1995. Tumor cell growth inhibition by several annonaceous acetogenins in vitro disk diffusion assay. *Cancer Letters,* 96, 55-62.

Oberlies, N. H.; Chang, C.; McLaughlin, J. L. 1997. Structure-activity relationships of diverse annonaceous acetogenins against multidrug resistant human mammary adenocarcinoma (MCF-7/Adr) cells. *Journal of Medical Chemical,* 40, 2102-2106.

Parellada, E. A.; Ramos, A. N.; Ferrero, M.; Cartagena, E.; Bardón, A.; Valdez, J. C.; Neske, A. 2011. Squamocin mode of action to stimulate biofilm formation of *Pseudomonas plecoglossicida* J26, a PAHs degrading bacterium. *Int. Biodeter. Biodegr.*, 65, 1066-1072.

Parellada, E. A.; Ramos, A. N.; Ferrero M.; Cartagena E.; Bardón, A.; Neske, A. 2012. Effect of the annonaceous acetogenins, squamocin and laherradurin, on the degradation of naphthalene mediated by *Pseudomonas plecoglossicida* J26. *Int. Biodeter. Biodegr.*, 72, 82-87.

Parellada, E. A.; Cartagena, E.; Bardón, A.; Neske, A. 2013. Laherradurin, a natural stressor, stimulates QS mechanism involved in biofilm formation of a PAHs degrading bacterium. *Int. Biodeter. Biodegr.*, 85, 78-84.

Pohle, W. and Selle, C. 1996. Fourier-transform infrared spectroscopic evidence for a novel lyotropic phase transition occurring in dioleoylphosphatidylethanolamine. *Chem. Phys. Lipids,* 82, 191-198.

Popova, A. V. and Hincha, D. K. 2003. Intermolecular interactions in dry and rehydrated pure and mixed bilayers of phosphatidylcholine and digalactosyldiacylglycerol: a Fourier-transform infrared spectroscopy study. *Biophys. J.,* 85, 1682-1690.

Rodney, M.; Donlan, L.; Costerton, J. W. 2002. Biofilms: Survival mechanisms of clinically relevant microorganisms. *Clin. Microbial. Rev.*, 15, 167-193.

Shimada, H.; Kozlowski, J. F.; McLaughlin, J. L. 1998a. The localisations in liposomal membranes of the tetrahydrofuran ring moieties of the annonaceous acetogenins, annonacin and sylvaticin, as determined by ^{1}H NMR spectroscopy. *Pharmacol. Res.,* 37, 358-364.

Shimada, H.; Grutzner, J. B.; Kozlowski, J. F.; McLaughlin, J. L. 1998b. Membrane Conformations and Their Relation to Cytotoxicity of Asimicin and Its Analogues. *Biochemistry,* 37, 854-866.

Smith, J. R.; Nakles D. V.; Sherman, D. F.; Neuhauser, E. F.; Loehr, R. C. 1989. Environmental fate mechanisms influencing biological degradation of coal-tar derived polynuclear aromatic hydrocarbons in soil systems. In: The Third International Conference on New Frontiers for Hazardous Waste Management, U.S. Environmental Protection, Agency Washington D.C., pp. 397-405.

Tormo, J. R.; Gallardo, T.; Peris, E.; Bermejo, A.; Cabedo, N.; Estornell, E.; Zafra-Polo, M. C. Cortes, D. 2003. Inhibitory effects on mitochondrial complex I of semisynthetic mono-tetrahydrofuran acetogenin derivatives. *Bioorg. Med. Chem. Lett.,* 13. 4101-4105.

Waters, C. M. and Bassler, B. L. 2005. Quorum sensing: Cell-to-cell communication in bacteria. *Annu. Rev. Cell. Dev. Biol.*, 21, 319-346.

Zafra-Polo, M. C., González, M. C.; Estornell, E.; Sahpaz, S.; Cortes, D. 1996. Acetogenins from Annonaceae, Inhibitors of Mitochondrial Complex I. *Phytochemistry,* 42, 253-271.

Zeng, L.; Ye, Q.; Oberlies, N. H.; Shi, G.; Gu, Z. M.; He, K.; McLaughlin, J. L. 1996. Recent Advances in Annonaceous Acetogenins. *Nat. Prod. Rep*., 13, 275-306.

In: Natural Heterocycles
Editors: K. L. Ameta and Saksit Chanthai
ISBN: 978-1-63463-424-3

Chapter 2

INDAZOLE MOTIFS IN HETROCYCLIC COMPOUNDS: A BIO-ACTIVE COMPONENT

Digambar D. Gaikwad[1], Archana D. Chapolikar[1], Chabutai S. Patil[2], Saroj R. Bembalkar[2], Maya V. Katariya[2], K. L. Ameta[3] and Rajendra P. Pawar[2*]

[1]Department of Chemistry, Govt. College of Arts & Science, Aurangabad, India
[2]Department of Chemistry, Deogiri College, Aurangabad, India
[3]Department of Chemistry, FASC, Mody Institute of Technology & Science, Lakshmangarh, Rajasthan, India

ABSTRACT

Naturally occurring class of heterocyclic compounds indazoles possesses a wide range of biological and pharmaceutical applications. These novel heterocyclic compounds were known for several pharmaceutical and medicinal applications. The present chapter focuses on various biological activities of indazole moiety.

ABBREVIATIONS

PI3K	phosphatidylinositol 3-kinase
PDK1	3-phosphoinositide-dependent protein kinase-1
RTK	receptor tyrosine kinases
AKT	protein kinase B
mTOR	mammalian target of rapamycin
PTEN	phosphatase and tensin homologue protein
PIP2	phosphatidylinositol 4,5-bisphosphate
PIP3	phosphatidylinositol 3,4,5-triphosphate

* E-mail: rppawar@yahoo.com.

PDCD4	programmed cell death protein
S6K1	S6 kinase β-1
GFP	green fluorescent protein
HAART	highly active antiretroviral therapy
HIV	human immunodeficiency virus
AIDS	acquired immunodeficiency syndrome
NNRTI	non-nucleoside reverse transcriptase inhibitor
wt	wild type
RT	reverse transcriptase
NRTI	nucleoside reverse transcriptase inhibitor
ΔE	change in energy
CIC95	cell inhibitory concentration that blocks 95% of cell replication
FBS	fetal bovine serum
NHS	normal human serum
EFV	efavirenz
iv	intravenous
nAUC	normalized area under the plasma concentration vs time curve
*V*d	volume of distribution
Clp	plasma clearance
*t*1/2	half-life
F	bioavailability
BER	base excision repair
BRCA	breast cancer susceptibility gene
DSB	double strand break
HLM	human liver microsomes
HR	homologous recombination
NAD þ	nicotinamide adenine dinucleotide
PAR	poly (ADP)ribose
PARP	poly (ADP-ribose)-polymerase
RLM	rat liver microsomes
SFC	supercritical fluid chromatography
shRNA	short hairpin interfering RNA
SSB	single strand break
TANK	tankyrase
V-PARP	vault PARP
WT	wild type

1. Introduction

Heterocyclic chemistry is an important branch of organic chemistry accounting for nearly one-third of modern publications. In fact, two third of organic compounds are heterocyclic compounds. A cyclic organic compound containing only carbon atoms in ring formation is referred as a carbocyclic compound. If at least one atom other than carbon forms a part of the ring system then it is designated as a heterocyclic compound. Nitrogen, oxygen and sulphur are the most common heteroatoms. But heterocyclic rings containing other hetero atoms are also widely known. An enormous number of heterocyclic compounds are known. This number is increasing rapidly. Accordingly, the literature on the subject is very vast. Heterocyclic compounds may be classified into aliphatic and aromatic. The aliphatic heterocyclics are the cyclic analogues of amines, ethers, thioethers, amides etc. Their properties are particularly influenced by the presence of strain in the ring. These compounds generally consist of small (3 and 4 membered) and common (5 to 7 membered) ring systems. In contrast the heterocyclic compounds are those which have a heteroatom in the ring and be have as benzene in some of their properties. Furthermore, these compounds also comply with the general rule proposed by Huckel. Besides the vast distribution of heterocycles in natural products, they are also the major components of biological molecules such as DNA and RNA. DNA is no doubt the most important macromolecule of life. Nucleotides, the building blocks of our genes are derivatives of pyrimidine and purine ring structures. Chlorophyll and haemoglobin are the oxygen carriers in plants and animals respectively. These are derivatives of large porphyrin rings. Heterocycles are present in a wide variety of drugs, vitamins, natural products, biomolecules and biologically active compounds, including antitumor, antibiotic, anti-inflammatory, antidepressant, antimalarial, anti-HIV, antimicrobial, antibacterial, antifungal, antiviral, antidiabetic, herbicidal, fungicidal, and insecticidal agents. Also, they have been frequently found as a key structural unit in synthetic pharmaceuticals and agrochemicals. Some of these compounds exhibit a significant photochromic and biochemi-luminescence properties. Most of the heterocycles possess important applications in materials science such as dyestuff, fluorescent sensor, brightening agents, information storage, plastics and analytical reagents.

Indazoles are rare in nature and so far only three natural products based on indazole ring have been isolated [1]. These are the indazole alkaloids nigellicine (**1**) [2], nigeglanine (**2**) [3] and nigellidine (**3**) [4]. The total synthesis of nigellicine and nigeglanine are also well documented.

Indazoles are naturally occurring alkaloids *Nigellicine*, *Nigeglanine* and *Nigellidine*. *Nigellicine* was isolated from widely distributed plant *Nigella sativa*. *Nigeglanine* was isolated from extracts of *Nigella glandulifera*. Only few of the alkaloids studied on isolation showed the presence of indazole ring system. The first member of this alkaloid family *Nigellicine* was isolated in 1985 from the plant *Nigella sativa*an annual flowering plant, native to Southwest Asia [5]. The seeds of this plant have been used for thousands of years as a spice and for the treatment of various diseases [6, 7]. The structure of *nigellicine* has an intramolecular hydrogen bond between the carboxylate oxygen atom and the hydroxyl group. The structure of *nigellicine* is a pseudo-cross-conjugated heterocyclic mesomeric betaine, which means it can be presented by dipolar canonical formula where both the positive and negative charge are delocalized in the structure [8].

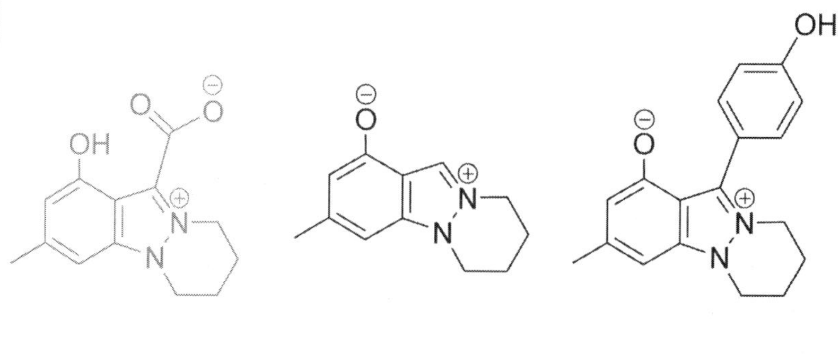

Nigellicine (1) Nigeglanine (2) Nigellidine (3)

Similarly other two alkaloids containing the same ring system, *Nigeglanine* and *Nigellidine* have been isolated from extracts of *Nigella glandulifera* and *Nigella sativa.* These two compounds can also be presented by their zwitterions formulae [9].

2. Extraction Methods of Medicinal Plants

Extraction, the term is used pharmaceutically, involves the separation of medicinally active portions of plant or animal tissues from inactive or inert components by using selective solvents by standard extraction procedures. The products so obtained from plants are relatively impure liquids, semisolids or powders intended only for oral or external use. These include classes of preparations known as decoctions, infusions, fluid extracts, tinctures, pilular (semisolid) extracts and powdered extracts. Such preparations popularly have been called galenicals, named after Galen, the second century Greek physician. The purposes of standardized extraction procedures for crude drugs are to attain the therapeutically desired portion and to eliminate the inert material by the treatment with a selective solvent known as *menstruum*.

The extract thus obtained may be ready for use as a medicinal agent in the form of tinctures and fluid extracts. It may be further processed to be incorporated in any dosage form such as tablets or capsules, or it may be fractionated to isolate individual chemical entities such as ajmalicine, hyoscine and vincristine modern drugs. Thus, standardization of extraction procedures contributes significantly to the final quality of the herbal drug.

Hot Extraction (Soxhlet) Process

In this method, finely ground crude drug is placed in a porous bag or "thimble" made of strong filter paper and placed in chamber E of the Soxhlet apparatus (Figure 1). The extracting solvent in flask **A** is heated and its vapors are condensed in condenser **D**. The condensed extract drips into the thimble containing the crude drug and extracts it by contact. When the level of liquid in chamber E rises to the top of siphon tube **C**, the liquid contents of chamber E siphon into flask **A**. This process is continuous and is carried out until a drop of solvent from the siphon tube does not leave residue when evaporated. The advantage of this method, compared to previously described methods, is that large amounts of drug can be extracted with a much smaller quantity of solvent. This effect is tremendous by economical in terms of time, energy and consequently financial inputs. At small scale, it is employed as a batch process only but it becomes much more economical and viable when converted into a continuous extraction procedure on medium or large scale.

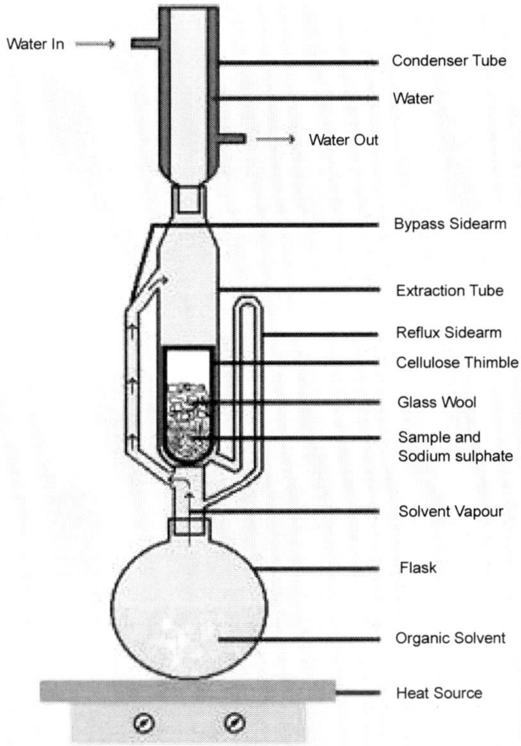

Figure 1. Soxhlet apparatus.

3. Chemistry of Indazoles

Indazole was first defined by scientist Emil Fisher as a "pyrazole ring fused with the benzene ring". It has been extensively studied due to its interesting chemical and biological properties. Indazole belongs to the azoles family containing carbon, hydrogen and nitrogen atoms. Indazoles are also called as benzpyrazole or isoindazolone heterocyclic organic compounds, possessing two nitrogen atoms. Indazole having ten π-electron aromatic systems as like of pyrazole molecule. Indazoles also resembles with pyridine and pyrrole. The structure of indazole is as follows.

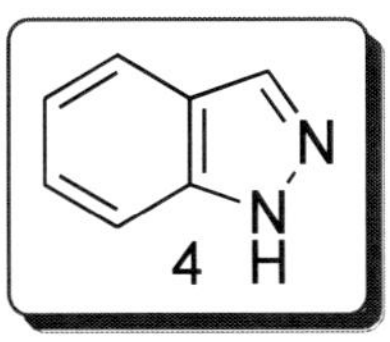

Indazole derivatives are pharmacologically important, they form the basic structure of several drug molecules, such as Granisetron, $5HT_3$ receptor antagonist used as an anti-emetic in cancer chemotherapy and Benzydamine an anti-inflammatory agent. Indazole ring has two nitrogen atoms and can be functionalized with high selectivity at different positions. Planarity of the indazole ring, side chain length and functionalizations at different positions can afford an enormous number of indazole derivatives, providing new molecules with biological and therapeutic properties.

1H-Indazoles represent an important class of heterocyclic compounds that exhibit a wide range of biological and pharmaceutical activities [10] including anti-inflammatory [11] antitumor [12, 13] and anti-HIV [14] activity among others. The same examples such as 1-alkyl-1H-indazoles with notable pharmacological activities include granisetron, a serotonin 5-HT_3 receptor antagonist used to treat nausea and vomiting after chemotherapy [15]. lonidamine used for the treatment of brain tumors [16] and CL-958 an antitumor agent, which is currently in clinical evaluation [17,18]. Recently, several methodologies have been reported that involve aryne intermediates in [3 + 2] cycloaddition reactions with diazocompounds [19, 20]. N-tosylhydrazones generatednitrile imines [21, 22, 23]. These methods afford 1H-indazoles, 1-acyl-1H-indazoles or 1-aryl-1H-indazoles under mild reaction conditions. Similarly for aryne-annulation approach to 1-alkyl-1H-indazoleshas been reported [24].

World's largest selling drugs are nitrogen containing heterocycles [25]. It is because of their ubiquitous nature which is the key scaffolds of many biological molecules and pharmaceutical products. Thus many chemists from all over the world were motivated and started developing different methods for the synthesis of these heterocycles. One of these which have biological, agricultural and an industrial application is indazole [26]. Indazole and its derivatives possess several activities such as anti-inflammatory, anti-tumor, anti-HIV, anti-platelet and serotonin 5-HT_3 receptor antagonist activities [27, 28]. Indazole has potent affinity for 5-HT_{1A} receptor β and high affinity for I_2 receptor. Thus, looking towards its important activities the methods for its synthesis should be developed [29]. This would help to explore the chemistry of 1H & 2H-Indazole.Ortho-hydrazine benzoic acid (**5**) on heating results in the formation of (**6**) which was later named as indazolone, as reported by Emil Fisher in 1800.

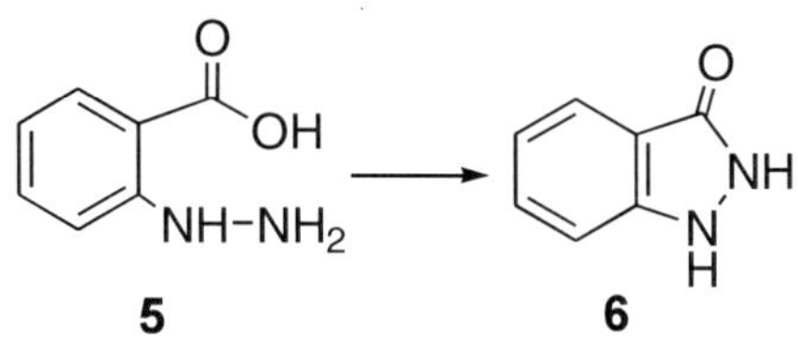

First reported synthesis of indazolone.

Indazolone also called as the anhydride of (**5**), Emil Fisher and Kuzel attempted to apply the same strategy to obtain the anhydride of ortho-hydrazino cinnamic acid (**7**). Among the other products, they isolated a molecule that did not contain any oxygen. According to Fisher and Kuzel, this was remarkable in the highest degree and they named it as indazole by analogy with the name of indole.

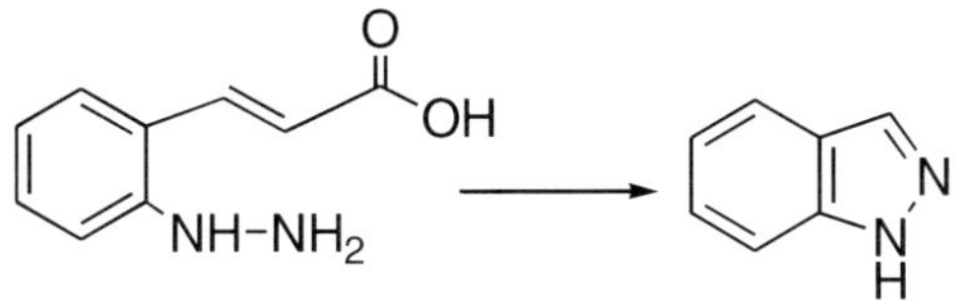

7 8

First reported synthesis of indazole.

Most of the studies on indazole have been carried out by Karl Von Auwers during 1920s. His work on the recent studies of indazole nucleus is reviewed in the following sections.

4. Tautomerism of Indazoles

To understand the chemical reactivity of indazoles, it is important to know its tautomerism and aromaticity. Indazole may exist in two forms which result from the movement of proton between two nitrogen atoms, a process described as proto-tropic annular tautomerism. Nonsubstituted indazole exist predominantly as 1H-tautomer (**9**) based on the results from molecular refractivity studies. This was supported by the finding that, the UV and Raman spectra of non substituted indazole were similar to the spectra obtained for 1H-substituted indazoles and different than the spectra of 2H-substituted indazole. Comparison between hydrogen and nitrogen NMR spectra of indazole with 1-methyl and 2-methylindazole and comprehensive study of ^{13}C and ^{1}HNMR studies provide the evidence for 1H-tautomer [30].

The indazole heterocycle is normally referred to as 1*H*-indazole, although it has two other potential tautomers 2*H*-indazole (**10**) and 3*H*-indazole (**11**). The 1*H*-indazole was found to be dominant tautomer in the gaseous state and in aqueous solution and this result is not reversed in excited state by a solvent effect. X-ray diffraction studies of *N*-substituted indazoles confirmed the general preference for 1*H*-tautomers in the solid state. 1*H*-Indazoles have benzenoid properties (are aromatic in nature), while 2-substituted 2*H*-indazoles have ortho-quinoid character. 3*H*-Indazoles lack heteroaromatic character and are very rare. There is evidence regarding the influence of the tautomeric equilibrium in indazoles on different biological properties. However, the effect of tautomer identity on the anti-proliferative activity, biological mechanisms involved and other physico-chemical properties, which can have an impact on pharmacokinetic and pharmacodynamic behaviors in the case of metal complexes with indazole remains unexplored.

The indazole ring has two nitrogen atoms and annular tautomerism with regards to the position of NH hydrogen atom. Due to the difference in energy between tautomers, the benzonoid form (**9**) predominates in the gas phase solution. Solid state derivatives are usually found thermodynamically more stable than the corresponding 2H forms, annular tautomerism of indazole benzenoid 1H-indazole tautomer (**9**) and quinonoid 2H-indazole tautomers (**10**) [31].

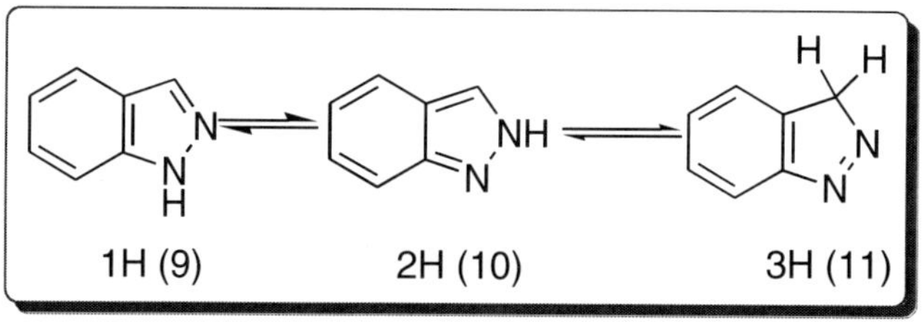

The tautomers of indazole.

Initially, tricyclic structure (**12**) was suggested due to evidence for the presence of benzene ring based on the reactivity of 2H-substituted indazoles in electrophilic aromatic substitution reactions and molecular refractivity measurements. However, a tricyclic formula creates a stereogenic center at C-3 and efforts to resolve the putative indazole-3-carboxylic acid (**13**) were unsuccessful. Eventually, Von Auwers proposed the quinonoid structure (**14**) for indazole.

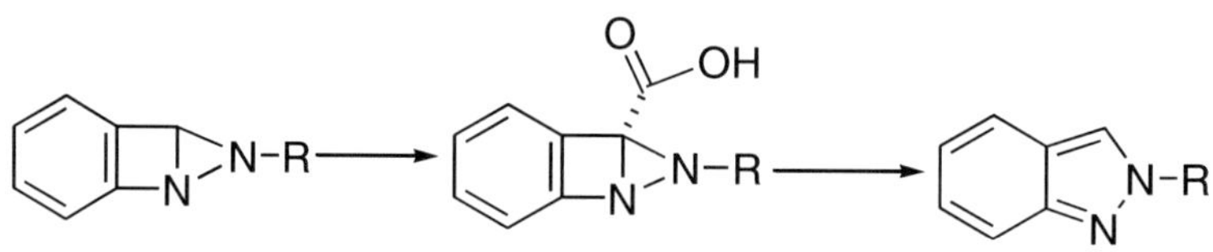

Another important observation reported by Von Auwers is related to the structure of 3-phenylindazole. 3-Phenylindazole is having a melting point 107-108^{0}C. On heating above the melting point and allowed to cool it resolidify to form fine crystalline product (M.P. 115-116^{0}C). Acetylation of both compounds results in the formation of the same product which is converted to the low melting 3-phenylindazole on hydrolysis.

5. General Importance of Indazoles

Indazole nucleus is present in naturally occurring alkaloids and biologically active molecules. *Nigellidine* is a natural product containing an indazole nucleus, isolated from plant *Nigella sativa* and used in the treatment of various diseases. Commonly have carminative, stimulatory and diaphoretic properties. Indazole based heterocycles like indazole pyrimidines and their derivatives are found to have a wide range of activities. Earlier findings on indazole derivatives are specifically known to be active as protein kinase inhibitors, cancer cell proliferative disorders, alzheimers disease, viral infections, auto immune and neuro degenerative disorders. In recent years, some of the indazole ring systems are being evaluated as potential drugs for variety of physiological activities with at least few compounds approved for clinical use. Further, pyrimidines ring in an organic molecule also showed prominent activity against several diseases.

In addition, fluorine or trifluoro-methyl group at an appropriate position of molecule promotes lipophilicity of the molecule and results in increased solubility and transport of the drug in lipid system. Therefore, indazole based compounds with fluorine or trifluoromethyl group at a specified position have gained significance as potent organic molecules, which ultimately serve like drugs. Some of the drug molecules containing indazole moieties are used as pharmaceutical agents.

However, the development of efficient and general methodologies for the synthesis of wide variety of indazoles and derivatives has given limited success and significant limitations such as the need for harsh reaction conditions and high temperatures often limits the scope and applications. Potentially hazardous and explosive intermediates [32] sometimes limit applications in large-scale processes. In addition, most of the present procedures involve several steps and of overall efficiency. Considering the limitations of these previous procedures and the importance of such indazole derivatives, a simple, efficient, and general method to synthesize indazoles and their derivatives would be quite attractive. Recently, we had considerable success using arynes prepared in situ from *o*- (trimethylsilyl)aryl triflates under mild reaction conditions in a variety of synthetic processes [33]. 3-aminoindazole derivatives have been identified as potent dopamine receptor antagonists, useful for antipsychotic treatment [34] as well as non-steroidal anti-inflammatory compounds with analgesic properties [35] for the preclinical evaluation of 3-piperazinylindazole derivative. The preparation of *N, N*-disubstituted 3-aminoindazoles usually involves the substitution of 2- or 3-nitro or 3-haloindazole [36] derivatives with the appropriate secondary amine, *N,N*-disubstituted 3-aminoindazoles [37]. The synthesis of 1H-indazoles have also been reported the reaction of ortho hydroxyl acetophenone on reaction with hydrazine hydrate in presence of silica sulphuric acid (SSA) [38]. The synthesis of indazoles from ortho alkoxy acetophenone, hydrazine hydrates using DMSO and molecular iodine has been reported [39].

6. Indazoles of Reverse Transcriptase (RT) Enzyme

An enzyme which is used to generate complementary DNA (cDNA) from an RNA template is termed as reverse transcription for replication of retroviruses (e.g., HIV) RT is needed. These RT inhibitors are widely used as antiretroviral drugs. The replication of chromosome ends (telomerase) and some mobile genetic elements is also associated with RT activity (retrotransposons). Retroviral RT has three sequential biochemical activities:

- RNA-dependent DNA polymerase,
- DNA-dependent DNA polymerase.
- Ribonuclease

The conversion of single-stranded genomic RNA into double-stranded cDNA these activities are used by retro virus which can integrate into the host genome, potentially generating a long-term infection that can be very difficult to eradicate. The same sequence of reactions is widely used in the laboratory to convert RNA to DNA for the use in molecular cloning, RNA sequencing, polymerase chain reactions (PCR), or genome analysis.

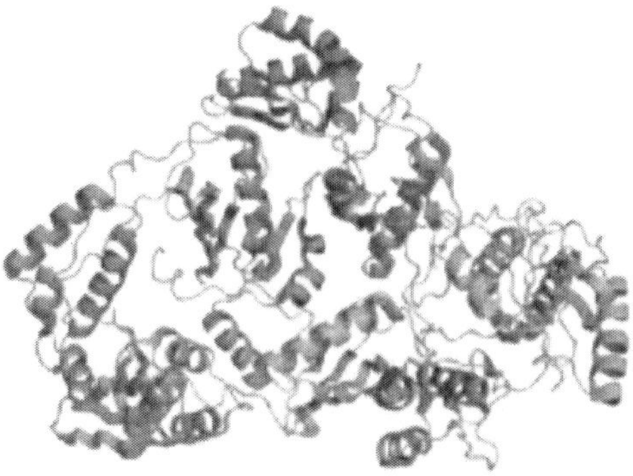

Figure 2. Reverse transcriptase (RNA-dependent DNA polymerase) [40].

Inhibition of the reverse transcriptase (RT) enzyme present in the human immunodeficiency virus (HIV) represents a key method for the treatment of HIV infection. Welcome boost against the resistance experienced by the older class of nucleoside reverse transcriptase inhibitors (NNRTIs) by viral mutation is provided by the emergence of nucleoside reverse transcriptase inhibitors (NRTIs) [41]. However, recently marketed NNRTI treatments such as efavirenz, nevirapine and delavirdine have themselves met with viral mutation and resistance. Second-generation NNRTIs, such as etravirene [42] have recently been developed specifically for treating experienced adult patients with HIV resistance to an NNRTI. Merck recently reported on the discovery and development of a novel class of second-generation NNRTIs containing the pyrazolo [3,4-b]pyridine and bi-aryl ether key structural features [43, 44].

HIV has a high rate of mutation due to its infidelity during replication. As a result, single drug therapy is not found to be durable because of the rapid selection of resistant mutant strains [45, 46, 47]. HAART significantly reduces HIV viral load and has led to a dramatic decrease in AIDS related mortality and a better quality of life for patients [48]. HAART typically combines at least two anti-HIV drugs simultaneously, as this greatly reduces the probability of virus developing drug resistance [49]. The emergence of HIV-1 strains resistant to at least one antiretroviral drug highlights the need for further development of antiviral agents with improved efficacy against the mutant strains that are observed clinically [50, 51]. Reverse transcriptase (RT) inhibitors block the conversion of single stranded viral RNA to double stranded proviral DNA, a prerequisite to integration into host DNA [52]. NNRTIs interact with RT enzyme at an allosteric site and induce a change in the substrate binding site which interferes with its DNA polymerase activity [53]. This class of inhibitor has gained an increasingly important role in HAART therapy because of its high specificity and lack of mitochondrial toxicity which characterizes the nucleoside inhibitors (NRTI's) [54]. However, because of their allosteric mechanism of action, a common consequence of NNRTI use is the emergence of drug resistant forms of HIV. K103N and Y181C variants are the most predominant mutants selected by first generation NNRTI therapy [55, 56]. Next generation NNRTIs are sought that possess broad spectrum antiviral activity against key mutant strains (especially K103N and Y181C) and that also present a high genetic barrier to the selection of new drug resistant viral strains. To facilitate patient compliance additionally a next generation NNRTI that is both efficacious with once daily dosing and amenable to fixed dose combination HAART regimens is highly desirable. The NNRTI TMC-125 [57] has been approved by the FDA for twice daily oral dosing and has demonstrated activity toward a number of key clinical mutations. More recently, the more potent follow-up compound TMC-278 [58] has been approved by the FDA to be administered orally once daily. Tucker et al. [59] have previously reported a series of diaryl ethers that feature a pyrazolopyridine, as exemplified by MK-1107. This compound has excellent potency versus wild type and key mutant viruses.

7. PI3K/AKT/MTORA Cascade Is an Important Cellular Signaling Pathway

The *PI3K/AKT/MTOR* pathway is an intracellular signalling pathway important in apoptosis and hence cancer (e.g. breast cancer and non-small-cell lung cancer) and longevity. mTOR signaling pathway. Hypertrophic muscle growth is contributed by the PI3K/AKT/MTOR pathway. It is activated by IGF1 and has number of downstream effects which either promote protein synthesis or inhibit protein breakdown. PI3K activation activates AKT which activates mTOR. ('PDK1' in the diagram is Phosphoinositide-dependent kinase-1.) In many cancers, this pathway is overactive, thus reducing apoptosis and allowing proliferation. Consequently, to inhibit the signalling sequence at some point is the aim of some experimental cancer drugs. The PI3K pathway may be overactive because *PTEN* is faulty or deficient.

The PI3K/AKT/mTORa cascade is an important cellular signaling pathway that regulates intersecting biological processes such as cell growth and proliferation, cell survival, protein synthesis and glycolysis metabolism [60] Physiological stimulation of this cascade occurs through the binding of growth factors (e.g., insulin or insulin-like growth factor IGF-1) to receptor tyrosine kinases (RTK) on the cell surface. Subsequent activation of the lipid phosphatidylinositol 3-kinase (PI3K) leads to the phosphorylation of phosphatidylinositol 4,5-bisphosphate (PIP2) to produce phosphatidylinositol 3,4,5-triphosphate (PIP3) [61], which in turn interacts with the pleckstrin homology (PH) domain of AKT and recruits the kinase to the plasma membrane. Full activation of AKT requires phosphorylation of two distinct residues: Thr308 by the upstream kinase PDK1 and Ser473 by the mammalian target of rapamycin complex 2 (mTORC2) [62]. ActivatedAKT, among its wide array of downstream effects, increases protein synthesis rate by phosphorylation at Thr246 of the proline-rich substrate of 40 kDa (PRAS40).

This substrate is further regulated through phosphorylation at Ser183 by the mTOR complex 1 (mTORC1). Downstream of mTORC1, signaling events can go through p70 S6 kinase (S6K1), which in turnphosphorylates the programmed cell death protein (PDCD4)at residue Ser457 (Figure 3). Because of its involvement in a variety of cellular events, such as survival, proliferation, cell motility and invasion [63], hyper activation of the PI3K signaling pathway is one of the most common molecular events in nearly every form of human cancer [64, 65]. Indeed, aberrant activation of the PI3K cascade, including PTEN in activation and PI3K activating mutations, is present in about 32% of colon cancers, 30% of breast cancers,30% of melanomas [66] 27% of brain cancers, and 25% of stomach cancers [67, 68].

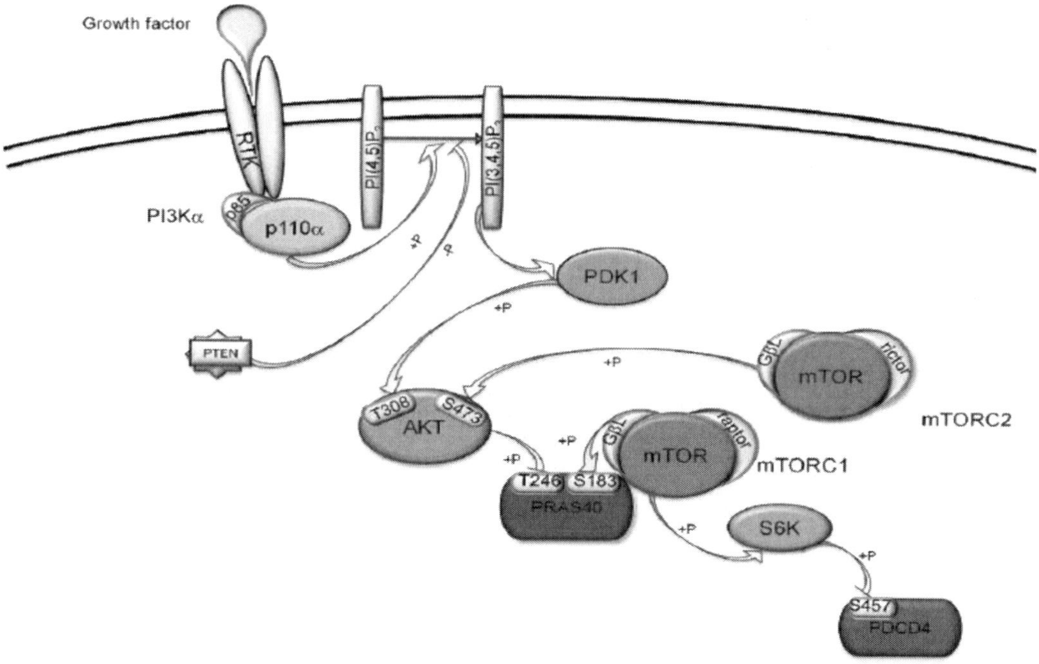

Figure 3. The schematic representation of the P13K/mTOR signaling pathway.

Recently, this data have emerged and suggested several DNA repairing pathways in targeting tumor cells. They are selective killing of BRCA-1 or BRCA-2 deficient tumor cell by PARP inhibitors. BRAC-1 and BRC-2 are known tumor suppressors and are key components in the repair of DNA double strand breaks by homologous recombination (HR) pathway and mutation in these genes predispose individuals to hereditary breast, ovarian cancer, also prostate and pancreatic cancer. The development of novel series of 2-phenyl-2H-indazole-7-carboxamide as PARP-1 and 2 inhibitors for the treatment of BARC-1 and 2 deficient cancers is reported. This indazole-7-carboxamide inhibits proliferation of cancer cell lines with mutant BRCA-1 and BRCA-2 in vitro and is efficacious in xenograft models oral dosing [69, 70]. It is for these reasons that considerable drug discovery efforts are ongoing, targeting several components of this signaling cascade using either single and multiple target strategies [71, 72]. Interestingly, clinical data suggest that multi kinase inhibitors (MKIs) produce greater benefit over single kinase inhibition, especially in solid tumors where different kinases and different pathways synergistically contribute to tumor proliferation [73, 74]. Indeed, emerging phase II clinical trial data reveal that dual PI3K/mTOR inhibitors may be more effective, having the advantage of being less susceptible toPI3K drug resistance owing to their preserved activity against mTOR. In our drug discovery program we interrogated the PI3Ksignaling pathway and report, the identification of a novel scaffold of the 3-ethynylindazole family found to be multiplePI3K/PDK1/mTOR inhibitors [75].

Such scaffold represents a novel structure for the PI3Kpathway, suitable for further fictionalization and drug development.

The importance of indazole nucleus has been discussed in biological and pharmacological properties. Our main aim is to explore the medicinal properties. Accordingly the following outline has been proposed

I. Medicinal importance of indazoles
II. Drug design
III. Miscellaneous

i. Medicinal Importance of Indazoles

The indazole ring system is of much current interest as partial structure of large number of biologically active compounds. Different aspects of pharmaceutical and other useful applications of indazoles have been reviewed. Some substituted indazoles exhibits relevant biological properties for the development as anticancer drugs [76]. One of the tetrasubstituted indazoles, namely, CL-958, entered clinical trials for prostate cancer treatment about a decade ago [77]. From unsubstituted indazole derivatives the most prominent example is the ruthenium (III) compound (H_2ind) [*trans*-$Ru^{III}Cl_4$ $(Hind)_2$] (KP1019, Hind = 1*H*-indazole), which is now in clinical trials as an anticancer agent against metastatic solid tumors [78]. Potentially interested complexes related to (H_2im) [*trans*-$Ru^{III}Cl_4$ (DMSO) (Him)] (NAMI-A, Him = imidazole) [79] are currently evaluated in a clinical phase II trial for its capacity of inhibiting the process of metastasis, namely (H_2ind) [*trans*-$Ru^{III}Cl_4$ (DMSO) (Hind)] [80] and its osmium counterpart [81].

Among the heterocyclic compounds the indazole nucleus is potentially valuable as building-blocks in medicinal chemistry. Even so, a large number of synthetically prepared indazoles derivatives have displayed various biological and pharmacological properties [82]. Bendazac (**15**) is a non-steroidal anti-inflammatory agent used as an anti-cataract drug. New YC-1 indazole derivative (**16**) were synthesized and evaluated for HIF-1 transcriptional activity, in vivo [83]. Furthermore, the mechanistic study of YC-1 derivatives, showed the anti-proliferative activity against human cancer cells and 1, 3, 5-trisustituted indazole derivative (**17**) were found to beextremely potent antioxidant [84]. Recently, it has been reported that a combination of lonidamine (**18**) was shown to have antitumor properties in phase-II and phase-III studies for the treatment of advanced breast, ovarian and lung cancers. 7-nitro indazoles derivative (**19**) was tested for the treatment of alcohol dependence, anti-mutagenesis and anti-oxidant properties [85]. Similarly piperidine derivative of indazole (**20**) has been patented as a non-narcotic, analgesic and antipsychotic drug [86] as shown below.

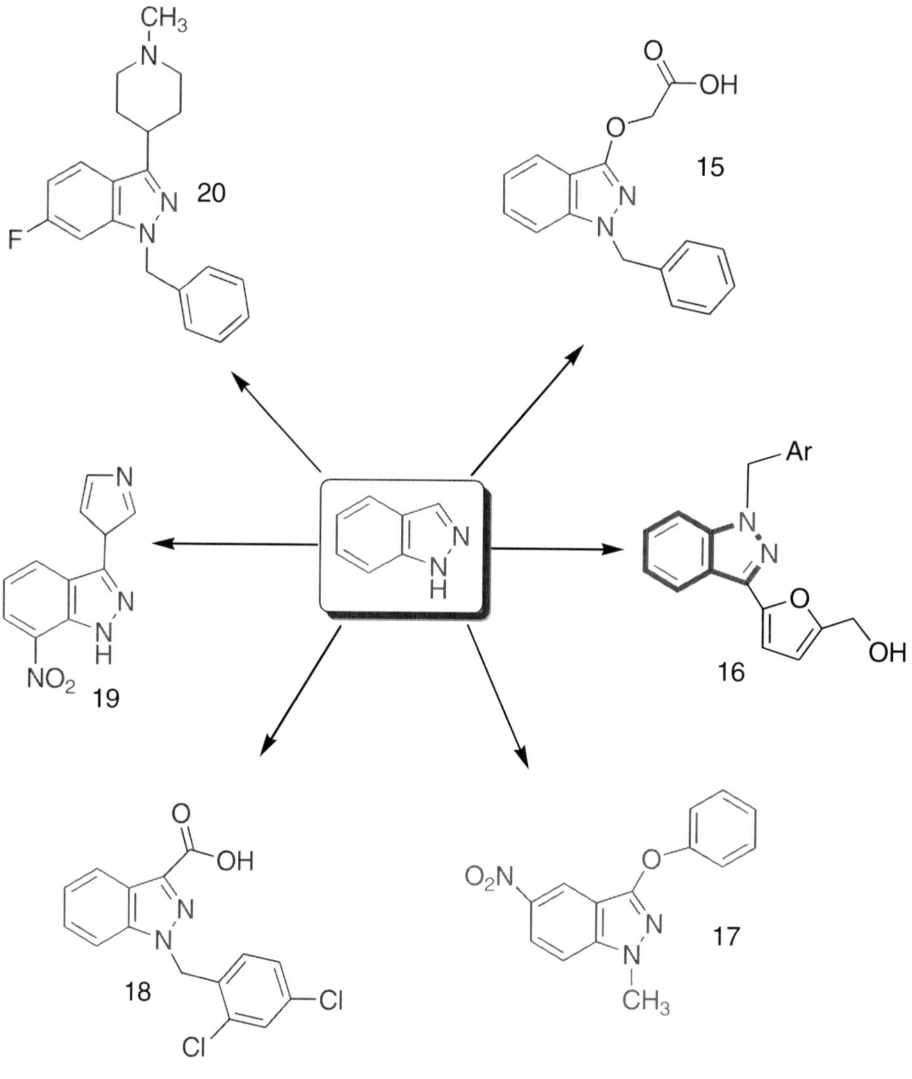

Drug molecules containing indazoles moiety.

Apart from this, indazoles constitute an important class of heterocycles that display interesting biological and pharmacological properties. Literature survey reveals that, various molecules containing indazoles moiety gives an anti-depressant, anti-inflammatory, analgesic, anti-pyretic, dopamine antagonistic, anti-tumor, anti-emetic and anti-HIV activities. The indazole ring system is also present in many other compounds such as herbicides, dyes and sweeteners. Granisetron (**21**), serotonin 5-HT_3 receptor antagonist used to treat and prevent nausea and vomiting induced by cancer chemotherapy [87]. The other compounds of interest are YC-1 (**22**) used [88] for the treatment of cardiovascular diseases or erectile dysfunction. Similarly, the melamine concentrating hormone receptor antagonist (**23**) is believed to have a number of biological functions, including in the regulation of appetite [89].

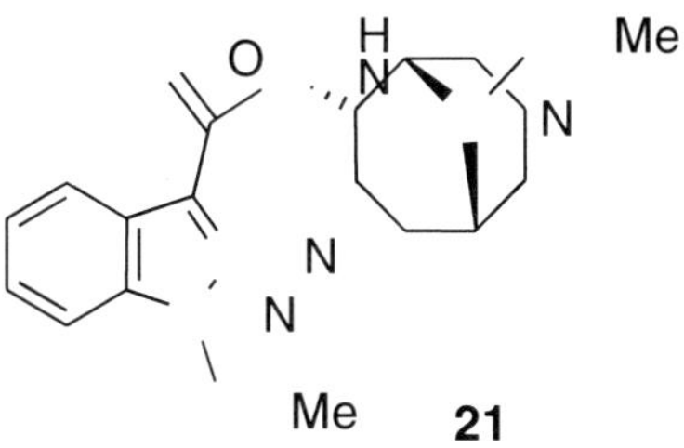

21

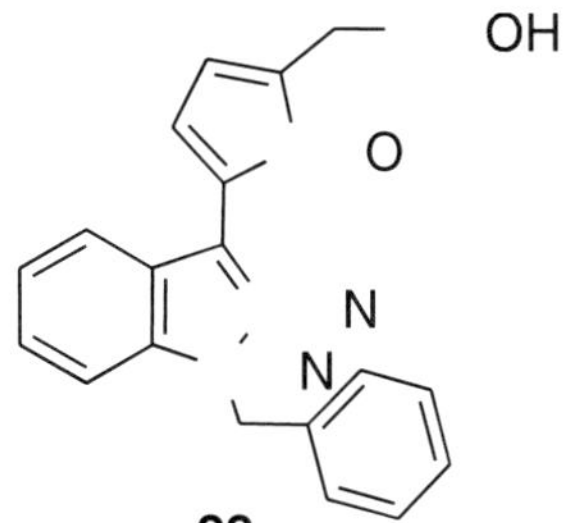

22

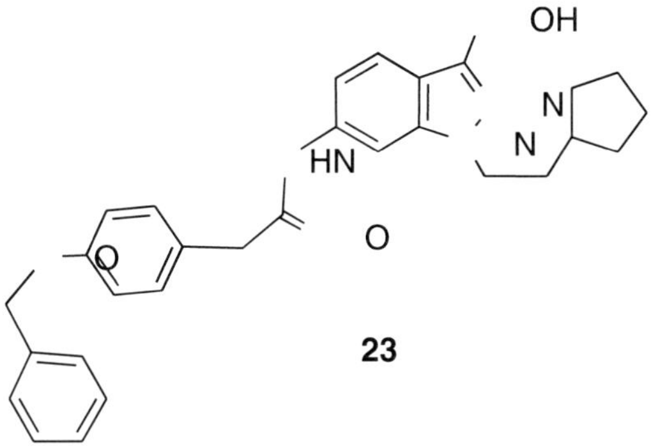

23

Compound (**24**) an indazolyl substituted urea derivatives is found to be extremely potent HIV protease inhibitors. Molecular modeling studies underline the biological importance of hydrogen bond donor properties of indazolyl nitrogen atoms [90, 91, 92].

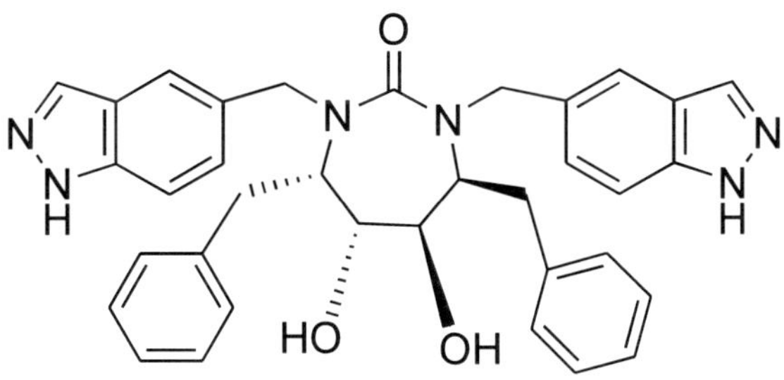

24

Indazole derivatives (**25**) have significant anti-inflammatory properties [93]. Metz, S. et al. [94] described fused pyrazole compounds (**26**) for the treatment of inflammation. Their physiologically acceptable salts also possess anti-inflammatory and analgesic properties. A (**27**) number of in vivo biological assays were used to identify novel anti-inflammatory agents among the class of 2,3-diaryl-4, 5, 6, 7-tetrahydro-2H-indazoles [95].

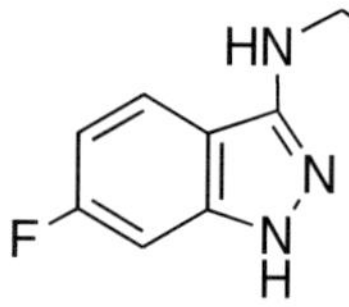

25

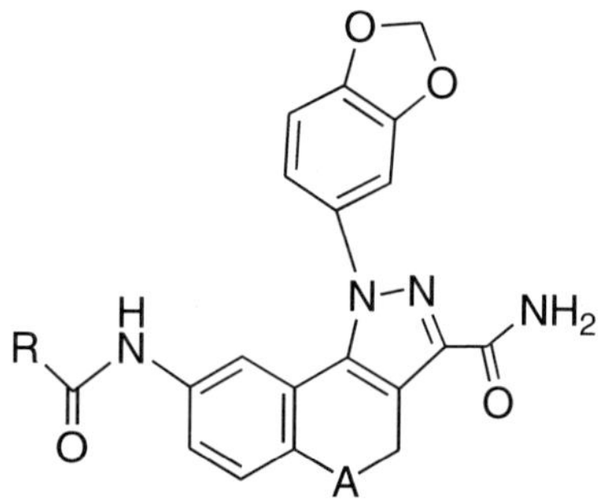

26

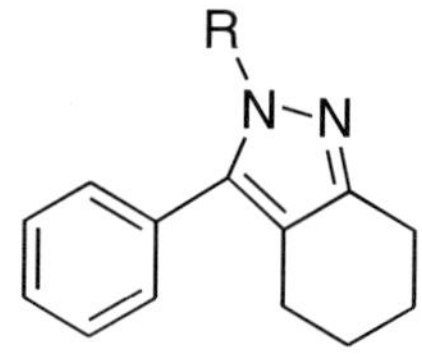

27

3- (4-Methylphenyl)-1*H*-indazole (**28**) of *N*-substituted derivatives are evaluated for anti-angiogenic activity. Synthesis of *N*- (substituted benzyl)-3- (4-methylphenyl) indazoles (**29**) has reffered as novel anti-angiogenic agents [96].

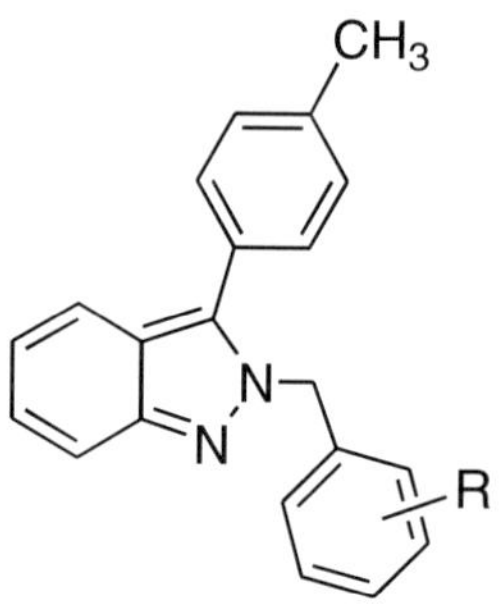

28

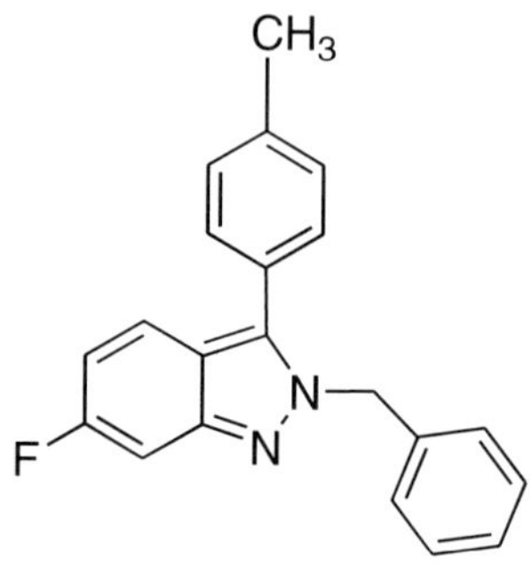

29

Indazole (**30, 31**) based series of glucocorticoid receptor antagonist reported by M. Bai et al. [97] Glucocorticoid the receptor antagonist. The SAR exploration of this scaffold yielded compounds with nanomolar affinity for the glucocorticoid receptor with indications of selectivity for the preferred transrepression mechanism.

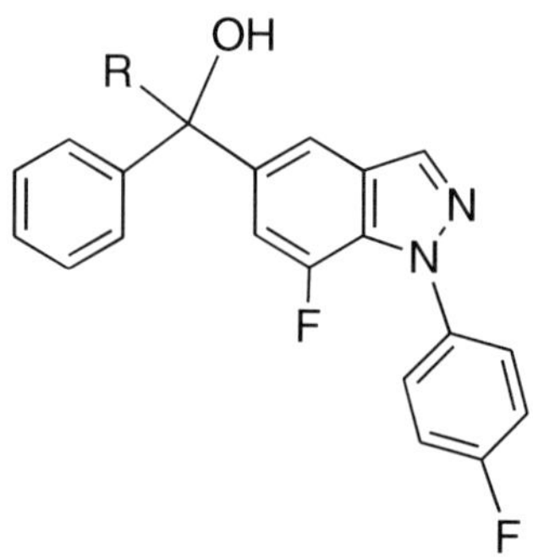

30

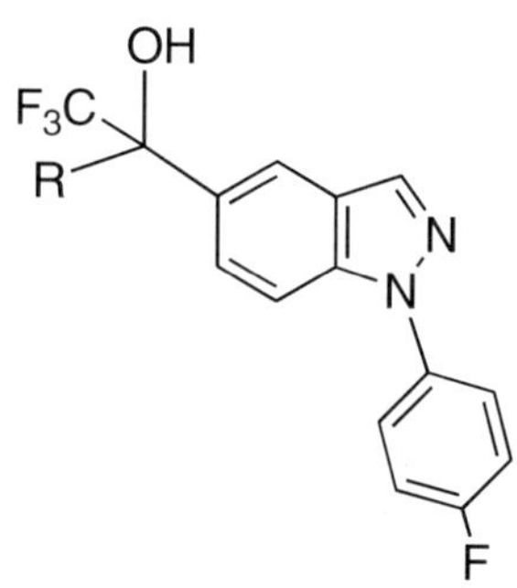

31

Brian S. Brown et al. [98] carried out the synthesis and SAR of a series of indazole (**32**) TRPV1 antagonists leading to the discovery is described. Biological studies demonstrated potent in vitro and in vivo activity for suitable physicochemical and pharmacokinetic.

Salvatore Plescia et al. [99] describes new indazole derivatives (3- (1-ethyl-1 H -indazol-3-yl)-2-phenylquinazolin-4 (3H)-one) (**33**) and tested to evaluate antimicrobial, anti-proliferative activity and showing scarce or moderately anti-proliferative activity and some inhibitory activity against COX-1 and COX-2.

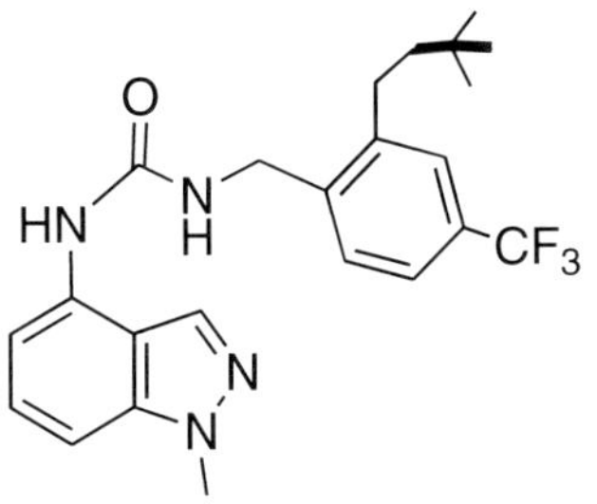

32

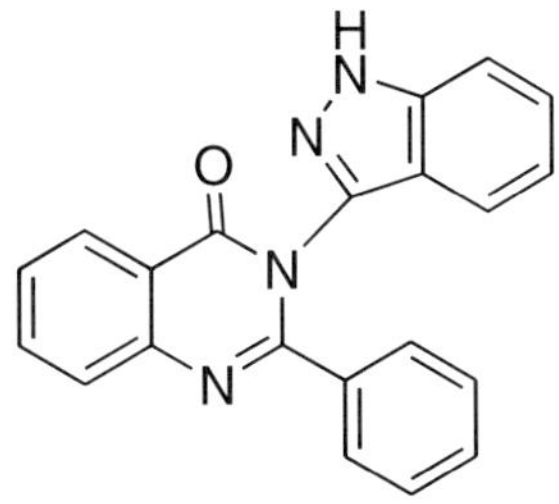

33

The synthesis and SAR of 5- (pyridinon-1-yl) indazoles (**34**) and 5- (furopyridinon-5-yl) indazoles (**35**) as potent MCH-1 antagonists are described

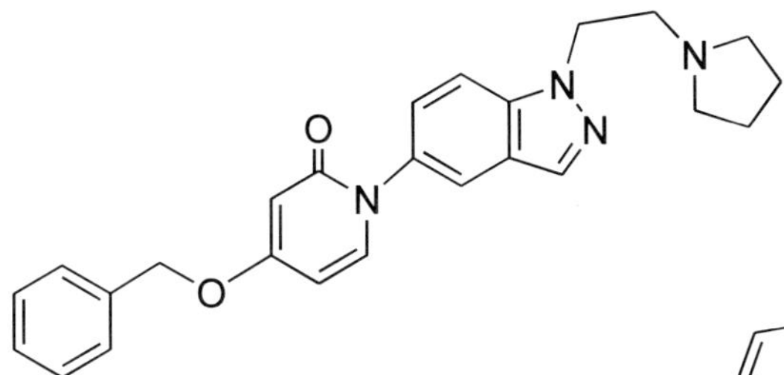

34

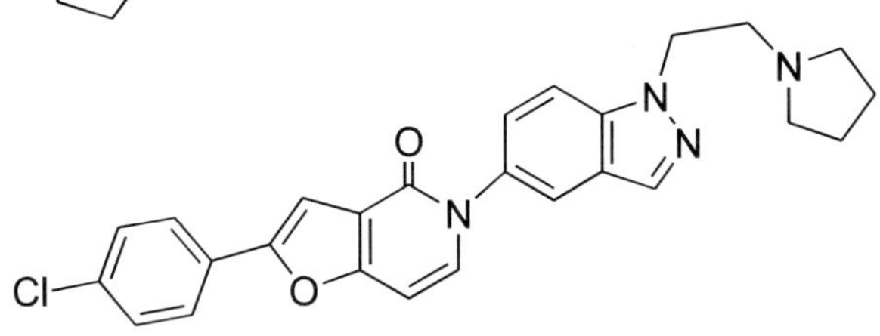

35

Maninder M. T. et al. [101]described the synthesis of new 2,3-disubstituted-3,3a,4,5,6,7-hexahydro-2H–indazole derivatives.3- (4-chlorophenyl)-2- (4-nitrophenyl sulfonyl) 3,3a,4,5,6,7-hexa hydro 2H-indazole (**36**) and 3- (4-fluorophenyl)-2- (4-nitrophenyl sulfonyl)-3,3a,4,5,6,7-hexahydro-2H–indazole (**37**) showed the prominent antimicrobial activity.

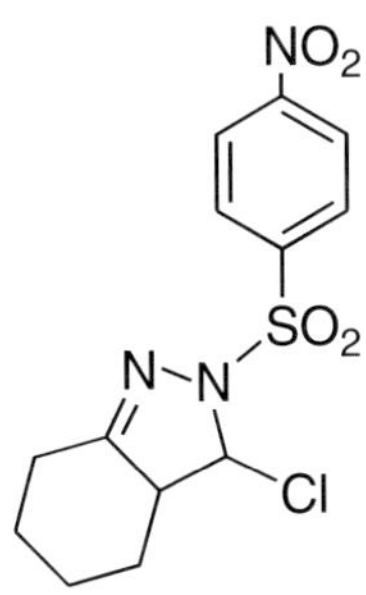

36

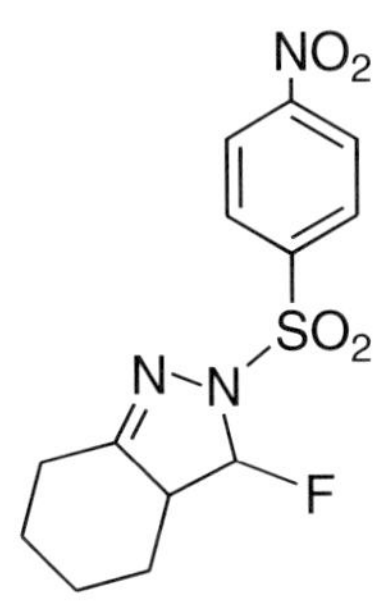

37

Synthesis, modeling and structure–activity relationship of indazole as inhibitors of Tpl2 kinase are described. Through SAR modifications at C3 and C5 positions of the indazole

(**38**), we discovered compound with good potency in LANCE assay (0.047 μM) and cell-based *p*-Erk assay (0.079 μM) [102]. A series of urea-based *N-1-* (2-aminoethyl)-indazoles (**39**) were synthesized and evaluated for melanin-concentrating hormone receptor 1 (MCHr1) antagonism. Several compounds that acted as MCHr1 antagonists were identified, and optimization afforded a compound with excellent binding affinity, good functional potency, and oral efficiency in a chronic model for weight loss in diet-induced obese mice [103].

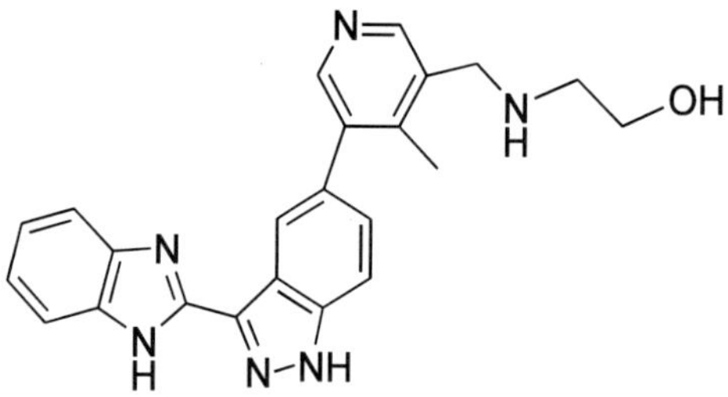

38

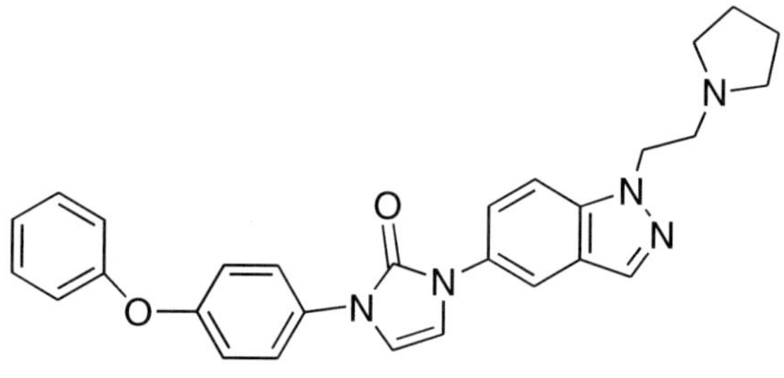

39

Design, synthesis and insight into the structure–activity relationship of 1, 3-disubstituted indazoles (**40**) were reported as novel HIF-1 inhibitors [104]. Anti-spermatogenic activity of 1-p-chlorobenzyl-1 H -indazole-3-carboxylic acid (**41**) has led to the synthesis of a large number of 1H -indazole-3-carboxylic acid derivatives [105]. The optimization of potency and selectivity for a 3- (indol-2-yl) indazole (**42**) class of Chek1 kinase inhibitors is described [106].

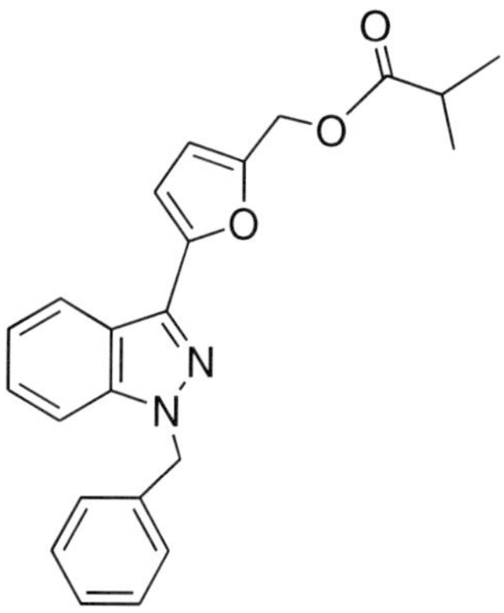

40

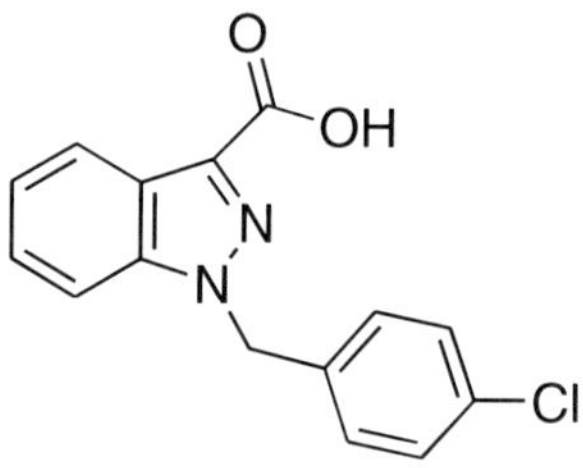

41

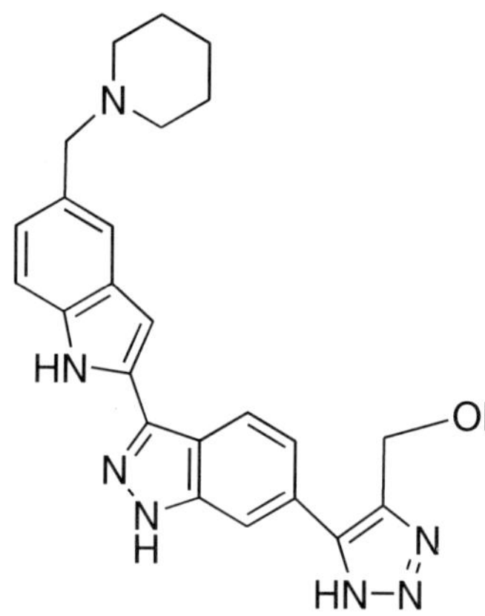

42

3-benzimidazol-2-yl-1H-indazole (**43**) scaffold was developed for receptor tyrosine kinase (RTK) inhibitor program. Exploring the SAR of this series, subsets of these compounds were potently inhibit the enzyme c-ABL [107]. Corbera A. and Esteve, S.A. et al. [108] had reported some tetrahydro indazole (**44**) and fused pyrazole derivatives having pharmacological activity towards the sigma receptor, and used in the treatment of psychosis or pain. [1, 2] oxazolo [5,4-*e*] indazoles (**45**) showed the growth-inhibitor activity against human tumor cell lies generally at low micromolar concentrations [109].

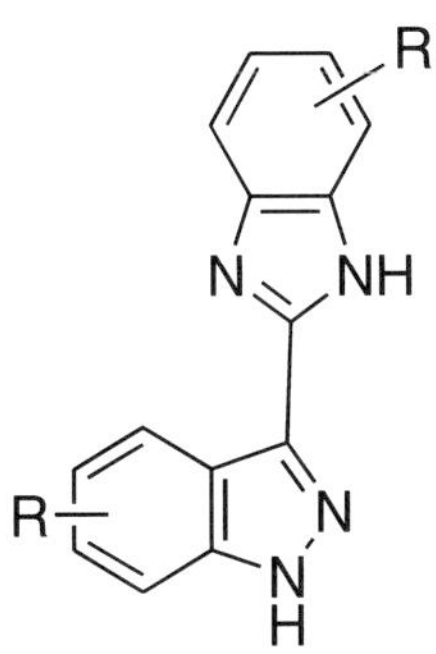

43

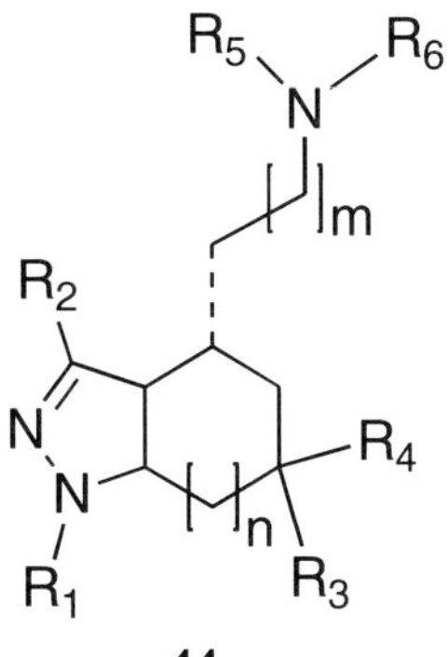

44

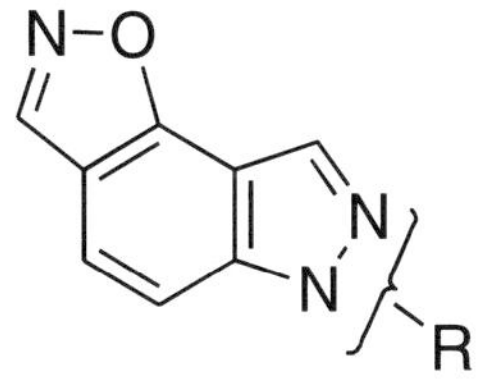

45

Pharmaceutical drug molecule containing indazoles nucleus which have different medicinal activity

Sr. No	Compounds	Activity	Published
01	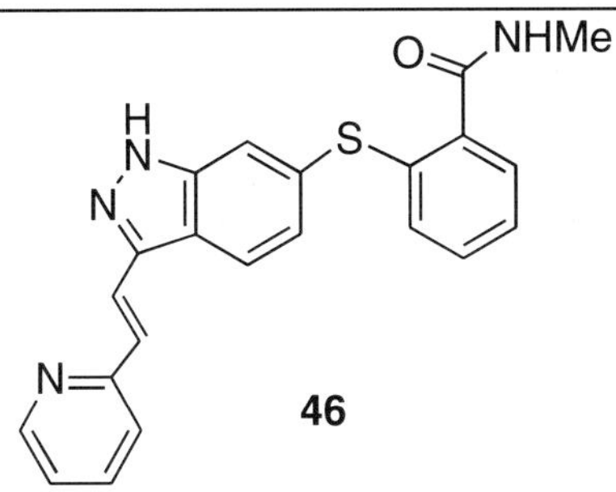 46	Axitinib	Ferrara N., [110]
02	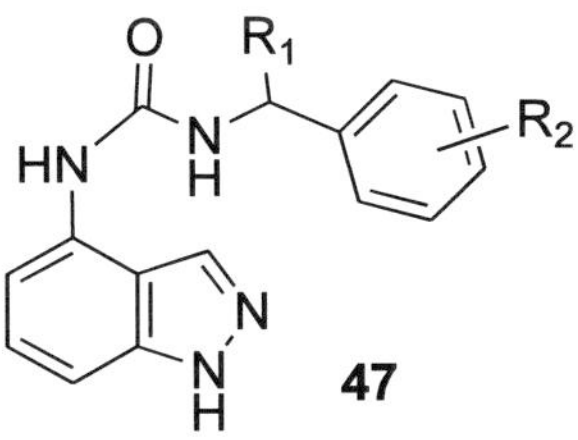 47	Chronic Pain	Kiril Lukin et al. [111]
03	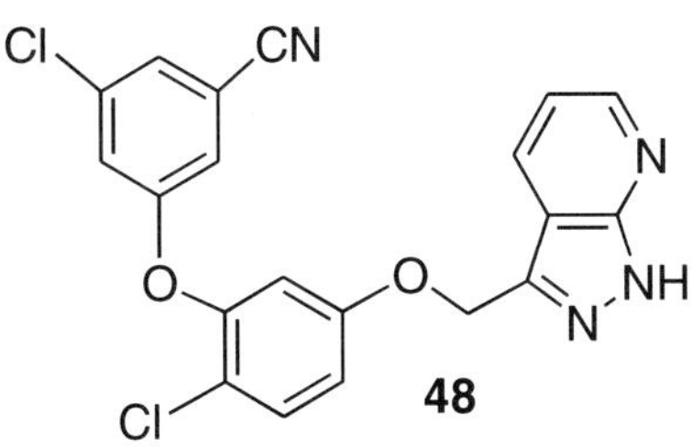 48	Antiretroviral	Robert Gomez et al. [112]
04	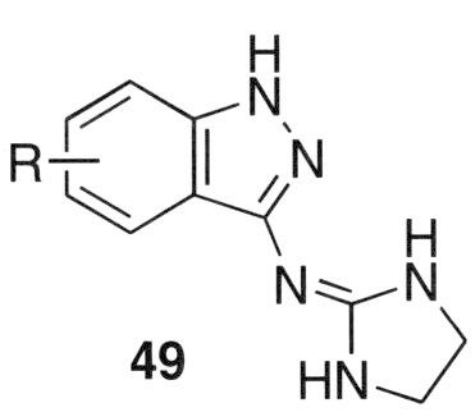 49	Hypertensive	Franciszek Saczewski [113]

ii. Drug Design

The drug is most commonly a small organic molecule that activates or inhibits the function of a bio-molecule such as a protein, which in turn results in a therapeutic benefit to the patient. In the most basic sense, drug design involves the designing of small molecules that are complementary in shape and charge to the bio-molecular target with which they interact and bind to it. Drug design frequently but not necessarily relies on computer modeling techniques. This type of modeling is often referred as computer-aided drug design. Finally, drug design that relies on the knowledge of the three-dimensional structure of bio-molecular target is known as structure-based drug design.

Pan-Kinase Inhibitor Anthrapyrazolone is rationally used for designing of indazoles based potent Cell-Active Mps1 Kinase Inhibitors (**50, 51**) [114].

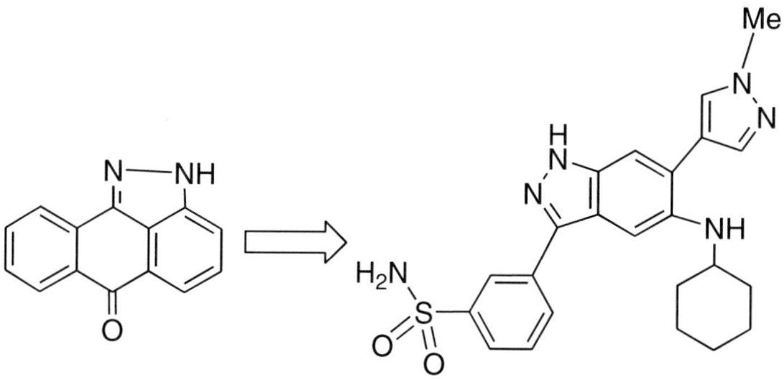

50 **51**

5-aryl substituent of the oxazole core is identified as a potent inhibitor. The analogues showed poor aqueous solubility. But *N*-acyloxymethyl analogue has good activity like aqueous solubility and susceptibility to enzymatic hydrolysis (**52, 53**) [115].

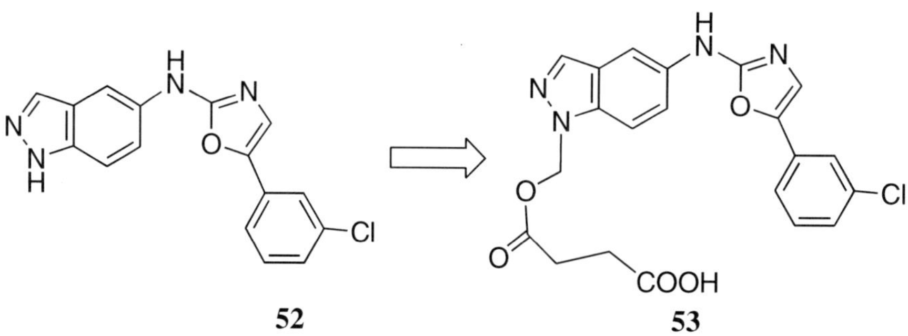

52 **53**

Members of this series of benzopyrazole analogues had demonstrated good oral bioavailability (**54, 55, 56**) [116]

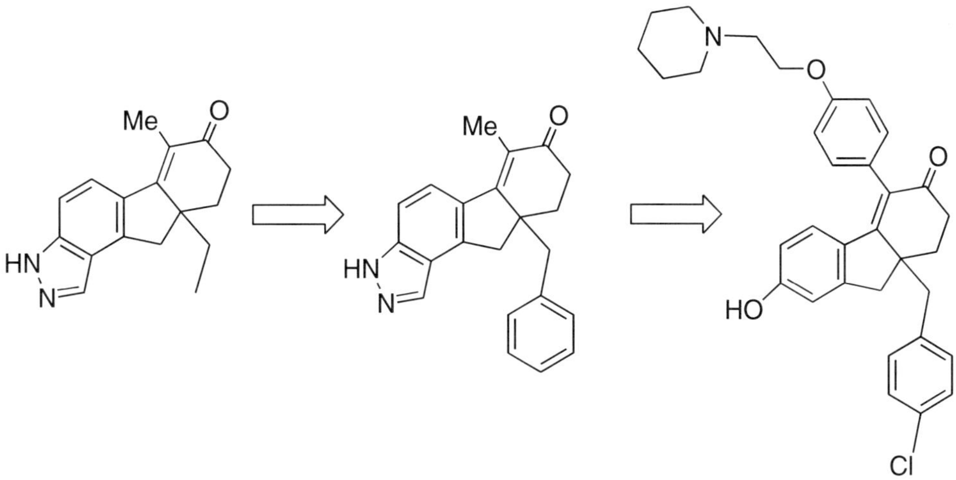

54 **55** **56**

Microwave synthesis, characterization and bio-efficiency evaluation of novel chalcone based 6-carbethoxy-2-cyclohexen-1-one and 2H-indazol-3-ol derivatives (**57, 58**) [117].

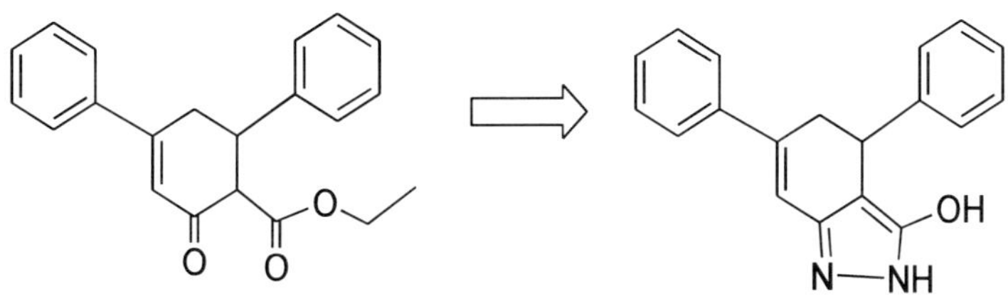

Antifungal and antibacterial agents **Antibacterial and antioxidant agents**

57 **58**

Convenient synthesis of novel N- (5-allyl-7,7-difluoro)-4,5,6,7-tetrahydro-2H-indazol-3-yl)-carboxymides has been reported (59, 60) [118].

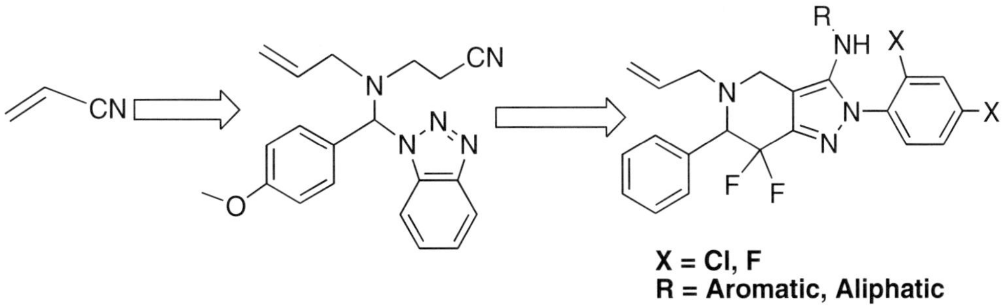

59 **60**

In this work, bioconjugates combine with acetylenic derivatives known to have anticancer activity and condensed anthraquinoid cycles, capable of intercalating DNA duplex (**61, 62, 63**) [119].

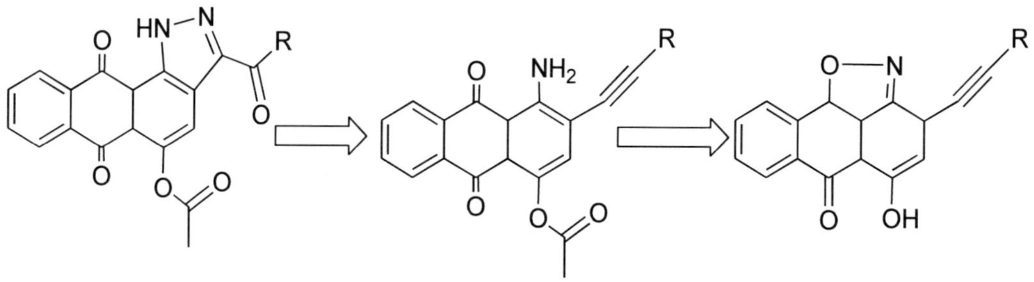

61 **62** **63**

Base-catalyzed rearrangement of 2*H*-indazoles 1-oxides, prepared by tandem carbon–carbon followed by nitrogen–nitrogen bond formations from easily accessible *N*-alkyl-2-nitro-*N*- (2-oxo-2-aryl-ethyl)-benzenesulfonamides using glycine, 2-nitrobenzene sulfonyl chlorides, and bromo ketones/acetates, yielded high purity quinazolines. After substitution of bromide with amine nuclophiles, acylation with carboxylic acid, and subsequent reduction of nitro group, the resulting precursors undergo cyclocondensation to quinazolinone derivatives (**64**) [120].

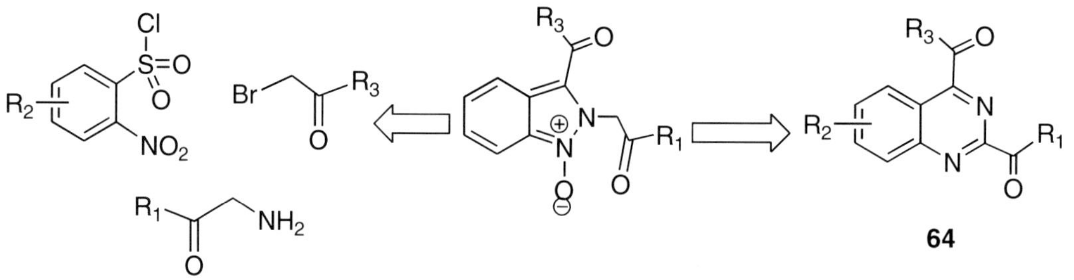

64

2,4-Dianilino pyrimidines with a phenolic group at 4-position are potent inhibitors of Lck tyrosine kinase enzyme activity. But they have poor pharmacokinetic properties. Analogues where the 4-position was replaced by 4-amino (5-methyl-1H-indazole) had comparable enzyme potency and improved pharmacokinetic properties (**65, 66**) [121].

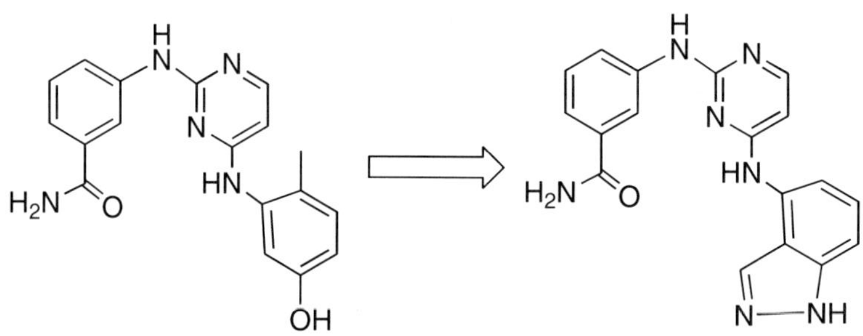

65 **66**

Studies on substituted benzo [*h*]quinazolines, benzo [*g*]indazoles, pyrazoles, 2,6-diarylpyridines as anti-tubercular agents has been reported (**67, 68, 69, 70**) [122]

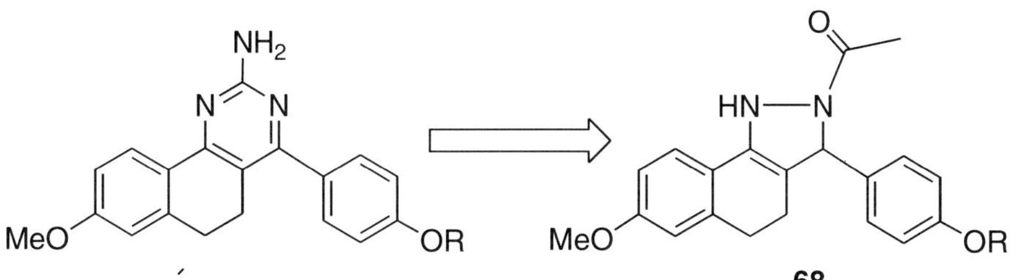

67 68

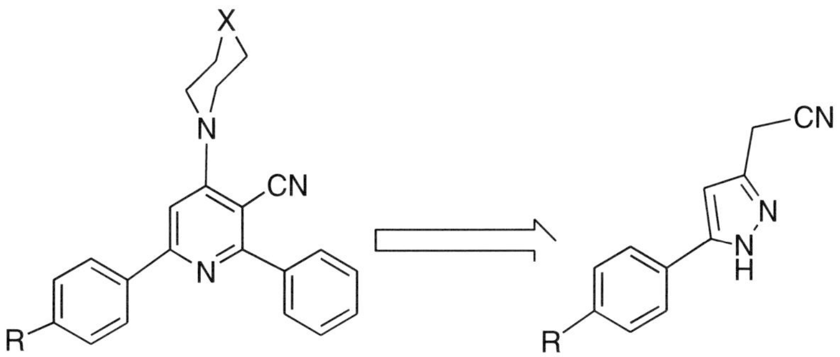

69 70

iii. Miscellaneous

Certain indazole derivatives provides useful new route to indazoles which should find application in the construction of biological properties and pharmaceutical potential.

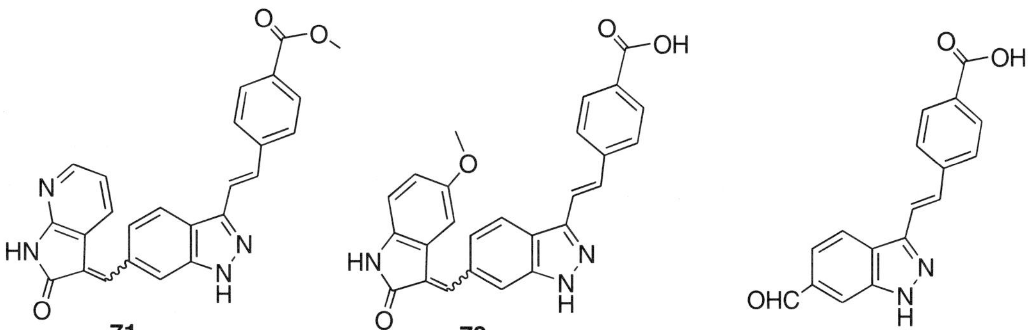

71 72 73

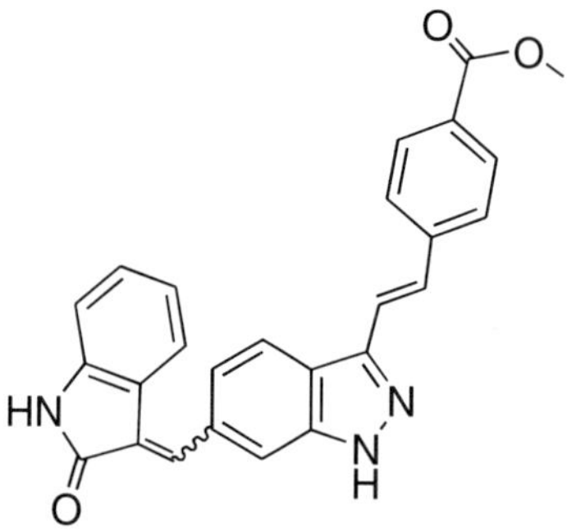

74

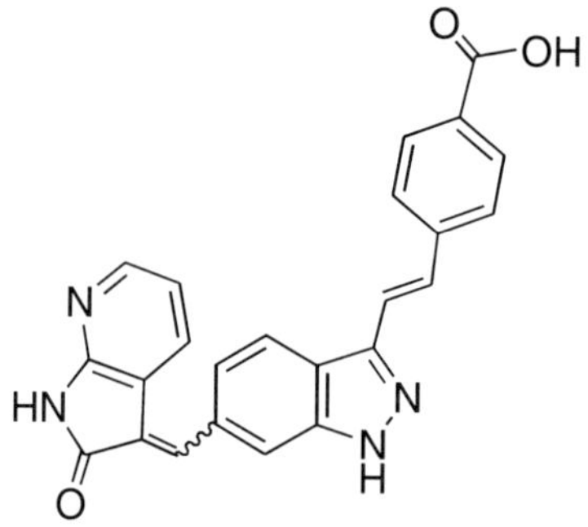

75

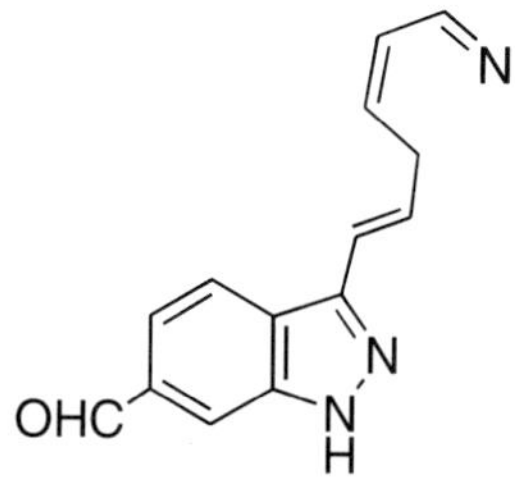

76

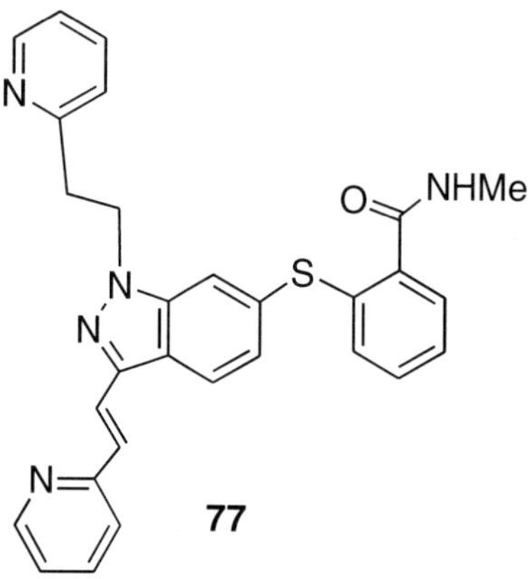

77

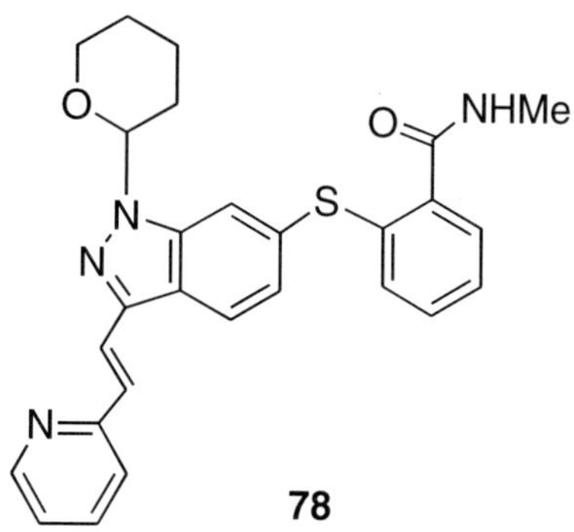

78

CONCLUSION

- The biological and pharmaceutical properties of substituted indazoles find a wide range of application in novel drug designing.
- Indazole derivatives have great relevance in many field of heterocyclic chemistry.
- Future developments in the field of drug designing will utilize the applications of indazole in target oriented drug synthesis.
- This chapter will trigger research in this area for the synthesis of new indazole derivatives.

REFERENCES

[1] Schamidt, A., Beutler, A., Snovydovych, B. (2008), Recent Advances in the Chemistry of Indazoles, *Eur. J. Org. Chem.* 4073-4095.

[2] Rahman, A., Malik, S., Cun-Heng, H., Clardy, J. (1985), Isolation and structure determination of nigellicine, a novel alkaloid from the seeds, *Tetrahedron Lett.* 26, 2759-2762.

[3] Liu, Y. M., Yang, J. S., Liu, Q. H. (2004), A new alkaloid and its artificial derivatives with an indazole ring from Nigella glandulifera, *Chem. Pharm. Bull.*52, 454-455.

[4] Ali, B. H., Gerald, B. (2003), Pharmacological and toxicological properties of Nigella Sativa, *Phytotherapy Research*, 17, 299-305.

[5] Ali, Z., Khan I. A. (2008), Pimarane type diterpene analogus from Nigella Sativa, *Planta Med.,* 74, 152.

[6] Eddouks, M., Maghrani, M., Lemhadri, A., Quahidi, M. L., Jouad H. (2002), Ethanopharmacological survey of medicinal plants used for the treatment of diabetes mellitus, hypertension and diseases in south-east region of Morocco, *Journal of Ethnopharmacology,*82, 97-103.

[7]] Said, O., Khali, K., Fulder, S., Azaizeh, H. (2002), Ethnopharmacological survey of medicinal herbs in Israel, the Golan Heights and the West Bank region, *Journal of Ethnopharmacology,* 83, 251-265.

[8] Schmidt, A., Merkel, L., Eisfeld, W., (2005), Nucleophilic carbine and pseudo-cross-conjugated mesomeric Betaines of indazole starting from analogus of the Alkaloid-Betaine Nigellicine, *Eur. J. Org. Chem.*10, 2124-2130.

[9] Attaur, A., Malik, S., Hasan, S. S., Choudhary, M. I., Ni, C. Z., Clardy, J. (1995), Nigellidine-A new indazole alkaloid from the seeds of Nigelle Sativa, *Tetrahedron Lett.* 36, 1993-1996.

[10] Brase, S., Gil, C., Knepper, K., (2002), The recent impact of solid-phase synthesis on medicinally relevant benzoannelated nitrogen heterocycles, *Bio-org. Med. Chem.* 10, 2415-2437.

[11] Guglielmotti, A., De Joannon, A. C., Cazzolla, N., Marchetti, M., Soldo, L., Cavallo G., Pinza, M. (1995), Radical scavenger activity of bendazac an anticataract non-steroidal anti-inflammatory agent. *Pharmacol. Res.*, 32, 369-373.

[12] Hartmann, M., Sommer, M. E., Keppler, B. K., Kratz F., Einhaüser, Th. J. (1995), New tumor-inhibiting ruthenium compounds: investigations into their mode of action in human blood plasma and binding towards DNA, *J. Inorg. Biochem.*, 59, 214.

[13] Clarke, M. J. (2003), Ruthenium metallopharmaceuticals, *Coord. Chem. Rev.,* 236, 209-233.

[14] Sun, J. H., Teleha, C. A., Yan, J. S., Rodgers, J. D., Nugiel, D. A. (1997), Efficient synthesis of 5- (Bromomethyl)-and 5- (Aminomethyl)-1-THP-Indazole, *J. Org. Chem.,* 62, 5627-5629.

[15] Hsu, E. S., (2010), A review of granisetron 5-hydroxytryptamine 3-receptor antagonists and other antiemetics, *Am. J. Ther.,* 17, 476-486.

[16] Di Cosimo, S., Ferretti, G., Papaldo, P., Carlini, P., Fabi, A., Cognetti, F. (2003), Lonidamine: Efficacy and safty in clinical trials for the treatment of solid tumors, *Drugs Today,* 39, 157-174.

[17] Krapcho, A. P., Haydar, S. N., (2001), Synthesis of regioisomeric 2,5-bis-substituted-aza-benzothiopyranoindazoles, *J. Heterocycl. Chem.,* 38, 1153-1166.

[18] Dees, E. C., Whitfield, L. R., Grove, W. R., Rummel, S., Grochow, L. B., Donehower, R. C., (2000), A phase I and pharmacologic evalution of the DNA intercalator CI-958 in patients with advanced solid tumors, *Clin. Cancer Res.,* 6, 3885-3894.

[19] Jin, T., Yamamoto, Y., (2007), An efficient, facile and general synthesis of 1H-indazoles by 1,3-Dipolar cycloaddition of Arynes with Diazomethane Derivatives, *Angew. Chem., Int. Ed.,* 46, 3323-3325.

[20] Liu, Z., Shi, F., Martinez, P. D. G., Raminelli, C., Larock, R. C., (2008), Synthesis of indazoles by the [3+2] cycloaddition of Diazo compounds with Arynes and Subsequent Acyl Migration, *J. Org. Chem.*,73, 219-226.

[21] Li, P., Zhao, J., Wu, C., Larock, R. C., Shi, F. (2011), Synthesis of 3-substituted indazoles from Arynes and N-Tosylhydrazones, *Org. Lett.,* 13,3340-3343

[22] McAusland, D., Seo, S., Pintori, D. G., Finlayson, J. (2011), Greaney, N-Tosylhydrazones have been alternatively employed as starting materials for the benzyne Fischer-indole reaction: *Org. Lett.,* 2011, 13, 3667.

[23] Spiretti, C. , Keeling, S., Moses, J. E., (2010), New synthesis of 1-Substituted-1H-indazoles via 1,3-Dipolar cycloaddition of in situ generated Nitrile imines and Benzyne, *Org. Lett.,* 12, 3368-3371.

[24] Nataliya, A. M., Anton V. D., Richard C. L., (2012), One-pot synthesis of 1-alkyl-1H-indazols from 1,1-dialkylhydrazones via aryne annulations, *Organic & Biomolecular chemistry,* 10, 2409-2412.

[25] Aaron D. Mills, Musa Z. Nazer, Makhluf J. haddadin, and Mark J Kurth. (2006), (N, N-Bond-Forming Heterocylization: Synthesis of 3-Alkoxy-2H-indazoles, *J. Org. Chem.*712687-2689.

[26] Tianshui Zhang and Weiliang Bao., (2013), Synthesis of 1H-indazoles and 1H-pyrazoles via $FeBr_3/O_2$ Mediated intramolecular C-H Amination, *J. Org. Chem.*78,1317-1322.

[27] Pan Li, Chunrui Wu, Jingjing Zhao, Donald C. Rogness, and Feng Shi. (2012), Synthesis of substituted 1H-indazoles from arynes and hydrazones, *J. Org. Chem.* 77, 3127-3133.

[28] Christian Spiteri, Steve Keeling, and John E. Moses., (2010), New synthesis of 1-substituted-1H-indazoles via 1,3-Dipolar cycloaddition of in situ generated Nitrile imines and Benzyne, *Org. Lett.*,12, 3368-3371.

[29] Manian Rajesh Kumar, Ahbyeol Park, and Sunwoo Lee., (2011), Consecutive Condensation, C–N and N–N Bond Formations: A Copper- Catalyzed One-Pot Three-Component Synthesis of 2*H*-Indazole, *Org. Lett.*,13,3542-3545.

[30] Elguero J., katritzky A. R., Denisko O. V. (2000), The tautomerism of heterocycles: Five-membered rings with two or more heteroatoms, *Adv. Heterocyclic Chem.,* 76, 157-323.

[31] Fatima C. T., Helene R., Ines F. A., Joao M., Curto M., Teresa D. M., Isabel Bento, (2006), Synthesis and Structural Characterization of 1- and 2-Substituted Indazoles: Ester and Carboxylic Acid Derivatives, *Molecules,* 11, 867-889.

[32] Andreas Unsinn, Paul Knochel, (2012), Regioselective zincation of indazoles using TMP_2Zn and *Negishi* cross-coupling with aryl and heteroaryl iodides*Chem. Commun.*, 48, 2680-2682.

[33] Liu, Z., Zhang, X., Larock, R. C., (2005), Synthesis of Fused Polycyclic Aromatics by Palladium-Catalyzed Annulation of Arynes Using 2-Halobiaryls, *J. Am. Chem.* Soc.,127, 15716-15717

[34] Strupczewski, J. T., Helsey, G. C., Grover, C., Glamkowski, E. J.,Bordeau, K. J., (1998),3- (heteroaryl)-1-› (2,3-dihydro-1h-isoindol-2-yl)alkyl!pyrrolidines and 3-(heteroaryl)-1-› (2,3-dihydro-1h-indol-1-yl)alkyl!pyrrolidines and related compounds and their therapeutic untility,*U.S. Patent* 5,776,963.

[35] Vincent, L.,George E. L.,Jiang L.,Sandra H. H.,Thomas B. L., (2001), Facile Preparation of 3- (1-Piperazinyl)-1H-indazoles, Org. Proc. Res. Dev., *5* (2), 179-183.
[36] Wrzeciono, U., Majewska, K., Dudzinska-Usarewicz J., Bernas, M., (1986), Synthesis and SAR of indazole-pyridine based protein kinase B/Akt., *Pharmazie 41*, 472-474.
[37] Strupczewski, J. T., Helsey, G. C., Grover, C.; Chiang, Y., Bordeau, K. J., Glamkowski, E. J., (1994),Heteroarylpiperidines, pyrrolidines and piperazines and their use as antipsychotics and analetics, *U.S. Patent* 5,364,866.
[38] Digambar, D. Gaikwad; Rajendra, P. Pawar, Silica sulphuric acid: An efficient catalyst for the synthesis of substituted indazoles, (2010), *J. Iran. Chem. Res.* 3, 191-194.
[39] Digambar, D. Gaikwad, Syed, Abed, Rajendra, P. Pawar, Molecular iodine as an efficient catalyst for the synthesis of indazole, (2009), *Int. J. ChemTech Res.,* 01, 442-445.
[40] URL: http://en. Wikipedia.org/wki/Reverse_trancriptase.
[41] Tarby, C. M., (2004), Recent Advances in the Development of Next Generation Non-nucleoside Reverse Transcriptase Inhibitors, *Curr. Top. Med. Chem.* 4 (13), 1045-1057.
[42] Tucker, T. J.; Saggar, S.; Sisko, J. T.; Tynebor, R. M.; Williams, T. M.; Felock, P. J.; Flynn, J. A.; Lai, M. T.; Liang, Y.; Poehnelt, R.; Prasad, S.; Sanchez, R.; Torrent, M.; Vacca, J. P.; Wan, B.L.; Yan, Y. (2008),The design and synthesis of diaryl ether second generation HIV-1 non-nucleoside reverse transcriptase inhibitors (NNRTIs) with enhanced potency versus key clinical mutations. *Bioorg. Med. Chem. Lett.* 18, 2959–2966.
[43] L.; Cai, D.; Fleitz, F. J.; Gibb, A. D.; Kassim, A.; Linn, K.; Mancheno,D.; Marcune, B.; Pye, P. J.; Scott, J. P.; Tellers, D. M.; Xiang, B.; Yasuda, N.; Yin, J.; Davies, I. W. (2009),Development of practical synthesis of potent non-nucloside reserve transcripts inhibiters, *Tetrahedron* 65, 5013–5023.
[44] Gomez, R.; Jolly, S. J.; Williams, T.; Vacca, J. P.; Torrent, M.;McGaughey, G.; Lai, M. T.; Felock, P.; Munshi, V.; DiStefano, D.;Flynn, J.; Miller, M.; Yan, Y.; Reid, J.; Sanchez, R.; Liang, Y.; Paton, B.; Wan, B. L.; Anthony, N. (2011), Design and synthesis of conformationally constrained inhibitors of non-nucleoside reverse transcriptase,*J. Med. Chem.* 54, 7920–7933.
[45] Coffin, J. M. (1986),Genetic variation in AIDS viruses. *Cell* ,*46* (1),1–4.
[46] Perrino, F. W.; Preston, B. D.; Sandell, L. L.; Loeb, L. A. (1989), Extension of mismatched 3′ termini of DNA is a major determinant of the infidelity of human immunodeficiency virus type 1 reverse transcriptase. *Proc. Natl. Acad. Sci. U.S.A. 86* (21), 8343–8347.
[47] Menendez-Arias, L. (2002) Molecular basis of fidelity of DNA synthesisand nucleotide specificity of retroviral reverse transcriptases. *Prog.Nucleic Acid Res. Mol. Biol. 71*, 91–147.
[48] Gulick, R. M.; Mellors, J. W.; Havlir, D; Eron, J. J.; Meibohm, A.;Condra, J. H.; Valentine, F. T.; McMahon, D.; Gonzalez, C.; Jonas, L.;Emini, E. A.; Chodakewitz, J. A.; Isaacs, R.; Richman, D. D. (2000), HIV viremia with indinavir, zidovudine, and lamivudine. *Ann. Intern. Med. 133* (1), 35–39.
[49] Cane, P. A. (2009), New developments in HIV drug resistance. *J. Antimicrob.Chemother.*, 64, 37-40.
[50] Little, S. J.; Holte, S.; Routy, J.-P.; Daar, E. S.; Markowitz, M.; Collier, A. C.; Koup, R. A.; Mellors, J. W.; Connick, E.; Conway, B.; Kilby, M.; Wang, L.; Whitcomb, J.

M.; Hellmann, N. S.; Richman, D. D. (2002), Antiretroviral-drug resistance among patients recently infected with HIV. *N. Engl. J. Med. 347* (6), 385–394.

[51] Richman, D. D.; Morton, S. C.; Wrin, T.; Hellmann, N.; Berry, S.; Shapiro, M. F.; Bozzette, S. A. (2004), The prevalence of antiretroviral drug resistance in the United States. *AIDS (London) 18* (10), 1393–1401.

[52] De Bethune, M.-P. (2010), Non-nucleoside reverse transcriptase inhibitors (NNRTIs), their discovery, development, and use in the treatment of HIV-1 infection: a review of the last 20 years (1989–2009). *Antiviral Res. 85* (1), 75–90.

[53] Sluis-Cremer, N.; Temiz, N. A.; Bahar, I. (2004), Conformational changes in HIV-1 reverse transcriptase induced by nonnucleoside reverse transcriptase inhibitor binding. *Curr. HIV Res. 2* (4), 323–232.

[54] Fokunang, C. N.; Hitchcock, J.; Spence, F.; Tembe-Fokunang, E. A.; Burkhardt, J.; Levy, L.; George, C. (2006), An overview of mitochondrialtoxicity of nucleoside reverse transcriptase inhibitors associated with HIV therapy. *Int. J. Pharmacol. 2* (1), 152–162.

[55] Cheung, P. K; Wynhoven, B.; Harrigan, P. R. (2004), Which HIV-1drug resistance mutations are common in clinical practice? *AIDS Rev.6* (2), 107–116.

[56] Tambuyzer, L.; Azijn, H.; Rimsky, L. T.; Vingerhoets, J.;Lecocq, P.; Kraus, G.; Picchio, G.; de Bethune, M.-P. (2009), Compilation and prevalence of mutations associated with resistance to non-nucleoside reverse transcriptase inhibitors. *Antiviral Ther. 14* (1), 103–109.

[57] Das, K., Clark, A. D. Jr., Lewi, P. J., Heeres, J., De Jonge, M. R., Koymans, L. M. H., Vinkers, H. M., Daeyaert, F., Ludovici, D. W., Kukla, M. J., De Corte, B., Kavash, R. W., Ho, C. Y., Ye, H.; Lichtenstein, M. A., Andries, K., Pauwels, R., De Bethune, M. P., Boyer, P. L., Clark, P., Hughes, S. H., Janssen, P. A. J., Arnold, E., (2004), Roles of conformational and positional adaptability in structure-based design of TMC125-R165335 (etravirine) and related non-nucleoside reverse transcriptase inhibitors that are highly potent and effective against wild-type and drug-resistant HIV-1 variants. *J. Med. Chem. 47* (10), 2550–2560.

[58] Arasteh, K., Riege, R. A., Yeni, P., Pozniak, A., Boogaerts, G., Van Heeswijk, R., de Bethune, M.P., Peeters, M., Woodfall, B., (2009), Shortterm randomized proof-of-principle trial of TMC278 in patients with HIV type-1 who have previously failed antiretroviral therapy. *Antiviral Ther., 14* (5), 713-722.

[59] Tucker, T. J.; Sisko, J. T.; Tynebor, R. M.; Williams, T. M.;Felock, P. J.; Flynn, J. A.; Lai, M.-T.; Liang, Y.; McGaughey, G.; Liu,M.; Miller, M.; Moyer, G.; Munshi, V.; Perlow-Poehnelt, R.; Prasad, S.; Reid, J. C.; Sanchez, R.; Torrent, M.; Vacca, J. P.; Wan, B.-L.; Yan, Y., (2008),Discovery of 3-{5- [(6-amino-1*H*-pyrazolo [3,4-*b*]pyridine-3-yl)-methoxy]-2-chlorophenoxy}-5-chlorobenzonitrile (MK-4965): a potent,orally bioavailable HIV-1 non-nucleoside reverse transcriptaseinhibitor with improved potency against key mutant viruses. *J. Med. Chem. 51* (20), 6503-6511.

[60] Datta, S. R.; Dudek, H.; Tao, X.; Masters, S.; Fu, H.; Gotoh, Y.; Greenberg, M. E. (1997), Akt phosphorylation of BAD couples survival signals to the cell-intrinsic death machinery. *Cell,* 91, 231–241.

[61] Frame, S.; Cohen, P. (2001), GSK3 takes centre stage more than 20 years after its discovery. *Biochem. J.* 359, 1–16.

[62] Kandel, E. S.; Skeen, J.; Majewski, N.; Di Cristofano, A.; Pandolfi, P. P.; Feliciano, C. S.; Gartel, A.; Hay, N. (2002), Activation of Akt/ protein kinase B overcomes a G (2)/m cell cycle checkpoint induced by DNA damage. *Mol. Cell. Biol.* 22, 7831–7841.

[63] Shaw, R. J.; Cantley, L. C. Ras, (2006), PI (3)K and mTOR signaling controls tumour cell growth. *Nature,* 441, 424–430.

[64] Carnero, A.; Blanco-Aparicio, C.; Renner, O.; Link, W.; Leal, J. F. M. (2008),The PTEN/PI3K/AKT signalling pathway in cancer, therapeutic implications. *Curr. Cancer Drug Targets* 8, 187–198.

[65] Chiang, G. G.; Abraham, R. T., (2007) Targeting the mTOR signaling network in cancer. *Trends Mol. Med.* 13, 433–442.

[66] Haluska, F. G.; Tsao, H.; Wu, H.; Haluska, F. S.; Lazar, A.; Goel, V., (2006), Genetic alterations in signaling pathways in melanoma. *Clin. Cancer Res.* 12, 2301–2307.

[67] Cully, M.; You, H.; Levine, A. J.; Mak, T. W., (2006), Beyond PTEN mutations: the PI3K pathway as an integrator of multiple inputs during tumorigenesis. *Nat. Rev. Cancer,* 6, 184–192.

[68] Samuels, Y.; Wang, Z.; Bardelli, A.; Silliman, N.; Ptak, J.; Szabo,S.; Yan, H.; Gazdar, A.; Powell, S. M.; Riggins, G. J.; Willson, J. K. V.; Markowitz, S.; Kinzler, K. W.; Vogelstein, B.; Velculescu, V. E. (2004), High frequency of mutations of the PIK3CA gene in human cancers. *Science,* 304, 554.

[69] Philip Jones, Sergio Altamura, Julia Boueres, Federica Ferrigno, Massimiliano Fonsi, Claudia Giomini, Stefania Lamartina, Edith Monteagudo, Jesus M. Ontoria, Maria Vittoria Orsale, Maria Cecilia Palumbi, Silvia Pesci, Giuseppe Roscilli, Rita Scarpelli, Carsten Schultz-Fademrecht, Carlo Toniatti, Michael Rowley, (2009), Discovery of 2-{4- [(3S)-Piperidin-3-yl]phenyl}-2H-indazole-7-carboxamide (MK-4827): A Novel Oral Poly (ADP-ribose)polymerase (PARP) Inhibitor Efficacious in BRCA-1 and -2 Mutant Tumors, *Medicinal Chemistry,* 52, 7170-7185.

[70] Maira, S.-M.; Stauffer,F.; Brueggen, J.; Furet, P.; Schnell,C.;Fritsch,C.; Brachmann, S.; Chene, P.; De Pover, A.; Schoemaker, K.; Fabbro, D.; Gabriel, D.; Simonen, M.; Murphy, L.; Finan, P.; Sellers, W.; Garcia-Echeverria, C., (2008), Identification and characterization of NVPBEZ235,a new orally available dual phosphatidylinositol 3-kinase/mammalian target of rapamycin inhibitor with potent in vivo antitumoractivity. *Mol. Cancer Ther.* 7, 1851–1863.

[71] Lindsley, C. W.; Barnett, S. F.; Layton, M. E.; Bilodeau, M. T. (2008), The PI3K/Akt pathway: recent progress in the development of ATP-competitive and allosteric Akt kinase inhibitors. *Curr. Cancer Drug Targets,* 8, 7–18.

[72] Engelman, J. A.; Chen, L.; Tan, X.; Crosby, K.; Guimaraes, A. R.;Upadhyay, R.; Maira, M.; McNamara, K.; Perera, S. A.; Song, Y.; Chirieac, L. R.; Kaur, R.; Lightbown, A.; Simendinger, J.; Li, T.; Padera, R. F.; Garcia-Echeverria, C.; Weissleder, R.; Mahmood, U.; Cantley, L. C.; Wong, K. K. (2008), Effective use of PI3K and MEK inhibitors to treat mutant Kras G12D and PIK3CA H1047Rmurine lung cancers. *Nat. Med.* 14, 1351–1356.

[73] Morphy, R., (2010), Selectively nonselective kinase inhibition: striking theright balance. *J. Med. Chem.* 53, 1413–1437.

[74] Serra, V.; Markman, B.; Scaltriti, M.; Eichhorn, P. J. A.; Valero, V.; Guzman, M.; Botero, M. L.; Llonch, E.; Atzori, F.; Di Cosimo, S.; Maira, M.; Garcia-Echeverria, C.; Parra, J. L.; Arribas, J.;Baselga, J., (2008), NVP-BEZ235, a dual PI3K/mTOR

inhibitor, preventsPI3K signaling and inhibits the growth of cancer cells with activatingPI3K mutations. *Cancer Res.* 68, 8022–8030.

[75] Cerecetto, H., Gerpe, A., González, M., Arán, V.J., Ocáriz, C.O., (2005) Pharmacological properties of indazole derivatives: recent developments, *Mini Rev. Med. Chem.* 5, 869–878.

[76] S. Huang, R. Li, P.J. Connolly, S. Emanuel, A. Fuentes-Pescuera, M. Adams, R.H. Gruninger, J. Seraj, S.A. Middleton, J.M. Davis, D.F.C. Moffat, (2007), Synthesis and biological study of 2-amino-5-chloropyridine analogous as inhibiter og VEGFR-2 and cycline dependent kinase 1 (CDK1), *Bio. org. Med. Chem. Lett.* 17 (8), 2179–2183.

[77] Krapcho, A. P., Haydar, S.N., (2001), Synthesis of regioisomeric 2, 5-bis-substituted-aza-benzothiopyranoindazoles, *J. Heterocycl. Chem.* 38, 1153–1166.

[78] Galanski, M., Arion, V. B., Jakupec, M. A., Keppler, B. K., (2003), Recent developments in the field of tumor-inhibiting metal complexes, *Curr. Pharm. Des.* 9, 2078–2089.

[79] Alessio, E., Mestroni, G., Bergamo, B., Sava, G., (2004), Ruthenium antimetastatic agents. *Curr. Top. Med. Chem.* 4 , 1525–1535.

[80] Reisner, E., Arion, V. B., Guedes M.F.C., Da Silva, R., Lichtenecker, A. Eichinger, B.K., Keppler, V. Yu., Kukushkin, A.J.L., (2004), Pombeiro, Tuning of Redox Potentials for the Design of Ruthenium Anticancer Drugs an Electrochemical Study of [*trans*-$RuCl_4L$ (DM*S*O)]$^-$ and [*trans*-$RuCl_4L_2$]$^-$ Complexes, where L = Imidazole, 1,2,4-Triazole, Indazole, *Inorg. Chem.* 43 7083–7093.

[81] B. Cebrián-Losantos, A.A. Krokhin, I.N. Stepanenko, R. Eichinger, M.A. Jakupec, V.B. Arion, B.K. Keppler, (2007), Osmium NAMI-A Analogues: Synthesis, Structural and Spectroscopic Characterization, and Antiproliferative Properties, *Inorg. Chem.* 46, 5023–5033.

[82] Attaur, R.; Malik, S.; Hasan, S. S.; Choudhary, M. I.; Ni, C. Z. Clardy, (1995), Nigellidine - A new indazole alkaloid from the seeds of Nigella sativa, *Tetrahedron Lett.,* 36 (12), 1993-1996.

[83] Ayano, Takeuchi; Makihito, Hori; Shinichi, Sato; Hyun, Seung Ban; Takahir, Kuchimaru; Shinae, Kizaka, Kondoh; Takao, Yamori, Hiroyuki Nakamura, (2012),Synthesis and biological activity of furanylindazoles as inhibitors of hypoxia inducible factor (HIF)-1 transcriptional activity, Med. Chem. Comm., 3, 1455-1461.

[84] Brian, S.; Brown, R. K.; Richard, J. P.; Stanley, D.; John, R. K.; Tammie, K. J.; Steven, M. H.; Heath, A. M.; Bruce, R. B.; Prisca, Honore P. S.; Puttfarcken, R. B. M.; Kennan, C.M.; Connie, R. F.; Chih, Hung, L., (2010), Discovery of TRPV1 antagonist, *Bio. Org. Med. Chem. Lett.,* 20, 3291-3294.

[85] Locke, K.W.; Dunn, R.W.; Hubbard, J.W.; Vanselous, C.L.; Cornfeldt, M.; Stuart, F.; Strupczewski, J.T., (1990),HP 818: A centrally acting analgesic with neuroleptic properties, *Drug Dev. Res.,* 19, 239-256.

[86] Matthew, S. S.; Mitchell, J. S., (2004), The renaissance of palladium (II)-catalyzed oxidation chemistry, *Org Bio. Mol Chem.,* 2, 2551-2554.

[87] Hesketh, P.I.; Gandara, D. R., (1991), Serotonin Antagonists: A New Class of Antiemetic Agents, *J. Natl. Cancer Inst.* 83 (9), 613-620.

[88] Nakane, M., *Clin. Chem. Lab. Med.,* 2003, 41, 865-870.

[89] Souers, A. J.; Gao, J.; Brune, M.; Bush, F.; Wodka, D.; Vasudevan, A.; Judd, A. S.;Mulhern, M.; Brodjian, S.; Dayton, B.; Shapiro, R.; Hernandez, L. E.; Marsh, K. C.;

Sham, H. L.; Collins, C. A.; Kym, P. R., (2005), Identification of 2- (4-Benzyloxyphenyl)-*N*- [1- (2-pyrrolidin-1-yl-ethyl)-1*H*-indazol-6-yl]acetamide, an Orally Efficacious Melanin-Concentrating Hormone Receptor 1 Antagonist for the Treatment of Obesity, *J. Med. Chem.*, 48 (5), 1318-1321.

[90] Mona, Patel; James, D. Rodgers; Robert, J. McHugh Jr.; Barry, L. Johnson; Beverly, C.; Cordova, Ronald M. Klabe; Lee, T. Bacheler; Susan, Erickson Viitanen; Soo, S. Ko., (1999), Unsymmetrical cyclic ureas as HIV-1 protease inhibitors: Novel biaryl indazoles as P2/P2′ substituents, *Bio. Org. Med. Chem. Lett.*, 9 (22), 3217-3220.

[91] Dell, Erba C.; Novi, M.; Petrillo, G.; Tavani, C., (1992), Behaviour of Arylazo *tert*-Butyl Sulfides with Ketone Enolates. Competition between $S_{RN}1$ α-arylation and Azocoupling Reactions., *Tetrahedron,* 48 (2), 325-334.

[92] Jung, Hui Sun; Christopher, A. Teleha; Jia, Sheng Yan; James, D. Rodgers; David, A. Nugiel,. (1997), Efficient Synthesis of 5- (Bromomethyl)- and 5- (Aminomethyl)-1-THP-Indazole*J. Org. Chem.* 62, 5627-5629.

[93] Kawakubo, H.; Sone, T.; Wakigawa, K.; Narita, T. Yakugaki Zasshi, (1987), Anti-inflammatory-analgesic propetrties and anti-pyratic action of 3-aminoindazole derivatives, *J. of the Pharmaceutical Society of Japan*, 107 28-36.

[94] Metz, S.; Clare, M.; Crich, J.; Hagen, T.; Hanson, G. J.; Huang, H.; Houdek, S. J.; Stealey, M. A.; Vazquez, M. L.; Weierr, M.; Xu, X. EP 1427706 B1, 2004.

[95] Ornelio, Rosati; Massimo, Curini; Maria, Carla; Marcotullio, Antonio Macchiarulo;Marina, Perfumi; Laura, Mattioli; Francesco, Rismondo; Giancarlo, Cravotto, (2007),Synthesis, docking studies and anti-inflammatory activity of 4,5,6,7-tetrahydro-2H-indazole derivatives, *Bio.Org. Med. Chem.,* 15 (10), 3463-3473.

[96] Li, Jiau Huang; Mei, Ling Shih; Hua, Sin Chen; Shiow, Lin Pan; Che, Ming Teng; Fang,Yu Lee; Sheng, Chu Kuo, (2006),Synthesis of N^2- (substituted benzyl)-3- (4-methylphenyl) indazoles as novel anti-angiogenic agents, *Bio. Org. Med. Chem.,* 14 (2), *528-536*

[97] Mei, Bai; Grant, Carr; Russell, J. De Orazio; Thomas, D. Friedrich; Svetlana, Dobritsa; Kevin, Fitzpatrick; Peter, R. Guzzo; Douglas, B. Kitchen; Michael, A. Lynch; Denise, Peace; Mohammed, Sajad; Alexander, Usyatinsky; Mark, A. Wolf, (2010), 5-Functionalized indazoles as glucocorticoid receptor agonists, *Bio. Org. Med. Chem. Lett.,* 20 (10), 3017-3020.

[98] Brian, S.; Brown, R. K.; Richard, J. P.; Stanley, D.; John, R. K.; Tammie, K. J.; Steven, M. H.; Heath, A. M.; Bruce, R. B.; Prisca, Honore, P. S.; Puttfarcken, R. B. M.; Kennan, C. M.; Connie, R. F.; Chih, Hung L., (2010), Discovery of TRPV1 antagonist ABT-116, *Bio. Org. Med. Chem. Lett.,* 20 (11), 3291-3294.

[99] Salvatore, P.; Demetrio, R.; Fabiana, P.; Giovanni, C.; Benedetta, M.; Giuseppe, D.; Maria, Valeria, R.; Maria, G.C.; Gabriella, B.; Fiorella, M., (2010), Synthesis and biological evaluation of new indazole derivatives, *ARKIVOC,* 10, 163-177.

[100] Matthew, D. Surman; Emily, E. Freeman; James, F. Grabowski; Mark, Hadden; Alan, J. Henderson; Guowei, Jiang; Xiaowu, Jiang; Michele, Luche; Yuri, Khmelnitsky; Steven, Vickers; Jean, Viggers; Sharon, Cheetham; Peter, R. Guzzo, (2010), 5-(Pyridinon-1-yl)indazoles and 5- (furopyridinon-5-yl)indazoles as MCH-1 antagonists, *Bio. Org. Med. Chem. Lett.,* 20 (23), 7015-7019.

[101] Maninder, M.T.; Sharad, R. W.; Shyam, S.A.; Balasubramanian, (2009), Synthesis, antimicrobial activity and QSAR studies of new 2,3-disubstituted-3,3a,4,5,6,7-hexahydro-2*H*-indazoles, *Bio. Org. Med. Chem. Lett.*,2009, 19 (11), 2960-2964.

[102] Yonghan, Hu Derek Cole; Rajiah, Aldrin Denny; David, R. Anderson; Manus, Ipek; Yike, Ni; Xiaolun, Wang; Suvit, Thaisrivongs; Timothy, Chamberlain; Perry Hall, Julie Liu; Michael, Luong; Lih, Ling Lin; Jean, Baptiste Telliez; Ariamala, Gopalsam, (2011),Discovery of indazoles as inhibitors of Tpl2 kinase, *Bio. Org. Med. Chem. Lett.,* 21 (16), 4758-4761.

[103] Andrew, J. Souers; Ju, Gao; Dariusz, Wodka; Andrew, S. Judd; Mathew, M. Mulhern; James, J. Napier; Michael, E. Brune; Eugene, N. Bush; Sevan, J. Brodjian; Brian, D. Dayton; Robin, Shapiro; Lisa, E. Hernandez; Kennan, C. Marsh; Hing, L. Sham; Christine, A. Collins; Philip, R. Kym, (2005), Synthesis and evaluation of urea-based indazoles as melanin-concentrating hormone receptor 1 antagonists for the treatment of obesity, *Bio. Org. Med. Chem. Lett.,* 15, 2752-2757.

[104] Hongchan An, Nam-Jung Kim, Jong-Wha Jung, Hannah Jang, Jong-Wan Park, Young-Ger Suh, (2011), Design, synthesis and insight into the structure–activity relationship of 1,3-disubstituted indazoles as novel HIF-1 inhibitors, *Bio. Org. Med. Chem. Lett.,* 21, 6297-6300.

[105] Caputo, A.; Silvestrini, B., (1984), Lonidamine, a new approach to cancer therapy. Oncology, 41, 2-6.

[106] Mark J. Kurth, Marilyn M. Olmstead, Makhluf J. Haddadin, (2005),Claimed 2,1-Benzisoxazoles Are Indazalones, *J. Org. Chem.,* 70, 1060-1062.

[107] Christopher M. McBride, Paul A. Renhowe, Thomas G. Gesner, Johanna M. Jansen, Julie Lin, Sylvia Ma, Yasheen Zhou, Cynthia M. Shafer, (2006),3-Benzimidazol-2-yl-1*H*-indazoles as potent c-ABL inhibitors, *Bio. Org. Med. Chem. Lett.,* 16, 3789-3792.

[108] Corbera, A., Esteve S. A., EP.1634873 Al. 2006.

[109] Paola Barraja, Virginia Spanò, Daniele Giallombardo, Patrizia Diana, Alessandra Montalbano, Anna Carbone, Barbara Parrino, Girolamo Cirrincione, (2013) Synthesis of [1,2]oxazolo [5,4-*e*]indazoles as antitumour agents, *Tetrahedron,* 69 (31), 6474-6477.

[110] Ferrara, N., (2004) Vascular Endothelial Growth Factor as a Target for Anticancer Therapy, *The Oncologist,* 9, (Suppl.1).2-10.

[111] Kirill, Lukin; Margaret, C. Hsu; Gilles, Chambournier; Brian, Kotecki, C. J.; Venkatramani, M.; Robert, Leanna, (2007),Development of a Large Scale Asymmetric Synthesis of Vanilloid Receptor (TRPV1) Antagonist ABT-102,*Org. Process Res. Dev.,* 11 (3), 578-584.

[112] Robert, Gomez; Samson, J. Jolly; Theresa, Williams; Joseph, P. Vacca; Marice,l Torrent; Georgia, Mc Gaughey; Ming,Tain Lai; Peter, Felock; Vandna, Munshi; Daniel, DiStefano; Jessica, Flynn; Mike, Miller; Youwei, Yan; John, Reid; Rosa, Sanchez; Yuexia, Liang; Brenda, Paton; Bang, Lin Wan; Neville, Anthony, (2011),Design and Synthesis of Conformationally Constrained Inhibitors of Non-Nucleoside Reverse Transcriptase, *J. Med. Chem.* 54 (22), 7920-7933.

[113] Franciszek, Sączewski; Anita, Kornicka; Alan, L. Hudson; Shayna, Laird; Mika, Scheinin; Jonne, M. Laurila; Apolonia, Rybczyńska; Konrad, Boblewski; Artur, Lehmann; Maria, Gdaniec, *Bio. Org. Med. Chem.,* 2011, 19, 321-329.

[114] Ananda, Thangadurai; Maninder, Minu; Sharad, Wakode; Shyam, Agrawal; Balasubramanian, Narasimhan, (2012),Indazole: a medicinally important heterocyclic moiety, *Med. Chem. Res.*, 21, 1509-1523.

[115] Se, Ho Kim; Benjamin, Markovitz; Richard, Trovato; Brett, R. Murphy; Harry,Austin; Adam, J. Willardsen; Vijay, Baichwal; Scott, Morham; Ashok, Bajji, (2013),Discovery of a new HIV-1 inhibitor scaffold and synthesis of potential prodrugs of indazoles, *Bio Org. Med. Chem. Lett.*, 23, 2888-2892.

[116] Olaf, Kinzel; Daniela, Fattori; Ester, Muraglia; Paola, Gallinari; Maria, Chiara Nardi; Chantal, Paolini; Giuseppe, Roscilli; Carlo, Toniatti; Odalys, Gonzalez Paz; Ralph, Laufer; Armin, Lahm; Anna, Tramontano; Riccardo, Cortes; Raffaele, De Francesco; Gennaro, Ciliberto; Uwe, Koch, (2006), A Structure-Guided Approach to an Orthogonal Estrogen-Receptor-Based Gene Switch Activated by Ligands Suitable for in Vivo Studies, *J. Med. Chem.*, 49 (18), 5404-5407

[117] Shakil, N. A.; Manish, K. Singh; Sathiyendiran, M.; Kumar, J. Jasdeep; Padaria, C.; (2013), Microwave synthesis, characterization and bio-efficacy evaluation of novel chalcone based 6-carbethoxy-2-cyclohexen-1-one and 2H-indazol-3-ol derivatives Eur. J. Med. Chem., 59, 120-131.

[118] Chigalli N. Revanna, Goravanahalli M. Raghavendra, Kebbahalli N. Nandeesh, Doddamedur G. Bhadregowda, Kanchugarakoppal S. Rangappa, Kempegowda Mantelingu, (2013), Convenient synthesis of novel *N*- (5-allyl-7,7-difluoro)-4,5,6,7-tetrahydro-2*H*-indazol-3-yl)-carboxymides, *Tetrahedron Letters,* 54 (38), 5224-5226.

[119] A. A. Stepanov, L. M. Gornostaev, S. F. Vasilevsky, E. V. Arnold, V. I. Mamatyuk, D. S. Fadeev, B. Gold, I. V. Alabugin, (2011),Chameleonic Reactivity of Vicinal Diazonium Salt of Acetylenyl-9,10-anthraquinones: Synthetic Application toward Two Heterocyclic Targets, *J. Org. Chem.,* 76 (21), 8737–8748.

[120] Sona Krupkova, Greg A. Slough and Viktor Krchnak, (2010), Syhtesis of Quinoazolines from N- (2-nitrophenyl sulfonyl) iminodiacetate & α- (2-nitrophenyl sulfonyl) amino ketones via 2H-indazole-1-oxides, *J. Org. Chem.,* 75 (13), 4562–4566.

[121] Paul, Bamborough; Richard, M. Angell; Inder, Bhamra; David, Brown; James, Bull; John, A. Christopher; Anthony, W.J.; Cooper, Lynsey H. Fazal; Ilaria, Giordano; Lucy, Hind; Vipulkumar, K. Patel; Lisa, E. Ranshaw; Martin, J. Sims; Philip, A. Skone; Kathryn, J. Smith; Emma, Vickerstaff; Melanie Washington, (2007), *N*-4-Pyrimidinyl-1H-indazol-4-amine inhibitors of Lck: Indazoles as phenol isosteres with improved pharmacokinetics, *Bio Org. Med. Chem. Lett.,* 17, 4363-4368.

[122] Hardesh, K. Maurya; Ruby, Verma; Saba, Alam; Shweta, Pandey; Vinay, Pathak; Sandeep, Sharm; Kishore, K. Srivastava; Arvind, S. Negi; Atul, Gupta, (2013), Studies on substituted benzo [h] quinazolines, benzo [q]indazoles, pyrazoles, 2,6-diarylpyridines as anti-tubercular agents, *Bio Org. Med. Chem. Lett.,* 23, 5844-5849.

In: Natural Heterocycles
Editors: K. L. Ameta and Saksit Chanthai
ISBN: 978-1-63463-424-3

Chapter 3

ESSENTIAL OILS, PHYTOCHEMICALS AND BIOACTIVITIES OF *PIPER CANINUM* BLUME

Wan Mohd Nuzul Hakimi Wan Salleh[1], Farediah Ahmad[1,*] and Khong Heng Yen[2]

[1]Department of Chemistry, Faculty of Science, Universiti Teknologi Malaysia, Skudai, Johor, Malaysia

[2]School of Chemistry and Environment Studies, Faculty of Applied Sciences, Universiti Teknologi Mara (UiTM) Sarawak, Sarawak, Malaysia

ABSTRACT

The chemical compositions of the essential oils and phytochemicals of *Piper caninum* have been studied. The essential oils were obtained by hydrodistillation technique using Dean-stark apparatus from fresh leaves and stems. The oils were analyzed by capillary GC and GC-MS. The chemical compositions were identified by comparison of the mass spectral data of Wiley Library and Kovats Indices with literatures. A total of 48 components were identified in the leaves (77.9%) and stems (87.0%) of *P. caninum* which were characterized by their high content of safrole (17.1-25.5%). Extractions by soxhlet apparatus were carried out on the dried samples of *P. caninum* to get the crude extracts. Purification of the stems and leaves extracts of *P. caninum* by chromatographic techniques yielded eight compounds. Their structures were elucidated spectroscopically as 5,7-dimethoxyflavone, 4',5,7-trimethoxyflavone, *N*-isobutyl-(2*E*,4*E*,14*Z*)-eicosatrienamide, *N*-isobutyl-15-(3',4'-methylenedioxyphenyl)-2*E*,4*E*,12*Z*-pentadecatrienamide, 5,7-dimethoxyflavanone, 4'-hydroxy-5,7-dimethoxyflavone, 5-hydroxy-7-methoxyflavone and 2'-hydroxy-4',6'-dimethoxychalcone. Antibacterial and antioxidant studies were carried out on the essential oils, crude extracts and isolated compounds. The antibacterial assay was done by using disc diffusion method, followed by minimum inhibition concentration (MIC) and minimum bactericidal concentration (MBC) determinations. The leaves oil of *P. caninum* showed the highest activity towards *Escherichia coli* with MBCvalue 125 μg/mL. *N*-isobutyl-15-(3',4'-methylenedioxyphenyl)-2*E*,4*E*,12*Z*-pentadecatrienamide showed the highest antibacterial

---

*Email: farediah@kimia.fs.utm.my

activity with MIC value of 125 μg/mL against *Bacillus subtilis*. The methanol extracts showed the highest value of gallic acid and (±)-catechin equivalents for stem extract. This results correlated with the antioxidant assays on 2,2-diphenyl-1-picrylhydrazyl (DPPH) radical which showed that the methanolic stems extract of *P. caninum* had the highest radical scavenging activity with IC_{50} values of 802.66 μg/mL, respectively. The isolated compounds, 4'-hydroxy-5,7-dimethoxyflavone and 5-hydroxy-7-methoxyflavone showed weak activity with IC_{50} values more than 1000 μg/mL.

1. INTRODUCTION

Plants have always been a part of human lives since early times. Besides being used as shelter and utensils, plants are also consumed as food, flavor, and coloring agents as well as in the treatment of ailments and diseases. In the last few decades, herbal medicine is gaining worldwide acceptance as a complementary or alternative medicine. With the increase of applications of plants as an alternative medicine, more herbal drugs are available in the market [1]. Higher plants have served humankind as source of natural products for medicinal agents since the earliest beginnings. In fact, natural products once served as the source of all drugs. Presently, natural products represent over 50% of all drugs in clinical use. In the ancient times, the chemical investigation and purification of extracts of plants with medicinal properties, have yielded several purified compounds which have demonstrated to be essential in the practice of modern medicine [2].

Piperaceae is a family of herbs, shrubs, small trees, and hanging vines. It can be recognized by their leaves, flowers, stems and fruits. The leaves are simple, alternate and stipulate while the blade is often cordate. The flowers are tiny without perianth and packed in dense and succulent spikes. The stems are articulate and the fruits are capsular or drupaceous [3]. Economically, the Piperaceae is important as an origin for the pepper in the worldwide spice markets. The most important of these is *P. nigrum* (Black pepper) the best known agricultural product in the genus. In addition, other Piperaceae of relative pharmaceutical value are *P. betle* (Sireh) and *P. methysticum* (Kava). Leaf of *P. betle* is chewed with the betle nut in Asia, whereas *P. methysticum*, is a traditional drink that has been used as a herbal medicine. Both of these species are used in traditional practices for their narcotic properties [4].

Piper is the genus of the family Piperaceae. The greatest diversity of *Piper* species is in the Neotropics, where about two thirdsof the described species are found. Some 300 species are endemic to SoutheastAsia, including the East Indian islands and Northern Australia [4]. The *Piper* species were mostly woody perennial climbers and rarely shrub with dilated or swollen nodes and stipule. The leaves of various *Piper* species were typically aromatic or had a pungent smell. The flowers are tiny, bisexual or unisexual, borne in compact spikes which are usually leaf-opposed. The ovary was superior and surrounded by the bracts which are variable in shape. The genus *Piper* could also be recognized by its pulpy fruit, 2-6 stamens, and small drupe. In some species, these bodies are used as food by resident mutualist ants [5].

The chemistry of *Piper* species has been widely investigated and the phytochemical investigations from all parts of the World have led to the isolation of a number of pharmacologically active compounds. They have been extensively investigated as a source of new natural products with potential antitumoral, antimicrobial, antifungal and insecticidal activities [6]. Table 1shows bioactivities of several *Piper* species.

Table 1. Bioactivities of Several *Piper* Species

Bioactivities	*Piper* Species	Part Tested	Ref
Antifungal	*P. arboretum*	Amide	[6]
Antiplasmodial	*P. sanguineispicum*	Lignans	[7]
Antimicrobial	*P. nigrum*	Essential oil	[8]
Cytotoxic	*P. gaudichaudianum*	Essential oil	[9]
Toxicity	*P. abyssinica*	Seeds extract	[10]
Antioxidant	*P. retrofractum*	Leaves extract	[11]
Anti-inflammatory	*P. sarmentosum*	Leaves extract	[12]
Antimalarial	*P. umbellatum*	Leaves extract	[13]
Antiviral	*P. aduncum*	Leaves extract	[14]
Antileishmanial	*P. elongatum*	Dihydrochalcone	[15]
Antitubercolosis	*P. hostmanianum*	Dihydrochalcone	[16]
Antidepressant	*P. laetispicum*	Amide, alkaloid	[17]
Antiparasitic	*P. heterophyllum*	Prenylated benzoic acid	[18]

Figure 1. *Piper caninum* Blume.

P. caninum Blume (Figure 1) is locally known as '*cabai hutan*' or '*lada hantu*' in Malaysia. It grows as a slender climber on a small tree in lowland and upland rain forest. The leaves are used by people in lieu of betel-leaves for chewing. They are added to the betel-quid for treating hoarseness. It was also reported that a wash made from the leaves was use after childbirth [19]. Essential oils are derived from leaves, stems, roots, bark, flowers, and other parts of aromatic plants. These oils are highly volatile and characterized by a strong odour and are formed by aromatic plants as secondary metabolites. Essential oil has been utilized since the beginning of civilization. One of the most well known was found in Egypt, where they used aromatic oils for embalming corpses in the preparation of mummies. The people of ancient Greek and Rome used the perfumed oils as sacraments in their worship rituals [9]. Essential oils can be extracted from plants either by steam distillation, enfleurage (cold extraction), maceration (hot extraction), solvent extraction, expression and supercritical fluid extraction (SFE). Steam distillation is the most common process for essential oil extraction due to the volatility of the oil.

Table 2. Properties and Uses of the Essential Oils [12]

Essential Oils	Biological Name	Properties	Uses
Jasmine	*Jasminum officinale*	antidepressant analgesic antiseptic antispasmodic	Depression, nervous exhaustion, and stress related conditions Used for coughs, laryngitis, dysmenorrhoea, uterine disorders
Lavender	*Lavendula officinalis*	antidepressant antimicrobial antiseptic hypertensive antiseptic	For nervous system disorders such as depression, headache, hypertension, insomnia and migraine Useful in treating skin conditions such as acne, allergies and eczema
Lemon	*Citrus limonum*	stimulating antirheumatic antitoxic hypertensive insecticide	Depression, acne and indigestion, arthritis, obesity, rheumatism, asthma, throat infections, bronchitis, fever, flu Treatment of anemia, mouth ulcers, greasy skin and varicose veins
Peppermint	*Menthe piperita*	digestive analgesic antimicrobial antiseptic antiviral	Muscle fatigue, bad breath, toothache, bronchitis, indigestion and travel sickness, muscular pains, asthma, skin problem such as acne, dermatitis, ringworm and scabies
Clory sage	*Salvia sclarea*	warming antiseptic antiphlogistic digestive antidepressant	Menstrual problems, anxiety, depression, high blood pressure, acne boils, oily skin and hair, cramp, migraine, the genitor-urinary system disorders such as wrinkles, ulcers
Eucalyptus	*Eucalyptus globulus*	antiseptic analgesic antirheumatic expectorant antiviral	Muscular aches and pains, poor circulation, asthma, bronchitis, flu, cold, headaches, throat infections, skin disorders such as burns, wounds, insect bites

The extraction product can vary in quality, quantity and in composition according to climate, soil composition, plant organ, age and vegetative cycle stage [10].

Currently, approximately 3000 essential oils are known, 300 of which are commercially significant especially for the pharmaceutical, agronomic, food, sanitary, cosmetic and perfume industries. Essential oils and their components are used in perfumes and cosmetic products, sanitary products, dentistry, agriculture, food preservatives and additives and as natural remedies [11]. Table 2 shows some properties and uses of common essential oils.

Chemically, most of the essential oils are classified as terpenes, a large group of unsaturated hydrocarbons with common building block of C_5H_8 (isoprene unit). The essential oils of *Piper* species compose of a numerous organic compounds which can be classified into three major groups, namely monoterpenes, sesquiterpenes and phenylpropanoids.

Numerous phytochemicals investigations have been carried out on the genus of *Piper* compared to other genera of Piperaceae family. The investigations have led to the isolations of various classes of phytochemicals such as flavonoids, alkaloids, lignans and neolignans, benzoic acid derivatives, hydroquinones and alkylphenols.

Phytochemical research on *Piper* species has been intensified in recent years, with the finding that several species have shown to have a number of biological activities such as antimicrobial, antifungal, antioxidant, antimalarial and anti-inflammatory [13]. Table 3 shows the bioactivity studies of the isolated compounds from *Piper* species.

Table 3. Bioactivity Studies of Isolated Compounds

Bioactivities	Compounds	Species	Ref
Antifungal	scutifoliamide A (1) scutifoliamide B (2) hoffmannseggiamide A (3) hoffmannseggiamide B (4) coruscanone A (5) coruscanone B (6)	*P. scutifolium* *P. hoffmannseggianum* *P. coruscans*	[14] [15]
Anti-inflammatory	piperovatine (7) piperlonguminine (8)	*P. ovatum*	[16]
Antitubercolosis	pellitorine (9) sarmentine (10) brachyamide B (11) 1-piperettyl pyrrolidine (12) sarmentosine (13)	*P. sarmentosum*	[17]
Antiplasmodial	sarmentine (10) 1-piperettyl pyrrolidine (12) (-)-methyllinderatin (14) linderatone (15) 2',6'-dihydroxy-4'-methoxydihydrochalcone (16)	*P. sarmentosum* *P. hostmannianum*	[17] [18]
Anti-leishmanial	2',6'-dihydroxy-4'-methoxydihydrochalcone (16) 2',6',4-trihydroxy-4'-methoxydihydrochalcone (17) cubebin (18)	*P. elongatum* *P. cubeba*	[19] [20]
Antioxidant	quercetin (19) quercitrin (20) *N-p*-coumaroyl tyramine (21) piperumbellactams B (22) piperumbellactams C (23)	*P. arboreum* *P. umbellatum*	[21] [22]
Antimicrobial	pellitorine (9) trachyone (24) pergumidiene (25) isopiperolein B (26)	*P. nigrum*	[23]

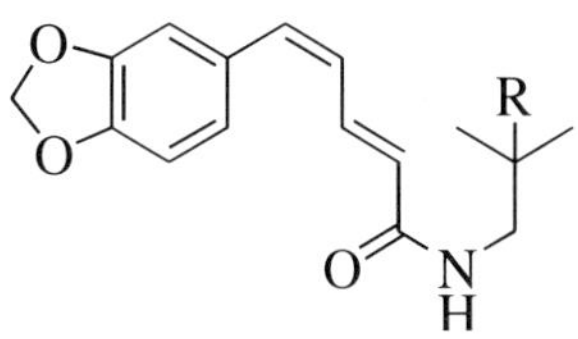

(1) R = H
(2) R = Ac

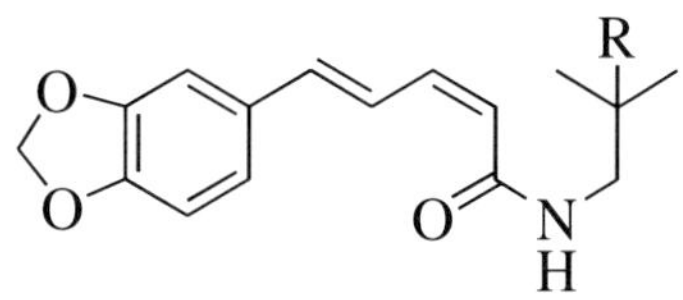

(3) R = H
(4) R = Ac

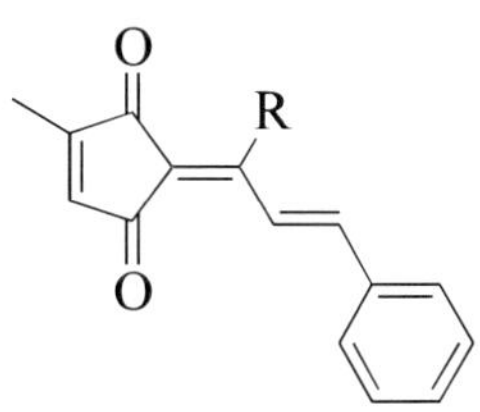

(5) R = OCH_3
(6) R = OH

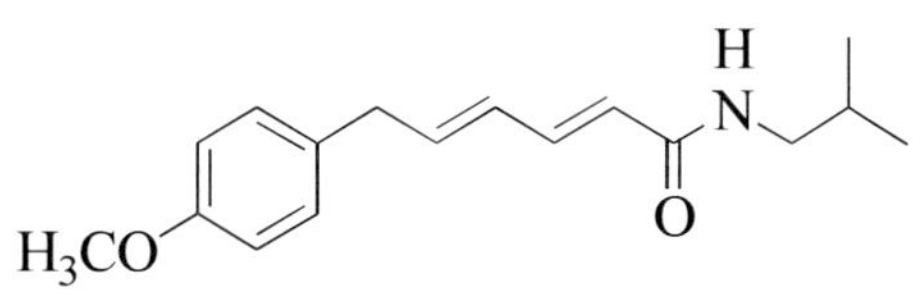

(7)

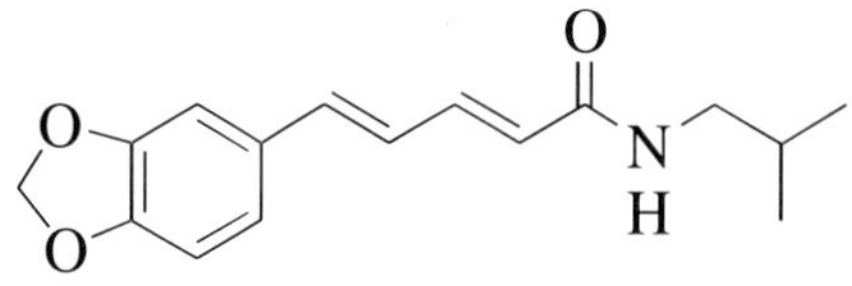

(8)

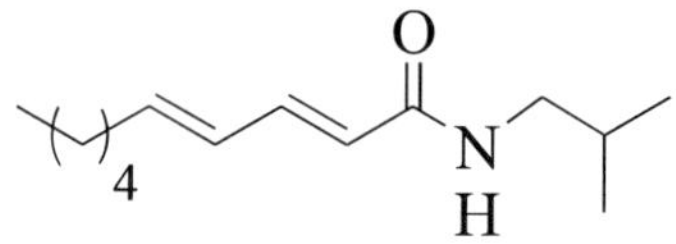

(9)

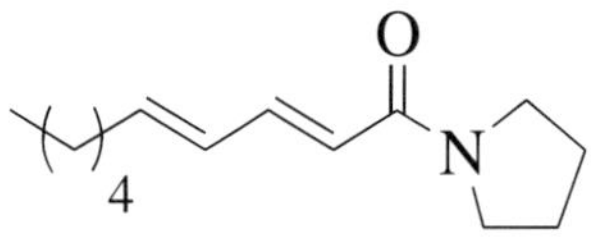

(10)

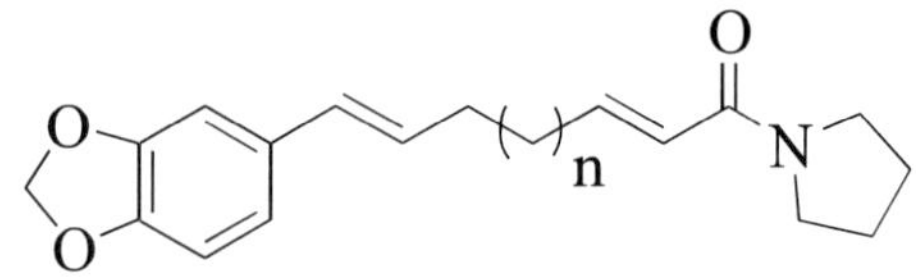

(11) n = 3
(12) n = 1

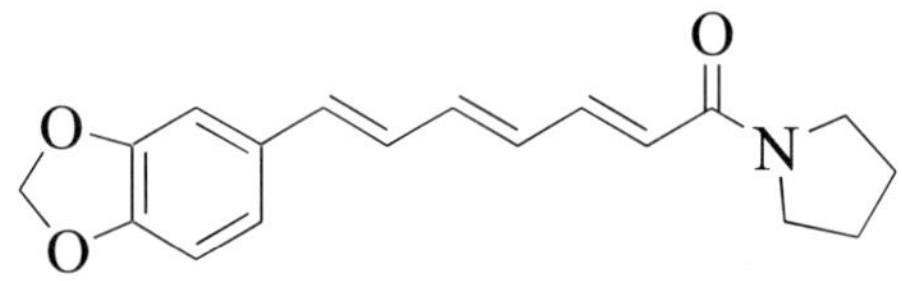

(13)

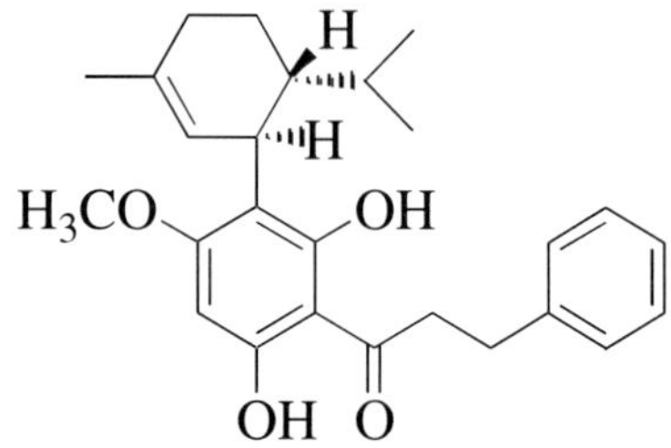

(14)

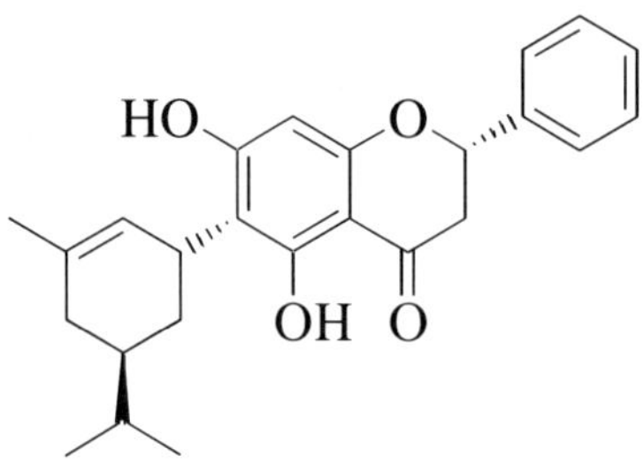

(15)

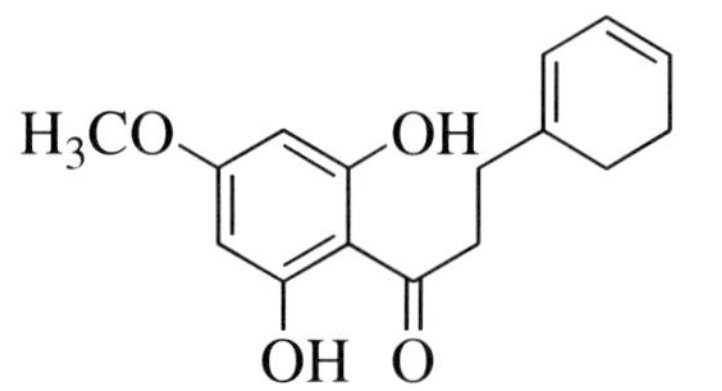

(16)

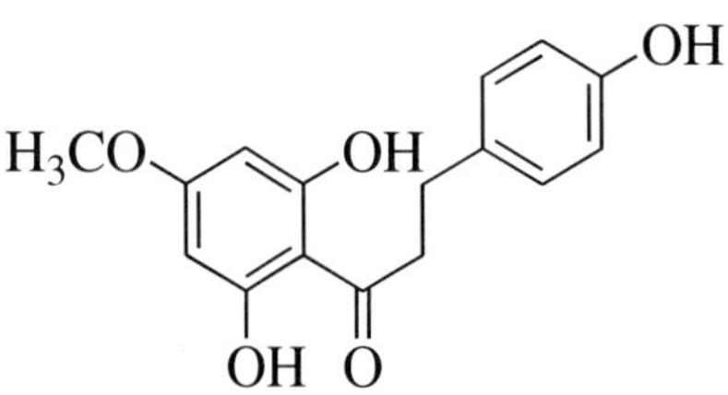

(17)

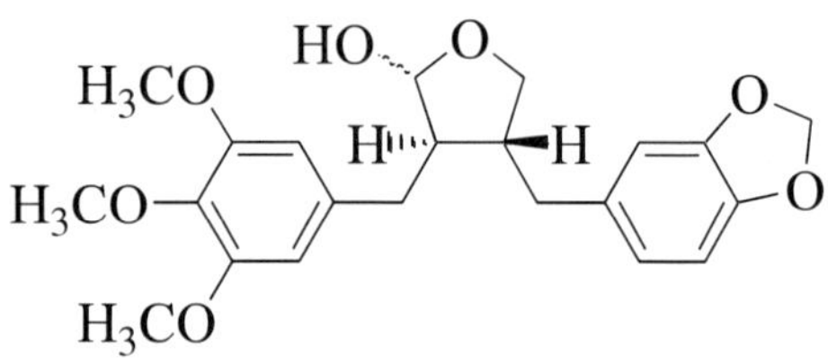

(18)

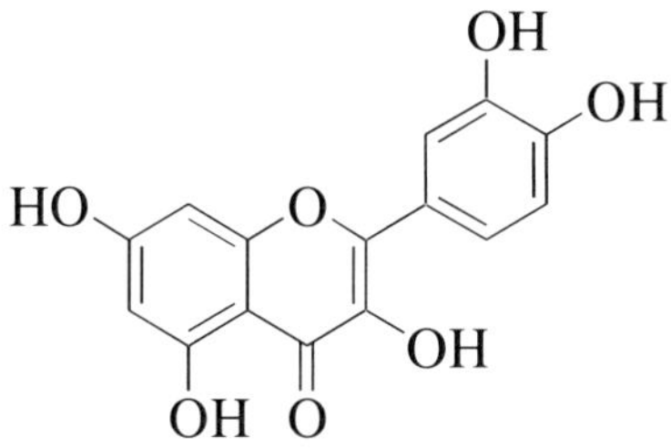

(19)

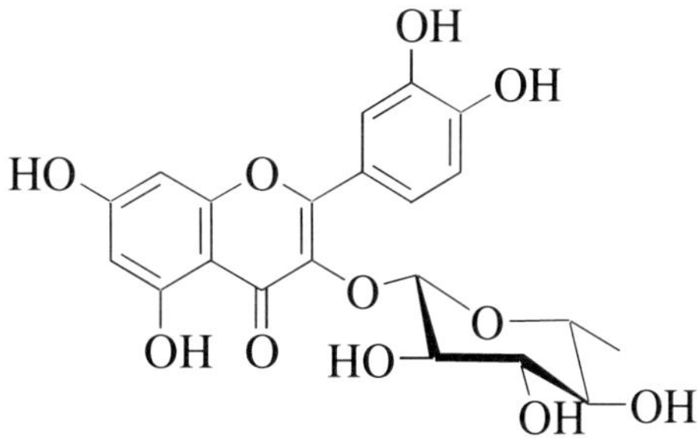

(20)

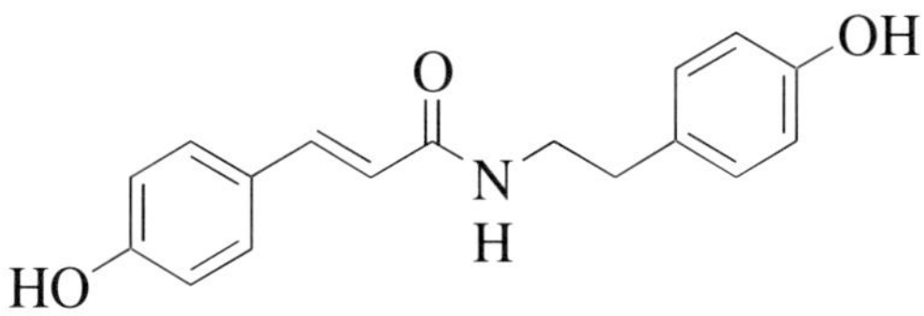

(21)

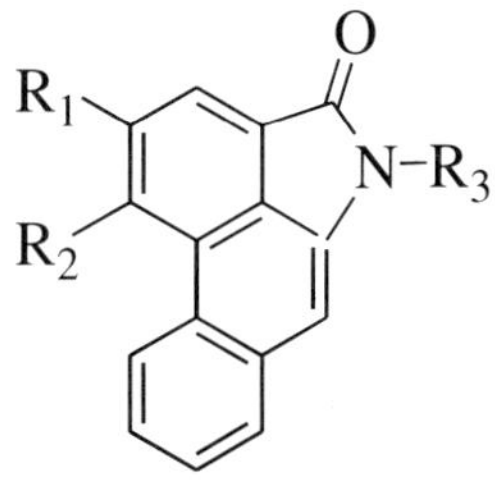

(22) $R_1 = OCH_3$; $R_2 = R_3 = OH$
(23) $R_1 = R_2 = OH$; $R_3 = OCH_3$

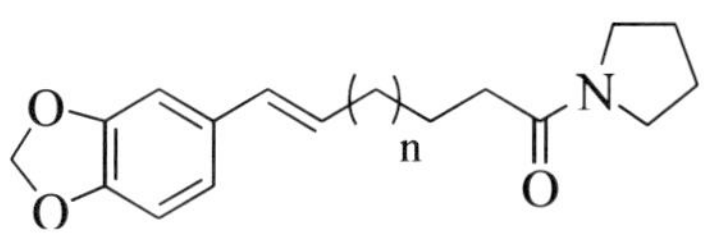

(24) n = 10
(25) n = 5

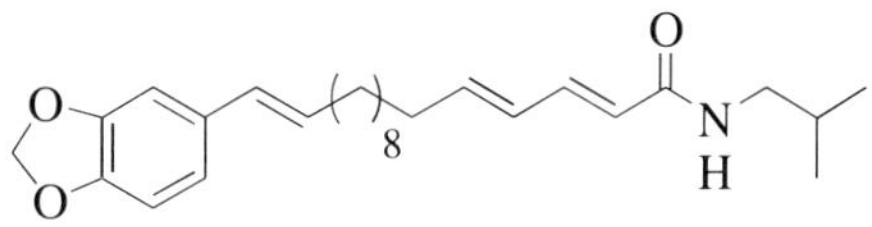

(26)

This chapter will discuss the essential oils, phytochemicals and bioactivities of *P. caninum* Blume.

2. Extraction and Analytical Methods

General Experimental Procedures

Petroleum ether (PE), *n*-hexane, diethyl ether (Et_2O), ethyl acetate (EtOAc), chloroform ($CHCl_3$) and methanol (MeOH) were used as solvent systems in chromatographic method. Petroleum ether (60-80°C) was distilled before used. Anhydrous $MgSO_4$ was used to dry the essential oils. A series of standard hydrocarbons (C_6-C_{30}) were used in Kovats Indices analysis. Soxhlet extraction technique was used to extract the phytochemicals from the dried samples, while hydrodistillation technique was applied to extract the essential oils of the fresh samples. Vacuum Liquid Chromatography (VLC) and gravity column chromatography were carried out by using Merck SiO_2 (230-400 Mesh) and Merck SiO_2 (70-230 Mesh) respectively. Thin layer chromatography (TLC) analysis was performed on 0.20 mm precoated silica gel aluminium sheets (Merck Kieselgel 60 F_{254}). The spots on TLC were detected by Ultraviolet, UV (254 nm and 365 nm) illumination after being sprayed with anilsaldehyde reagent. Melting points were measured by using melting point apparatus equipped with microscope, Leica Gallen III and were uncorrected. Preparative thin layer chromatography (TLC) was conducted by using 1 mm thin glass plate of Merck silica gel $60F_{254}$ (20 x 20 cm).

Chemical composition analysis of the essential oil was carried out by using a Hewlett Packard 6890 series II A. Gas chromatography (GC) analysis was equipped with FID detector (25.0 m long x 200 um i.d. film thickness x 0.11 um) with Ultra 1 column. The samples were injected by splitter by using Helium as a carrier gas. The column oven were programmed from 40°C (after 5 min) to 300°C at 4°C/min and the final temperature was held for 3 min. The chromatograms of gas chromatography-mass spectrometry (GC-MS) were recorded using a Hewlett Packard Model 5989A mass spectrometer. GC-MS was equipped with Wiley Library Software. Capillary GC-MS condition was performed as above except for the oven temperature program. The oven temperature programmed was 50°C (5 min) then 4°C/min to 300°C (5 min).

The ^{1}H (400 MHz) and ^{13}C NMR (100 MHz) spectra were recorded on a Bruker Avance 400 Spectrophotometer. Chemical shifts were reported in ppm and $CDCl_3$ as the solvent. Residual solvent was used as an internal standard. The IR spectra were recorded on Perkin Elmer series 1600 spectrophotometer as KBr disc or thin film on NaCl discs. The Ultraviolet (UV) spectra were recorded on Shimadzu UV 1601PC spectrophotometer. High resolution mass spectral data were obtained from UCL Spectrometry Service, London, UK.

Extraction of Essential Oil

Samples of *P. caninum* were collected in July 2010 from Bau and Lundu, Sarawak. The species were identified by Mrs. Mohizar Mohamad from Forest Research Centre, Kuching, Sarawak and the voucher specimen of *P. caninum* (UiTMKS3003) was deposited at Natural Products Research and Development Centre (NPRDC), UiTM Sarawak. The fresh samples of *P. caninum* were chopped into small pieces (500 g) and loaded in a round bottom flask (5 L). Distilled water was then added until the water covered the entire samples. The flask was equipped with a Dean-stark apparatus and water condenser and the mixture was hydrodistilled for 10 hours. The mixture of oil and water was extracted with Et_2O (3 × 10 mL), dried over anhydrous $MgSO_4$ and filtered. The filtrate was evaporated at room temperature to afford the leaves oil (2.30 g, 0.46%) and the stems oil (1.56 g, 0.31%) both were in the formed of pale yellow oils.

The composition of the essential oil was analysed by using GCand GC-MS (Figure 3-4) in Ultra-1 column. Identification of the constituents was carried out bycomparison of the mass spectra from the Wiley Library in the GC-MS and Kovats Indices [24]. The Kovats Indices was calculated using the equation below.

$$KI = 100\left[\frac{(\log t_{sample}) - (\log t_x)}{(\log t_{x+1}) - (\log t_x)}\right] + 100x$$

where,

t_{sample} = the retention time of the sample component

t_x	=	the retention time of the saturated hydrocarbons elute before the sample component
t_{x+1}	=	the retention time of the saturated hydrocarbon which contains x+1 carbon and that just elutes after the sample component
x	=	number of carbon atoms in saturated hydrocarbons

Extraction and Isolation of Phytochemicals

The dried stems and leaves of *P. caninum* was ground into powder and extracted with *n*-hexane, EtOAc and MeOH (4.5 L each) sequentially by soxhlet extraction for 15 hours. The extracts were concentrated using rotary evaporator to afford the crude extracts as summarized in Table 4.

The *n*-hexane stems extract (PCH) (7.00 g) was fractionated by VLC and eluted with *n*-hexane:$CHCl_3$:EtOAc to afford 6 major fractions (PCH1-6). Fractions PCH5 and PCH6 were purified by CC and recrystallized using cold *n*-hexane to yield 5,7-dimethoxyflavone (27) and 4',5,7-trimethoxyflavone (28), respectively. Fractions PCH2 and PCH3 were subjected to multiple CC to give *N*-isobutyl-(2*E*,4*E*,14*Z*)-eicosatrienamide (29) and *N*-isobutyl-15-(3',4'-methylene-dioxyphenyl)-2*E*,4*E*,12*Z*-pentadecatrienamide (30), respectively. The EtOAc stems extract (PCE) (8.00 g) was fractionated by VLC and eluted with *n*-hexane:EtOAc to afford 5 major fractions (PCE1-5). Fractions PCE3 were subjected to CC to afford 5,7-dimethoxyflavanone (31).

The MeOH stems extract (PCM) (8.00 g) was subjected to VLC and eluted with *n*-hexane:$CHCl_3$:EtOAc to form four fractions (PCM1-4). Purification of PCM4 by CC and recrystallized using cold *n*-hexane yield 4'-hydroxy-5,7-dimethoxyflavone (32). The MeOH leaves extract (PCLM) (6.00 g) was subjected to VLC and eluted with *n*-hexane:$CHCl_3$:EtOAc to afford 5 major fraction. Fraction PCM2 was further subjected to CC and washed using cold *n*-hexane to yield5-hydroxy-7-methoxyflavone (33). Fraction PCM5 was purified by CC and followed by preparative TLC to give2'-hydroxy-4',6'-dimethoxychalcone (34).

Table 4. *Piper caninum* Extract

Species	Sample (g)	Solvent	Yield (g, %)	Colour
P.caninum Stems	300	*n*-Hexane (PCH)	7.12 g, 2.37%	Green gum
		EtOAc (PCE)	9.24 g, 3.70%	Green gum
		MeOH (PCM)	8.65 g, 3.46%	Brown gum
P.caninum Leaves	250	MeOH (PCLM)	6.50 g, 3.25%	Green gum

5,7-Dimethoxyflavone (27)

PCH 5 (400 mg) was purified by using CC packed with SiO_2 and eluted with *n*-hexane:EtOAc to give 200 fractions. Fractions 70-110 were combined and recrystallized from cold *n*-hexane to give 5,7-dimethoxyflavone (**27**) (350.1 mg, 4.92%) as white crystalline needles; m.p 158-159°C (lit. [25] 159-160°C); R_f = 0.47 (*n*-hexane:EtOAc = 4:1); IR (KBr) υ_{max} cm^{-1}: 2921, 1643, 1618, 1603, 1452, 1121; ^1H NMR (400 MHz, $CDCl_3$): δ 3.92 (3H, s, 7-OCH_3), 3.96 (3H, s, 5-OCH_3), 6.38 (1H, d, *J* = 2.0 Hz, H-8), 6.58 (1H, d, *J* = 2.0 Hz, H-6), 6.68 (1H, s, H-3), 7.50 (3H, m, H-3', H-4', H-5'), 7.87 (2H, dd, *J* = 8.0 and 3.6 Hz, H-2', H-6'); ^{13}C NMR (100 MHz, $CDCl_3$): δ 55.7 (5-OCH_3), 56.4 (7-OCH_3), 92.8 (C-6), 96.2 (C-8), 109.1 (C-3), 109.3 (C-4a), 125.9 (C-2', C-6'), 128.9 (C-3', C-5'), 131.2 (C-4'), 131.5 (C-1'), 159.9 (C-2), 160.6 (C-5), 160.9 (C-7), 164.1 (C-8a), 177.6 (C-4); EIMS (% rel int.): *m/z* 282 [M^+, $C_{17}H_{14}O_4$] (100), 253 (38), 224 (14), 209 (27), 150 (21), 137 (12).

4',5,7-Trimethoxyflavone (28)

PCH 6 (400 mg) was purified by using CC packed with SiO_2 and eluted with *n*-hexane:EtOAc to give 200 fractions. Fraction 154-200 were combined and recrystallized from cold *n*-hexane to give 4',5,7-dimethoxyflavone (28) (220.58 mg, 3.08%) as colourless crystalline needles; m.p 154-155°C (lit. [25] 156-157°C); R_f = 0.40 (*n*-hexane:EtOAc = 3:2); IR (KBr) υ_{max} cm^{-1}: 3019, 2923, 1642, 1599, 1567, 1258; ^1H NMR (400 MHz, $CDCl_3$): δ 3.89 (3H, s, 4'-OCH_3), 3.92 (3H, s, 7-OCH_3), 3.96 (3H, s, 5-OCH_3), 6.38 (1H, d, *J* = 2.4 Hz, H-8), 6.56 (1H, d, *J* = 2.4 Hz, H-6), 6.62 (1H, s, H-3), 7.02 (2H, d, *J* = 8.8 Hz, H-3' and H-5'), 7.84 (2H, d, *J* = 8.8 Hz, H-2' and H-6'); ^{13}C NMR (100 MHz, $CDCl_3$): δ 55.4 (5-OCH_3), 55.7 (7-OCH_3), 56.4 (4'-OCH_3), 92.8 (C-6), 96.1 (C-8), 107.6 (C-3), 109.2 (C-4a), 114.3 (C-3', C-5'), 123.8 (C-1'), 127.6 (C-2', C-6'), 159.8 (C-4'), 160.7 (C-8a), 160.9 (C-5), 162.1 (C-2), 163.9 (C-7), 177.6 (C-4); EIMS (% rel int.): *m/z* 312 [M^+, $C_{18}H_{16}O_5$] (100), 283 (35), 266 (42), 239 (16), 180 (2), 167 (8), 149 (20), 132 (33), 117 (10).

N-isobutyl-(2*E*,4*E*,14*Z*)-Eicosatrienamide (29)

PCH 2 (250 mg) was purified by using CC on SiO_2 and eluted with *n*-hexane:EtOAc to give 200 fractions. Fraction 111-120 were combined to give *N*-isobutyl-(2*E*,4*E*,14*Z*)-eicosatrienamide (**29**) ascolourless crystalline needles (6.50 mg, 2.60%); m.p 60-61°C (lit. [26] 61-62°C); R_f = 0.3 (*n*-hexane:EtOAc = 4:1); IR (KBr) υ_{max} cm^{-1}: 3301, 2917, 2851, 1654, 1609;^1H NMR (400 MHz, $CDCl_3$): δ0.92 (3H, t, *J* = 6.8 Hz, H-20), 0.87 (6H, d, *J* = 6.8 Hz, H-3', H-4'), 1.26-1.33 (16H, m, H-7 – H-12, H-17 – H-19), 1.81 (1H, septet, *J* = 6.8 Hz, H-2'), 2.03 (2H, m, H-13, H-16), 3.17 (2H, t, *J* = 6.4 Hz, H-1'), 5.35 (2H, dt, *J* = 9.2 and 5.6 Hz, H-14, H-15), 5.47 (1H, br. s, N-H), 5.77 (1H, d, *J* = 15.2 Hz, H-2), 6.07 (1H, dt, *J* = 16.4 and 6.0 Hz, H-5), 6.13 (1H, dd, *J* = 16.4 and 9.6 Hz, H-4), 7.20 (1H, dd, *J* = 15.2 and 9.6 Hz, H-3); ^{13}C NMR (100 MHz, $CDCl_3$): δ14.1 (C-20), 20.1 (C-3', C-4'), 22.6 (C-19), 26.8 (C-12), 26.8 (C-16), 27.1 (C-11), 28.6 (C-2'), 28.6 (C-10), 28.7 (C-9), 29.2 (C-8), 29.3 (C-7), 29.6 (C-17), 31.9 (C-18), 32.9 (C-6), 46.9 (C-1'), 121.6 (C-2), 128.2 (C-4), 129.7 (C-14), 129.8 (C-15), 141.3 (C-5), 143.2 (C-3), 166.3 (C-1); EIMS (% rel int.): *m/z* 361 [M^+, $C_{24}H_{43}NO$] (11), 333 (18), 203 (35), 202 (100), 152 (32), 149 (20).

N-isobutyl-15-(3',4'-methylenedioxyphenyl)-2*E*,4*E*,12*Z*-pentadecatrien-amide (30)

PCH 3 (280 mg) was purified by using CC packed with SiO_2 and eluted with *n*-hexane:EtOAc to give 150 fractions. Fraction 61-80 were combined to give to give *N*-isobutyl-15-(3',4'-methylenedioxyphenyl)-2*E*,4*E*,12*Z*-pentadecatrienamide (**30**) ascolourless crystalline needles (7.20 mg, 2.57%); m.p 95-96°C (lit. [27] 95-98°C); R_f = 0.3 (*n*-hexane:EtOAc = 4:1); IR (KBr) υ_{max} cm^{-1}: 3300, 1655, 1625, 1611, 1505, 1487, 1250, 1190;^{1}H NMR (400 MHz, $CDCl_3$): δ0.87 (6H, d, *J* = 6.8 Hz, H-3", H-4"), 1.28 (4H, m, H-8, H-9), 1.42 (4H, m, H-7, H-10), 1.79 (1H, septet, *J* = 6.4 Hz, H-2"), 1.96 (2H, m, H-11), 2.15 (2H, m, H-6), 2.25 (2H, quartet, *J* = 6.8 Hz, H-14), 2.59 (2H, t, *J* = 6.8 Hz, H-15), 3.17 (2H, t, *J* = 6.4 Hz, H-1"), 5.42 (2H, dt, *J* = 7.2 Hz, H-12, H-13), 5.48 (1H, br.s, N-H), 5.73 (1H, d, *J* = 15.2 Hz, H-2), 5.92 (2H, s, OCH_2O),6.09 (1H, dt, *J* = 13.6 and 7.2 Hz, H-5), 6.13 (1H, dd, *J* = 13.6 and 10.0 Hz, H-4), 6.61 (1H, d, *J* = 8.0 Hz, H-5'),6.68 (1H, d, *J* = 1.2 Hz, H-2'), 6.71 (1H, dd, *J* = 8.0 and 1.2 Hz, H-6'), 7.19 (1H, dd, *J* = 15.2 and 10.0 Hz, H-3); ^{13}C NMR (100 MHz, $CDCl_3$): δ20.1 (C-4"), 20.1 (C-3"), 28.6 (C-2"), 28.7 (C-9), 28.8 (C-10), 28.9 (C-7), 29.4 (C-8), 32.5 (C-6), 32.9 (C-11), 34.6 (C-14), 35.8 (C-15), 46.91, 100. 6 (OCH_2O), (C-1"), 108.0 (C-2'), 108.9 (C-5'), 121.1 (C-6'), 121.7 (C-2), 128.2 (C-4), 129.2 (C-12), 131.1 (C-13), 136.1 (C-1'), 141.3 (C-5), 143.2 (C-3), 145.4 (C-3'), 147.4 (C-4'), 166.5 (C-1); EIMS (% rel int.): *m/z* 411 [M^+, $C_{26}H_{37}NO_3$] (7), 276 (5), 152 (8), 136 (9), 135 (100), 121 (11).

5,7-Dimethoxyflavanone (31)

Fraction PCE 3 (400 mg) was subjected to CC and eluted with *n*-hexane:$CHCl_3$ to give 100 fractions. Fraction 25-50 were combined to give 5,7-dimethoxyflavanone (**31**) (10.32 mg, 2.58%) as white crystals; m.p 98-99°C (lit. [28] 98-100°C); R_f = 0.28 (*n*-hexane:$CHCl_3$ = 4:1); IR (KBr) υ_{max} cm^{-1}: 2937, 1673, 1602, 1570, 1486, 1241; ^{1}H NMR (400 MHz, $CDCl_3$): δ 2.84 (1H, dd, *J* = 13.0 and 3.3 Hz, H-3a), 3.07 (1H, dd, *J* = 13.0 and 10.3 Hz, H-3b), 3.83 (3H, s, 7-OCH_3), 3.91 (3H, s, 5-OCH_3), 5.44 (1H, dd, *J* = 10.3 and 3.3 Hz, H-2), 6.11 (1H, d, *J* = 2.4 Hz, H-6), 6.18 (1H, d, *J* = 2.4 Hz, H-8), 7.36 (5H, m, Ph); ^{13}C NMR (100 MHz, $CDCl_3$): δ 45.6 (C-3), 55.6 (7-OCH_3), 56.2 (5-OCH_3), 79.2 (C-2), 93.2 (C-8), 93.6 (C-6), 106.0 (C-10), 126.1 (C-2', C-6'), 128.7 (C-4'), 128.7 (C-3', C-5'), 138.7 (C-1'), 162.3 (C-9), 164.9 (C-5), 166.0 (C-7), 189.2 (C-4); EIMS *m/z* (rel.int.): 284 [M^+, $C_{17}H_{16}O_4$] (11), 207 (5), 195 (100), 180 (16), 149 (37), 121 (51), 86 (20), 84 (39).

4'-Hydroxy-5,7-dimethoxyflavone (32)

PCM 4 (350 mg) was purified by using CC packed with SiO_2 and eluted with *n*-hexane:EtOAc to give 300 fractions. Fraction 254-280 were combined and recrystallized from cold *n*-hexane to give 4'-hydroxy-5,7-dimethoxyflavone (32) (15.24 mg, 4.35%) as white crystalline needles; m.p 288-289°C (lit. [29] 289-290°C); R_f = 0.40 (*n*-hexane:EtOAc = 1:4); IR (KBr) υ_{max} cm^{-1}: 3084, 2947, 1679, 1584, 1508, 1266; UV (MeOH) λ_{max} nm (45 to 65 nm): 328 (band II), 370 (band I), 45-65 nm (addition $AlCl_3$/HCl); ^{1}H NMR (400 MHz, CD_3OD): δ

3.92 (3H, s, 5-OCH_3), 3.95 (3H, s, 7-OCH_3), 6.53 (1H, d, J = 2.4 Hz, H-8), 6.79 (1H, d, J = 2.4 Hz, H-6), 6.59 (1H, s, H-3), 6.92 (2H, d, J = 8.8 Hz, H-3' and H-5'), 7.85 (2H, d, J = 8.8 Hz, H-2' and H-6'); ^{13}C NMR (100 MHz, CD_3OD): δ 55.1 (2 x OCH_3), 92.8 (C-6), 95.9 (C-8), 105.3 (C-3), 105.3 (C-4a), 115.6 (C-3', C-5'), 121.7 (C-1'), 127.7 (C-2', C-6'), 159.8 (C-4'), 160.6 (C-8a), 162.5 (C-5), 161.1 (C-2), 165.0 (C-7), 178.7 (C-4); EIMS *m/z* (rel.int.): 298 [M^+, $C_{17}H_{14}O_5$] (8), 218 (5), 149 (12), 109 (11), 97 (13), 95 (218), 86 (50), 84 (100).

5-Hydroxy-7-methoxyflavone (33)

PCLM 2 (450 mg) was purified by using CC packed with SiO_2 and eluted with *n*-hexane:$CHCl_3$ to give 200 fractions. Fraction 92-130 were combined and recrystallized from cold *n*-hexane to give 5-hydroxy-7-methoxyflavone (33) (12.12 mg, 2.69%) as colourless crystalline needles; m.p 162-163°C (lit. [30] 163-164°C); R_f = 0.40 (*n*-hexane:$CHCl_3$= 2:3); IR (KBr) υ_{max} cm^{-1}: 3445, 2973, 1614, 1541, 1448, 1252; UV (MeOH) λ_{max} nm (46 to 68 nm): 270 (band II) and 340 (band I), 46-68 nm (addition $AlCl_3$/HCl); ^1H NMR (400 MHz, $CDCl_3$): δ 3.90 (3H, s, 7-OCH_3), 6.40 (1H, d, J = 2.0 Hz, H-8), 6.53 (1H, d, J = 2.0 Hz, H-6), 6.69 (1H, s, H-3), 7.56 (3H, m, H-3', H-4', H5'), 7.90 (2H, dd, J = 2.0 and 1.6 Hz, H-2', H-6'), 12.74 (1H. s, 5-OH); ^{13}C NMR (100 MHz, $CDCl_3$): δ 55.8 (7-OCH_3), 92.7 (C-6), 98.2 (C-8), 105.7 (C-3), 105.9 (C-4a), 126.3 (C-2', C-6'), 129.1 (C-3', C-5'), 131.4 (C-4'), 131.8 (C-1'), 157.8 (C-2), 162.2 (C-5), 164.0 (C-7), 165.6 (C-8a), 182.5 (C-4); EIMS (% rel int.): *m/z* 268 [M^+, $C_{16}H_{12}O_4$] (100), 239 (16), 225 (13), 149 (27), 111 (10), 97 (17).

2'-Hydroxy-4',6'-dimethoxychalcone (34)

PCLM 5 (450 mg) was purified by CC with *n*-hexane:$CHCl_3$ as eluents to give 150 fractions. Fractions 65-100 (50.20 g) were combined and purified by preparative TLC to get 2'-hydroxy-4',6'-dimethoxychalcone (34) as yellow rhombics (10.83 mg, 2.40%); m.p 87-89°C (lit. [28] 89-90°C); R_f= 0.38 (*n*-hexane:$CHCl_3$ = 1:4); IR (KBr) υ_{max} cm^{-1}: 3445, 2920, 2850, 1623, 1508, 1440, 1260; UV (MeOH) λ_{max} nm: 338 nm; ^1H NMR (400 MHz, $CDCl_3$): δ 3.85 (3H, s, 4'-OCH_3), 3.93 (3H, s, 6'-OCH_3), 5.98 (1H, d, J = 2.4 Hz, H-5'), 6.12 (1H, d, J = 2.4 Hz, H-3'), 7.41 (3H, m, H-3, H-4, H-5), 7.61 (2H, m, H-2, H-6), 7.78 (1H, d, J = 15.6 Hz, H-7), 7.90 (1H, d, J = 15.6 Hz, H-8), 14.30 (1H, s, 2'-OH); ^{13}C NMR (100 MHz, $CDCl_3$): δ 55.5 (4'-OCH_3), 55.8 (6'-OCH_3), 91.0 (C-5'), 94.2 (C-3'), 105.6 (C-7), 126.3 (C-1), 127.5 (C-3, C-5), 128.4 (C-2, C-6), 128.9 (C-1'), 130.0 (C-4), 142.5 (C-8), 161.2 (C-6'), 166.6 (C-4'), 169.0 (C-2'), δ 193.1 (C-2); EIMS *m/z* (rel. int.): 284 [M^+, $C_{17}H_{16}O_4$] (57), 283 (50), 207 (5), 208 (12), 207 (100), 181 (34), 149 (20).

Antibacterial Activity

Chemicals, Microorganisms and Culture Media

Nutrient agar (NA), nutrient broth (NB) and standard of streptomysin sulphate antimicrobial disc (10 ug/disc) were obtained from Oxoid. All bacteria i.e. *Staphylococcus aureus*, *Bacillus subtilis* (Gram-positive bacteria), *Pseudomonas aeruginosa*, *Pseudomonas putida* and *Escherichia coli* (Gram-negative bacteria) were obtained from Department of Biology, Faculty of Bioscience and Bioengineering, Universiti Teknologi Malaysia. The bacteria were cultured in an appropriate nutrient broth (NB) at 37°C overnight. The concentrations of the cultures were adjusted by comparison with the McFarland solution. The Mc Farland solution was prepared by mixing the H_2SO_4 solution (1% in broth) and $BaCl_2$ solution (1% in broth). The H_2SO_4 solution was prepared by dissolving H_2SO_4 (1.02 mL) in broth (100 mL), while $BaCl_2$ solution was prepared by dissolving $BaCl_2$ (1 g) in broth (100 mL). The mixture of H_2SO_4 solution (9.95 mL) and $BaCl_2$ solution (0.05 mL) were equivalent with 150×10^6 coloni/unit. Nutrient agar (NA) (20 g/L) and nutrient broth (NB) (8 g/L) were dissolved in distilled water. All solutions were sterilized by autoclave for 15 minutes at 120ºC.

Disc Diffusion Method

The antibacterial assay was conducted by using disc diffusion method as described by Mackeen et al. [31]. The NA solution (17 mL) was placed into petri dishes. The agar plates were kept in refrigerator for overnight. Whatman filter paper discs of 6 mm diameter were impregnated with 10 µL of the solution of crude extract or essential oil (at 4 mg/mL) or isolated compound (at 1 mg/mL) prepared using DMSO. Standard disc of streptomycin sulphate (10 µg/disc) was used as positive control, while DMSO was used as a negative control.

The bacteria stock (400 µL) was spread onto the surface of agar in the petri dish by using a glass rod. Then, the sample discs and the control discs were placed on the agar surface. Figure 2 shows the arrangement of the discs. The petri dishes were inverted and incubated for 24 hours at 37°C. Clear inhibition zones around the discs indicated the presence of antimicrobial activity. All tests and analyses were carried out in triplicates.

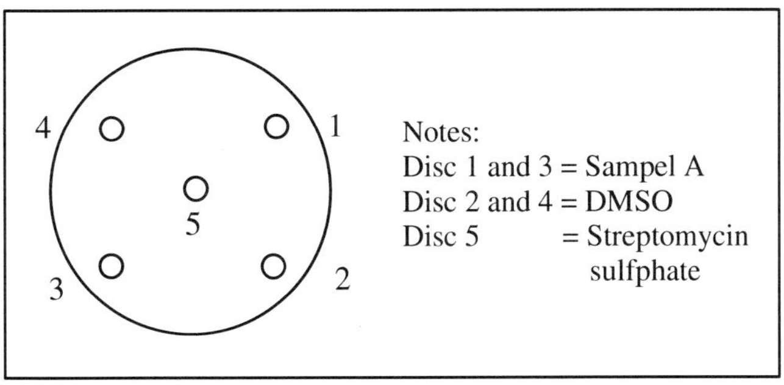

Figure 2. Arrangement of Discs in the Agar Plate.

Minimum Inhibition Concentration (MIC) and Minimum Bactericidal Concentration (MBC)

Minimum inhibitory concentration was determined by the micro-titer broth dilution method [32]. This test was performed in a sterile 96-well micro titer plates. The test samples

(1 mg each) were dissolved in methanol to obtain 1000 μg/mL stock solution. Each methanolic stock samples (10 μL) was transferred to micro titer plate well in duplicate at row A. A number of wells were reserved in each plate for positive and negative controls. Sterile broth (100 μL) was added to each micro-titer plate well from row B to row H. Then, the suspensions of microorganisms (200 μL) were added to the samples at row A. Mixture from row A (100 μL) was transferred to each micro titer plate well in order to obtain a two fold serial dilution of stock samples (concentration of 500 μg/mL to 3.9 μg/mL) plates were then incubated for 18 hours at 37°C. Bacterial growth was indicated by the presence of turbidity and a pellet at the bottom of the well. The lowest concentrations, which did not show any growth of tested microorganisms after macroscopic evaluation were determined as MIC values. MBC was determined according to method by Arias et al. [33]. All wells in the MIC study, which did not show any growth of bacteria after incubation period were first diluted in fresh nutrient broth (1:4) and then sub-cultured onto the surface of freshly prepared nutrient agar plates (θ, 15 mm). The plates were incubated for 18 hours at 37°C. The MBC were recorded as the lowest concentration of the sample that did not permit any visible bacteria colony growth on the appropriate agar plate after the incubation period

Antioxidant Activity

Chemicals and Instrumentation

Folin-Ciocalteau phenolic reagent, gallic acid, (±)-catechin, sodium phosphate monobasic monohydrate ($NaH_2PO_4.H_2O$), ammonium molybdate tetrahydrate, and butylated hydroxyanisole (BHA), were of Sigma brand. 2,2-Diphenyl-1-picrylhydrazyl (DPPH), sodium carbonate (Na_2CO_3) and dimethylsulfoxide (DMSO) were purchased from Fluka; L-ascorbic acid (vitamin C), butylated hydroxytoluene (BHT) and sulphuric acid (H_2SO_4) were purchased from Merck.

All absorbance were measured on Ultraviolet Perkin Elmer Lambda 25 Spectrophotometer, while the IC_{50} measurements were run on EXCEL program.

Free Radical Scavenging Activity (DPPH)

The free radical scavenging activity was measured by the DPPH method according to Loo et al. [33] with minor modification. Each stock solution (1.0 mg/mL) was diluted to final concentration of 1000, 500, 250, 125 and 62.5 μg/mL. Then, a total of 3.8 mL of 50 μM DPPH methanolic solution (1 mg/50 mL) was added to 0.2 mL of each sample solution and allowed to react at room temperature for 30 min.

The absorbance of the mixtures was measured at 517 nm. A control was prepared without sample or standard and measured immediately at 0 min. MeOH was used for the baseline correction. Lower absorbance of the reaction mixture indicates higher free radical scavenging activity, and vice versa. The percentage inhibition of radical scavenging activity was calculated using the following formula:

$$\text{Percentage Inhibition}(\%) = \frac{\text{Abs (DPPH)} - \text{Abs (DPPH + sample)}}{\text{Abs (DPPH)}} \times 100$$

The IC_{50} value, defined as the amount of antioxidant necessary to decrease the initial DPPH concentration by 50% was calculated from the results. Ascorbic acid and BHT were used as standards and were diluted into the same concentration as samples. All tests were carried out in triplicate.

Chemical Composition Analysis of *Piper caninum* Oils

The chemical compositions of the leaves and stems of *P. caninum* oils are listed in Table 5. The analysis of the oils successfully identified thirty six components (77.9%), and thirty seven components (87.0%), for the leaves and stems oil, respectively. The major fraction in leaves and stems oil were monoterpene hydrocarbons and sesquiterpene hydrocarbons, respectively. Both oils were found to possess lower percentage of oxygenated monoterpenes (4.7 - 7.5%) and oxygenated sesquiterpenes (5.2 - 6.9%). The phenylpropanoids, safrole was the major components in the leaves (17.1%) and stems (25.5%) oils.

Most of the components were similar in both oils. However, camphor (0.3%), α-cubebene (0.5%), eugenol(2.4%), β-cubebene (0.3%), aromadendrene (0.8%), *allo*-aromadendrene (0.3%), α-bisabolene (0.4%), germacrene B (1.1%), globulol (0.3%), t-muurolol (2.4%), α-cadinol (1.0%) and (*E,E*)-farnesyl acetate (1.2%) were found to be absent in the leaves oil. In addition, myrcene (0.9%), γ-terpinene (0.1%), α-terpinolene (0.4%), 2-nonanone (1.2%), cyclosativene (0.2%), α-ylangene (0.2%), β-bourbonene (1.1%), β-gurjunene (0.2%), (*E,E*)-α-farnesene (0.9%), chavibetol (0.9%), elemol (0.2%) and β-eudesmol (0.9%) were absent in the stems oil. Figure 3(a) and 3(b) are the GC-MS chromatograms of the leaves and stems oils of *P. caninum*.

Table 5. Chemical Compositions Analysis of *P. caninum* Oils

No	Components	Kovats Indices	Percentage	
		Ultra-1	Leaves	Stems
1	α-Pinene	932	4.0 ± 0.2	1.6 ± 0.1
2	Camphene	946	0.2 ± 0.1	0.6 ± 0.1
3	β-Pinene	974	8.9 ± 0.1	4.9 ± 0.1
4	Myrcene	988	0.9 ± 0.1	–
5	Limonene	1024	3.9 ± 0.2	2.7 ± 0.1
6	(*Z*)-β-Ocimene	1032	3.4 ± 0.2	0.2 ± 0.1
7	γ-Terpinene	1054	0.1 ± 0.2	–
8	α-Terpinolene	1086	0.4 ± 0.2	–
9	2-Nonanone	1087	1.2 ± 0.1	–
10	Linalool	1095	7.0 ± 0.1	2.9 ± 0.2
11	Camphor	1141	–	0.3 ± 0.1
12	Pinocarvone	1160	0.1 ± 0.2	–
13	Terpinen-4-ol	1174	0.2 ± 0.1	0.5 ± 0.1
14	α-Terpineol	1186	0.3 ± 0.1	1.0 ± 0.1
15	*n*-Decanal	1201	0.2 ± 0.2	0.8 ± 0.1
16	Safrole	1285	17.1 ± 0.1	25.5 ± 0.1

Table 5. (Continued)

No	Components	Kovats Indices	Percentage	
		Ultra-1	Leaves	Stems
17	2-Undecanone	1293	2.0 ± 0.1	1.1 ± 0.1
18	δ-Elemene	1335	1.8 ± 0.1	4.1 ± 0.2
19	α-Cubebene	1345	–	0.5 ± 0.2
20	Eugenol	1356	–	2.4 ± 0.1
21	Cyclosativene	1369	0.2 ± 0.1	–
22	α-Ylangene	1373	0.2 ± 0.1	–
23	α-Copaene	1374	0.5 ± 0.1	0.9 ± 0.1
24	β-Cubebene	1387	–	0.3 ± 0.1
25	β-Bourbonene	1387	1.1 ± 0.1	–
26	β-Elemene	1389	2.1 ± 0.1	2.4 ± 0.1
27	α-Gurjunene	1409	0.3 ± 0.1	0.9 ± 0.1
28	β-Caryophyllene	1417	6.7 ± 0.1	9.8 ± 0.1
29	β-Gurjunene	1431	0.2 ± 0.1	–
30	Aromadendrene	1439	–	0.8 ± 0.2
31	α-Humulene	1452	1.0 ± 0.1	1.6 ± 0.1
32	*allo*-Aromadendrene	1458	–	0.3 ± 0.1
33	Germacrene D	1484	4.9 ± 0.1	7.8 ± 0.1
34	Zingiberene	1493	0.4 ± 0.1	0.6 ± 0.2
35	Bicyclogermacrene	1502	1.1 ± 0.1	2.3 ± 0.1
36	(*E,E*)-α-Farnesene	1505	0.9 ± 0.1	–
37	α-Bisabolene	1506	–	0.4 ± 0.1
38	δ-Cadinene	1522	0.4 ± 0.1	1.0 ± 0.1
39	Chavibetol	1524	0.9 ± 0.1	–
40	Elemol	1548	0.2 ± 0.1	–
41	Germacrene B	1559	–	1.1 ± 0.1
42	(*E*)-Nerolidol	1561	3.9 ± 0.1	1.6 ± 0.2
43	Caryophyllene oxide	1582	0.2 ± 0.1	0.4 ± 0.1
44	Globulol	1590	–	0.3 ± 0.1
45	t-Muurolol	1644	–	2.4 ± 0.1
46	β-Eudesmol	1649	0.9 ± 0.1	–
No	Components	Kovats Indices	Percentage	
47	α-Cadinol	1652	–	1.0 ± 0.1
48	(*E,E*)-Farnesyl acetate	1845	–	1.2 ± 0.2
Group Components (%)				
	Phenylpropanoids		18.0 ± 0.2	27.9 ± 0.2
	Monoterpene Hydrocarbons		21.8 ± 0.2	10.0 ± 0.2
	Oxygenated Monoterpenes		7.5 ± 0.2	4.7 ± 0.2
	Sesquiterpene Hydrocarbons		21.9 ± 0.2	35.2 ± 0.3
	Oxygenated Sesquiterpenes		5.2 ± 0.2	6.9 ± 0.2
	Others		3.5 ± 0.2	2.3 ± 0.1
Identified Components			77.9 ± 0.2	87.0 ± 0.3

Each value is expressed as means ± SD of three injections.

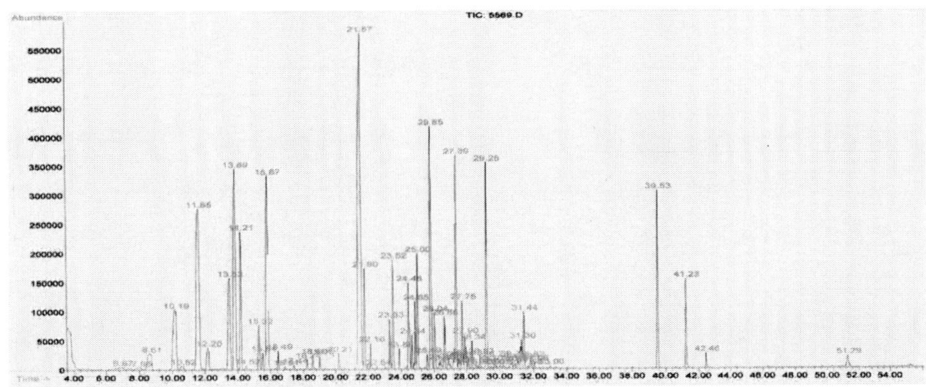

(a)

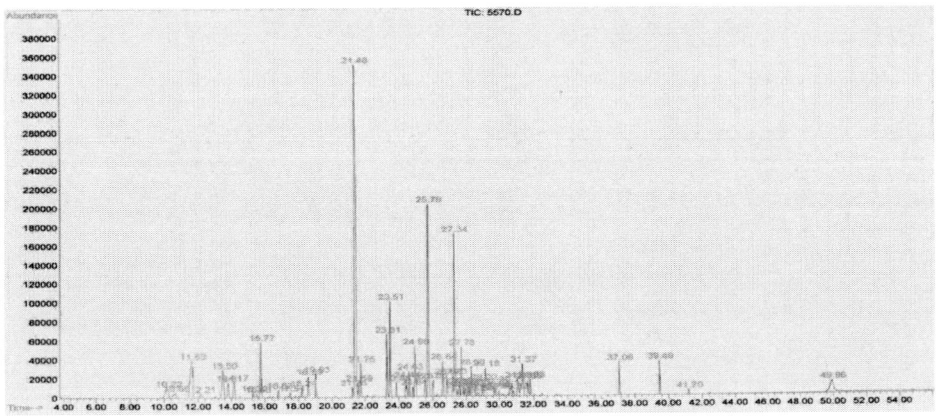

(b)

Figure 3. (a) GC-MS chromatogram of the leaves oil of *P. caninum*; 3(b) GC-MS chromatogram of the stems oil of *P. caninum*.

Compounds of *Piper caninum*

Soxhlet extraction of the powdered stems of *P. caninum* (300 g), sequentially with *n*-hexane, EtOAc and MeOH had yielded the crude *n*-hexane, PCH (7.12 g, 2.37%), EtOAc, PCE (9.24 g, 3.70%) and MeOH, PCM (8.65 g, 3.46%) extracts, while extraction of the powdered leaves of *P. caninum* (250 g) with MeOH yielded the crude MeOH, PCLM (6.50 g, 3.25%) extracts. Purification of the extracts were carried out by fractionation using vacuum liquid chromatography (VLC), followed by gravity column chromatography (CC), preparative thin layer chromatography and recrystallization techniques. Six compounds were isolated from the *n*-hexane, EtOAc and MeOH stems extracts.

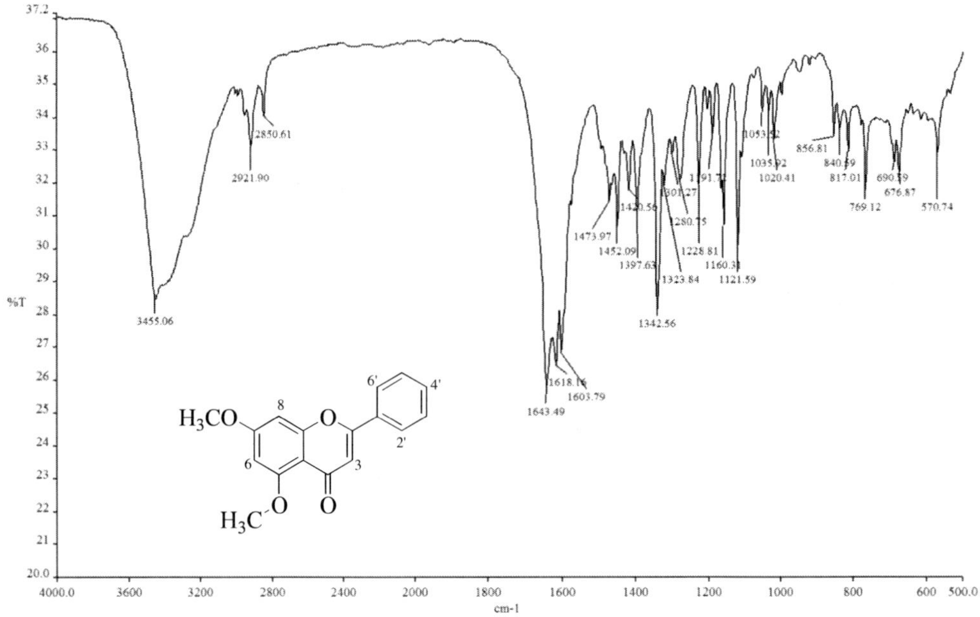

(a)

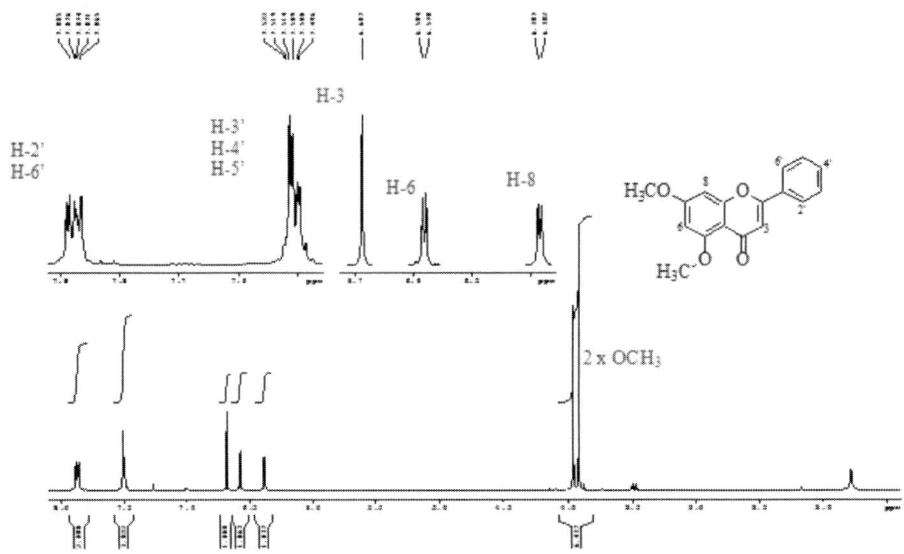

(b)

Figure 4(a). (Continued).

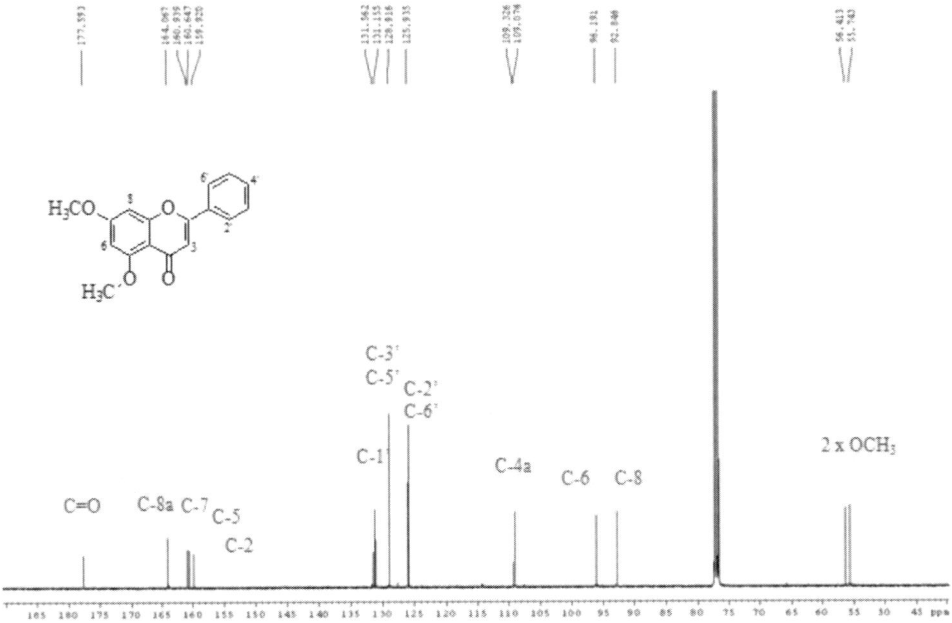

(c)

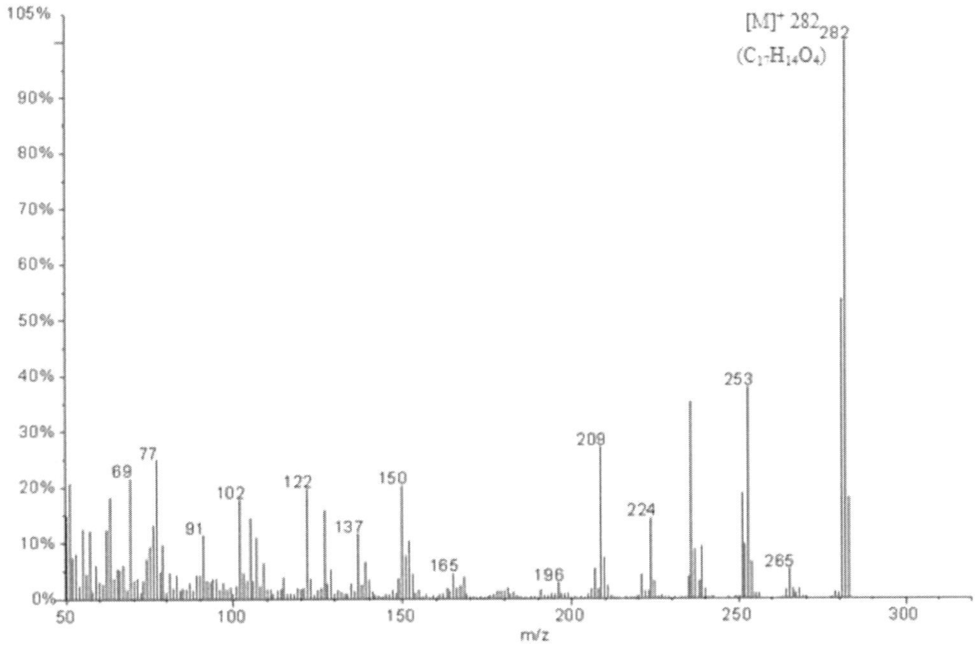

(d)

Figure 4(a). IR spectrum of compound (27); (b)^{1}H NMR spectrum of compound (27); (c)^{13}C NMR spectrum of compound (27);(d)MS spectrum of compound (27).

These compounds were identified as 5,7-dimethoxyflavone (27), 4',5,7-trimethoxyflavone (28), *N*-isobutyl-(2*E*,4*E*,14*Z*)-eicosatrienamide (29), *N*-isobutyl-15-(3',4'-methylene-dioxyphenyl)-2*E*,4*E*,12*Z*-pentadecatrienamide (30), 5,7-dimethoxyflavanone (31) and 4'-hydroxy-5,7-dimethoxyflavone (32). Purification of the MeOH leaves extract have yielded two compounds identified as 5-hydroxy-7-methoxyflavone (33) and 2'-hydroxy-4',6'-dimethoxychalcone (34). Structural elucidations of all phytochemicals were based on spectral analyses and comparison of the spectral and physical data with reported values. Below are examples of IR, NMR (^{1}H and ^{13}C) and MS spectra (Figure 4 (a-e)) for compound (27).

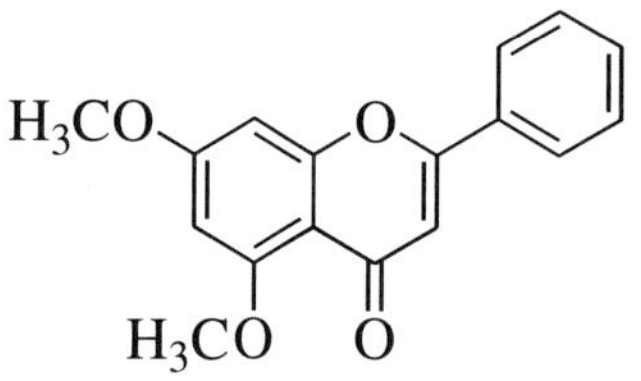

(27)

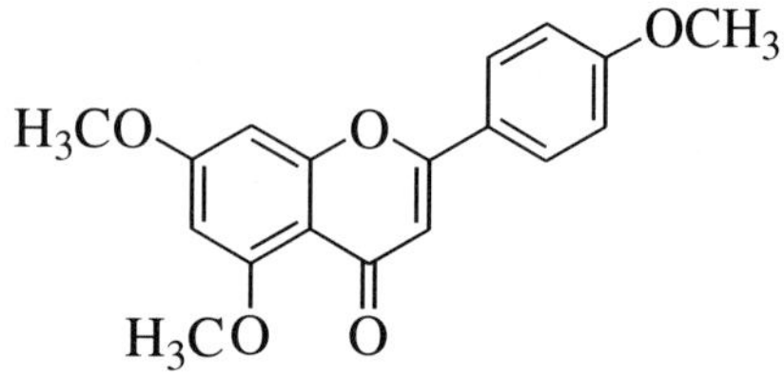

(28)

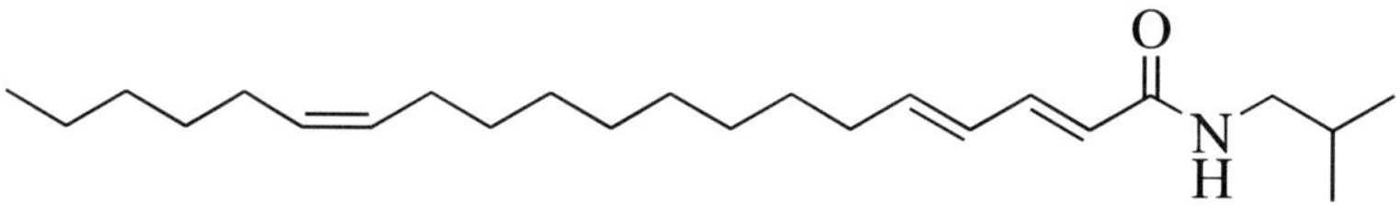

(29)

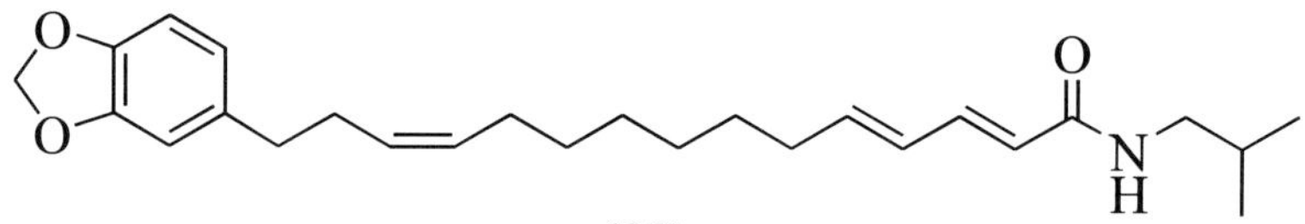

(30)

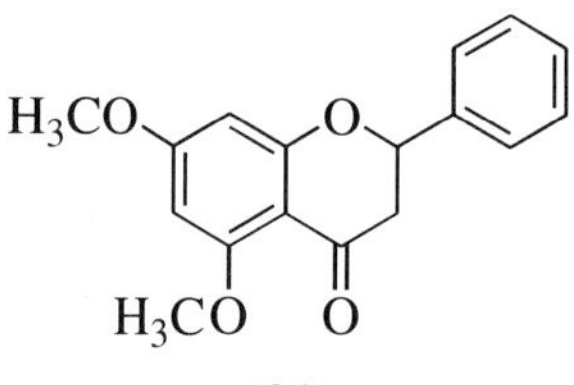

(31)

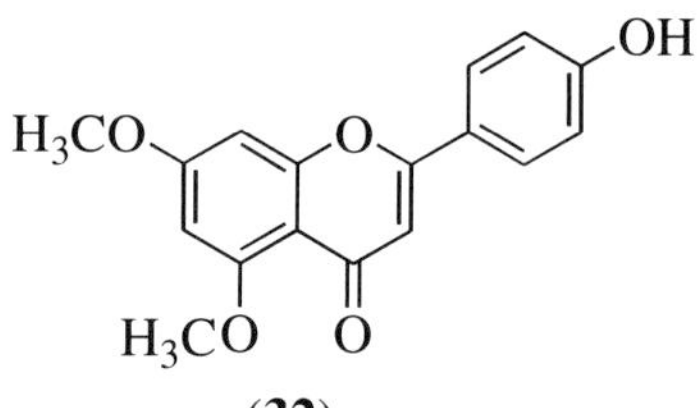

(32)

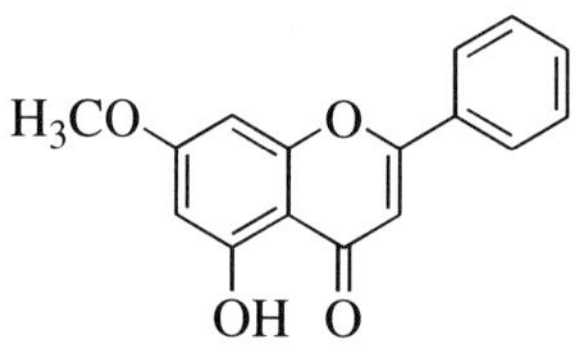

(33)

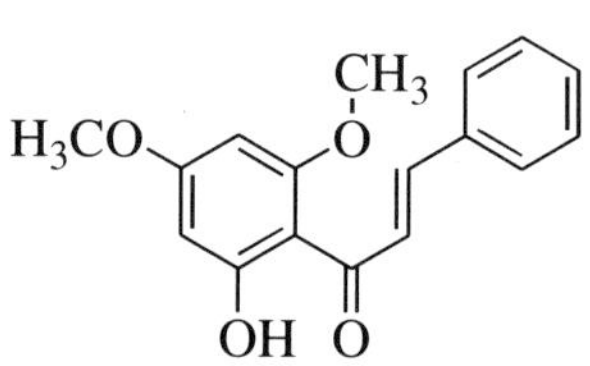

(34)

Antibacterial Activity

Evaluations of the antibacterial activity have been done on the essential oils, crude extracts and isolated compounds from *P. caninum*by using five bacteria, of two Gram-positive (*Staphylococcus aureus* and *Bacillus subtilis*) and three Gram-negative (*Pseudomonas aeruginosa*, *Pseudomonasputida* and *Escherichia coli*). The assay was carried out by using disc diffusion method and followed by the MIC and MBC. All tests were performed in triplicate and the antibacterial activity was expressed as the mean of inhibition zone diameter (mm) produced. Table 6 shows the results of antibacterial activity of the essential oils, crude extracts and pure compounds of the species. The inhibitions of the extracts against all the tested bacteria were in the range between 7 to 12 mm of inhibition zone diameter.

Table 6. Diameter of inhibition zone of the essential oils, crude extracts and isolated compounds

Samples	Inhibition Zone (mm)				
	Gram-positive bacteria		Gram-negative Bacteria		
	B. s	*S. a*	*P. a*	*P. p*	*E .c*
Essential Oils					
PCLO	12.2 ± 0.5	7.2 ± 0.5	8.2 ± 0.4	7.8 ± 0.5	7.2 ± 0.5
PCSO	10.4 ± 0.5	7.0 ± 0.5	8.8 ± 0.5	7.5 ± 0.4	7.8 ± 0.5
Crude Extracts					
PCH	-	-	-	-	-
PCE	-	-	-	-	-
PCM	8.4 ± 0.5	8.5 ± 0.4	-	-	7.5 ± 0.4
PCLM	-	-	-	-	-
Compounds					
(27)	11.2 ± 0.3	-	-	-	11.8 ± 0.5
(28)	10.1 ± 0.5	-	-	-	-
(29)	-	-	10.2 ± 0.6	-	9.4 ± 0.4
(30)	10.4 ± 0.5	-	-	-	-
(31)	-	-	10.1 ± 0.4	-	-
(32)	10.5 ± 0.5	-	10.2 ± 0.3	9.5±0.3	9.8 ± 0.4
(33)	8.2 ± 0.4	-	-	-	-
(34)	9.1 ± 0.5	9.0 ± 0.4	8.2 ± 0.5	7.8 ± 0.3	-
Standard					
SS	17.6 ± 0.2	17.8 ± 0.2	17.2 ± 0.2	17.3 ± 0.2	17.5 ± 0.2

All the essential oils showed significant activity towards all the tested bacteria. The leaves (PCLO) and stems (PCSO) oil of *P. caninum* showed the best activity towards *B. subtilis* and *P. aeruginosa*. The presence of safrole as the major components in the oils of *P. caninum* may contribute to theantimicrobial activity. Studies on the antimicrobial activity of single aroma compoundsfound that safrole inhibited the Gram negative, *E. coli* [40]. The extracts which showed 7.0-10.0 mm inhibition zone were considered weak towards all the tested bacteria. The MeOH stems extract of *P. caninum* (PCM) was active against *B. subtilis*, *S. aureus* and *E. coli*. In general, the Gram-positive bacteria showed better antimicrobial activity rather than the Gram-negative bacteria. Gram-positive bacteria, *B. subtilis* and *S. aureus*, were the most susceptible to this extracts, with inhibition zone of 8-10 mm. Gram-negative bacteria were the most resistant to this extract with inhibition zones between 7-8 mm diameters. These results are similar to those of previous reports in the literature, indicating that Gram-negative bacteria are more resistant than Gram-positive bacteria due to the thickness of its outer membrane [34].

Zone of inhibition including the diameter of the filter paper disc (6 mm); -, no activity; *B.s* = *Bacillus subtilis*, *S.a* = *Stapylococcus aureus*, *P.a* = *Pseudomonas aeruginosa*, *P.p* = *Pseudomonas putida*, *E.c* = *Escherichia coli*; SS = Streptomycin sulphate; PCLO = *P. caninum* Leaves Oil, PCSO = *P. caninum* Stems Oil, PCH = *P. caninum* Stems Hexane, PCE = *P. caninum* Stems Ethyl acetate, PCM = *P. caninum* Stems Methanol, PCLM = *P. caninum* Leaves Methanol; Data represent mean ± standard deviation of three independent experiments performed in triplicates.

Antibacterial screening of the pure compounds showed that they gave greater inhibition zonetowards *B. subtilis* except *N*-isobutyl-(2*E*,4*E*,14*Z*)-eicosatrienamide (29) and 5,7-dimethoxyflavanone (31), which showed no inhibition towards this bacteria. 4'-Hydroxy-5,7-dimethoxyflavone (32), and 2'-hydroxy-4',6'-dimethoxy-chalcone (34) inhibited four out of five tested microorganisms. 4'-Hydroxy-5,7-dimethoxyflavone (32) was active against the Gram-positive bacteria, *B. subtilis* and all the Gram-negative bacteria whereas 2'-hydroxy-4',6'-dimethoxychalcone (34) was found to be active towards *B. subtilis*, *S. aureus*, *P. aeruginosa* and *P. putida*.

Further tests using broth micro dilution method were carried out to determine the MIC and the MBC values of the active samples. MIC was defined as the lowest concentration of the sample that restricted the growth of the microbe, while MBC was defined as the lowest concentration of sample at which 99.9% of the bacteria were killed. Based on the MIC values, the proposed classification for antibacterial activity from plant product can be defined as strong inhibitors (<100 μg/mL), moderate inhibitors (100-500 μg/mL), weak inhibitors (500-1000 μg/mL) and inactive (>1000 μg/mL) [35].

The complete MIC and MBC values of the essential oils, crude extracts and isolated compounds are shown in Table 7.

The leaves oil of *P. caninum* (PCLO) displayed good activity towards *E. coli* with MBC value of 125 μg/mL, while the stems oil (PCSO) was found to be active towards *B. subtilis*, *S. aureus* and *E. coli*, all with MBC value 250 μg/mL. The MeOH stems extract of *P. caninum* (PCM) showed weak activity towards *B. subtilis* and *S. aureus*. The MIC values of the isolated compounds were in the range of 125-1000 μg/mL. The results showed that good antibacterial activity was shown by *N*-isobutyl-15-(3',4'-methylenedioxyphenyl)-2*E*,4*E*,12*Z*-pentadecatrienamide (30) towards *B. subtilis* with MBC value 125 μg/mL.

Table 7. MIC and MBC of the Essential Oils, Crude Extracts and Isolated Compounds

Samples	MIC (μg/mL)					MBC (μg/mL)				
	Gram-positive bacteria		Gram-negative bacteria			Gram-positive bacteria		Gram-negative bacteria		
	B. s	*S. a*	*P. a*	*P. p*	*E .c*	*B. s*	*S. a*	*P. a*	*P. p*	*E .c*
Essential Oils										
PCLO	250	250	250	125	62.5	250	500	500	250	125
PCSO	250	125	250	250	125	250	250	500	500	250
Crude Extracts										
PCH	-	-	-	-	-	-	-	-	-	-
PCE	-	-	-	-	-	-	-	-	-	-
PCM	1000	1000	-	-	1000	1000	-	-	-	-

Samples	MIC (μg/mL)					MBC (μg/mL)				
	Gram-positive bacteria		Gram-negative bacteria			Gram-positive bacteria		Gram-negative bacteria		
	B. s	*S. a*	*P. a*	*P. p*	*E .c*	*B. s*	*S. a*	*P. a*	*P. p*	*E .c*
PCLM	-	-	-	-	-	-	-	-	-	-
Compounds										
(27)	1000	-	-	-	500	-	-	-	-	-
(28)	1000	-	-	-	-	-	-	-	-	-
(29)	-	500	250	-	250	-	1000	250	-	250
(30)	125	500	-	1000	1000	125	1000	-	-	-
(31)	-	-	1000	-	-	-	-	1000	-	-
(32)	500	-	1000	1000	1000	1000	-	-	1000	-
(33)	250	-	-	-	500	250	-	-	-	500
(34)	250	250	250	250	-	500	250	250	500	-
Standard										
SS	7.81	7.81	7.81	7.81	7.81	7.81	7.81	7.81	7.81	7.81

Zone of inhibition including the diameter of the filter paper disc (6 mm); -, no activity; *B.s* = *Bacillus subtilis*, *S.a* = *Stapylococcus aureus*, *P.a* = *Pseudomonas aeruginosa*, *P.p* = *Pseudomonas putida*, *E.c* = *Escherichia coli*; SS = Streptomycin sulphate; PCLO = *P. caninum* Leaves Oil, PCSO = *P. caninum* Stems Oil; PCH = *P. caninum* Stems Hexane, PCE = *P. caninum* Stems Ethyl acetate, PCM = *P. caninum* Stems Methanol, PCLM = *P. caninum* Leaves Methanol; Data represent mean ± standard deviation of three independent experiments performed in triplicates.

Antioxidant Activity

In this study, the radical scavenging activities of the essential oils, crude extracts, and pure compounds were conducted using UV spectrophotometry assay. Table 8 summarized the DPPH radical scavenging activity of four essential oils, eight crude extracts, eight tested compounds and standards. Ascorbic acid and BHT were used as standards.

Among the oils, the leaves oil of *P. caninum* (PCLO) showed significant free radical scavenging activities at concentration 62.5 μg/mL with 16.5%, compared to BHT which gave 29.3% scavenging. However, it showed IC_{50} values more than 1000 μg/mL. Monoterpene hydrocarbons also have antioxidant activity due to the presence of strongly activated methylene groups, so the antioxidant activity of the essential oil might be mainly attributed to monoterpene hydrocarbons [36].

The MeOH stems extract of *P. caninum* (PCM) showed potent activity as free radical scavenger with IC_{50} value of 802.7 μg/mL. The free radical scavenging activity of the crude extracts may due to the presence of polyphenol compounds in the respective extracts [37]. Among the eight isolated compounds, 4'-hydroxy-5,7-dimethoxyflavone (**32**) and 5-hydroxy-7-methoxyflavone (**33**) showed the highest antioxidant activity at lowest concentration 62.5 μg/mL with IC_{50} value more than 1000 μg/mL. The other isolated compounds showed lower or less activity as free radical scavengers due to lack of hydroxyl groups in their structure.

Table 8. DPPH Radical Scavenging Activity of the Essential Oils, Crude Extracts and Isolated Compounds

Samples	Percent Inhibitionat 62.5 μg/mL	IC_{50} (μg/mL)
Essential Oils		
PCLO	16.5 ± 0.3	> 1000
PCSO	7.0 ± 0.4	-
Crude Extracts		
PCH	15.2 ± 0.2	> 1000
PCE	31.5 ± 0.4	> 1000
PCM	36.5 ± 0.3	802.7
PCLM	9.6 ± 0.3	-
Isolated Compounds		
(27)	14.1 ± 0.3	-
(28)	12.7 ± 0.7	-
(29)	6.1 ± 0.4	-
(30)	19.0 ± 0.5	> 1000
(31)	18.5 ± 0.6	> 1000
(32)	10.5 ± 0.3	-
Standard		
Ascorbic acid	37.5 ± 0.3	50.65
BHT	29.3 ± 0.2	124.38

BHT = Butylatedhydroxy toluene; PCLO = *P. caninum* Leaves Oil, PCSO = *P. caninum* Stems Oil, PCH = *P. caninum* Stem Hexane, PCE = *P. caninum* Stem Ethyl acetate, PCM = *P. caninum* Stem Methanol, PCLM = *P. caninum* Leaf Methanol, Data represent mean ± standard deviation of three independent experiments performed in triplicates.

CONCLUSION

Hydrodistillation of fresh leaf and stem of *P. caninum* gave 0.46% and 0.31% of the respective oils. The major fraction in leaf and stem oils were monoterpene hydrocarbons (21.8%) and sesquiterpene hydrocarbons (35.2%), respectively. Safrole was found to be the major component in both leaves and stems oils of *P. caninum*. Chemical investigations of the leaves and stems of *P. caninum* have successfully afforded ten compounds, which comprised of four flavones (5,7-dimethoxyflavone (27), 4',5,7-trimethoxyflavone (28), 4'-hydroxy-5,7-dimethoxyflavone (32) and 5-hydroxy-7-methoxyflavone (33)), two amides (*N*-isobutyl-(2*E*,4*E*,14*Z*)-eicosatrienamide (29), *N*-isobutyl-15-(3',4'-methylene-dioxyphenyl)-2*E*,4*E*,12*Z*-pentadecatrienamide (30)), and a flavanone (5,7-dimethoxyflavanone (31)) and a chalcone (2'-hydroxy-4',6'-dimethoxychalcone (34)).

The essential oils, crude extracts and isolated compounds were screened for antibacterial and antioxidant activities. The results for the antibacterial activity showed that all the crude extracts were found weak towards all the tested bacteria. As for the essential oils, strong antibacterial activity was shown by the leaves oil of *P. Caninum* (PCLO) towards *E. coli* with MBC value 125 μg/mL.

The antibacterial results of the isolated compounds showed that good antibacterial activity was shown by *N*-isobutyl-15-(3',4'-methylenedioxyphenyl)-2*E*,4*E*,12*Z*-pentadecatrienamide (30) towards *B. subtilis* with MBC value 125 μg/mL. The MIC values for all the tested isolated compounds were in the range 125-1000 μg/mL. The free radical scavenging activity was measured by using radical DPPH. *P. caninum* (PCLO) showed significant free radical scavenging activities at concentration 62.5 μg/mL with 16.5%. The MeOH stems extract of *P. caninum* (PCM) gave the strongest free radical scavenger with IC_{50} value of 802.66 μg/mL. 4'-Hydroxy-5,7-dimethoxyflavone (32) and 5-hydroxy-7-methoxyflavone (33) showed the highest antioxidant activity at lowest concentration 62.5 μg/mL with IC_{50} value more than 1000 μg/mL.

ACKNOWLEDGMENTS

The financial support from GUP with vote no QJ 130000.7126.02H30 and The Chemistry Department, Faculty of Science, Universiti Teknologi Malaysia are gratefully acknowledged.

REFERENCES

[1] Kinghorn, A.D., and Balandrin, M.F. (1993). *Human Medicinal Agents from Plants*. American Chemical Society, Washington DC Markakis, P. (1982). *Anthocyanins as Food Colors*. New York: Academic Press.

[2] Balunas, M.J., and Kinghorn, A.D. (2005). *Drug Discovery from Medicinal Plants. Life Sci.* 78, 431-441.

[3] Wiart, C. (2006). *Medicinal Plants of Asia and the Pacific*. Boca Raton, London, New York: Taylor & Francis, p. 15.

[4] Kato, M.J., and Furlan, M. (2007). Chemistry and Evolution of the Piperaceae. *Pure Appl. Chem.* 79, 529-538.

[5] Wiart, C. (2006). *Ethnopharmacology of Medicinal Plants: Asia and The Pacific*. Humana Press Inc. New Jersey, p. 122-123.

[6] Guerrini, A., Sacchetti, G., Rossi,. D., Paganetto, G., Muzzoli, M., Andreotti, E., Tognolini, M., Maldonado, M. E., and Bruni, R. (2009). Bioactivities of *Piper aduncum* L. and *Piper obliquum* Ruiz & Pavon (Piperaceae): Essential Oils from Eastern Ecuador. *Environ. Toxicol. Pharmacol.* 27, 39-48.

[7] Cabanillas, B.J., Lamer, A.C.L., Castillo, D., Arevalo, J., Rojas, R., Odonne, G., Bourdy, G., Moukarzel, B., Sauvain, M., and Fabre, N. (2010). Caffeic Acid Esters and Lignans from *Piper sanguineispicum.J. Nat. Prod.* 73, 1884-90.

[8] Raju, G., and Maridass, M. (2011). Evaluation of Antimicrobial Potential of *Piper* Species. *Nat. Pharm. Tech.***1**, 19-22.

[9] Bakkali, F., Averbeck, S., Averbeck, D., and Idaomar, M. (2008). Biological Effects of Essential Oils - A Review. *Food Chem. Toxicol.* **46**, 446-475.

[10] Zellner, B. A., Dugo, P., Dugo, G., and Mondello, L. (2010). *Analysis of Essential Oils. Handbook of Essential Oils; Science, Technology, and Applications*. Taylor & Francis Group, p. 85-88.

[11] Ladan, Z., Amupitan, J. O., Oyewale, O. A., Okonkwo, E. M., Ladan, E. O., Odjobo, B., and Habila, N. (2011). Chemical Composition and Biological Activity of the Volatile Oils of *Hyptis spicigera* against *Trypanosoma brucei brucei, (Tbb)* found in Northern Nigeria. *African J. Pure Appl. Chem.* **5**, 53-58.

[12] Thavaselvan a/l Superamaniam. (2008). *Extraction of Essential Oils from Jasmine Flower Using Solvent Extraction Method: A Study of Feed Ratio Effects*. Thesis Bachelor of Chemical Engineering, University Malaysia Pahang.

[13] Hussin, K., and Talip, N. (2006). *Anatomical Atlas of Malaysian Medicinal Plants* Volume 1. Penerbit Universiti Kebangsaan Malaysia, Bangi, Selangor, p. 2-4.

[14] Joaquim, V.M., Rodrigo, O.S., Joao, H. G., Maria, C.M., Elsie, F. and Massuo, J.K. (2007). Antifungal Amides from *Piper scutifolium* and *Piper hoffmanseggianum. J. Nat. Prod.* 70, 2036-2039.

[15] Xing, C.L., Daneel, F., Melissa, R.J., Qifeng, Z., Shabana, I.K, Hala, N.E, Dale, G., Troy, J.S., Ikhlas, A.K.,, Larry, A.W., and Alice, M.C. (2004). Antifungal Cyclopentenediones from *Piper coruscans. J. Am. Chem. Soc.* 126, 6872-6873.

[16] Silva, D.R., Baroni, S., Svidzinski, A.E., Bersani-Amado, C.A., and Cortez, D.A.G. (2008). Anti-inflammatory Activity of the Extract, Fractions and Amides from the Leaves of *Piper ovatum* Vahl (Piperaceae). *J. Ethnopharmacol.* 116, 569-573.

[17] Zakaria, Z.A., Patahuddin, H., Mohamad, A.S., Israf, D.A., and Sulaiman, M.R. (2010).In vivo Anti-nociceptive and Anti-inflammatory Activities of the Aqueous Extract of the Leaves of *Piper sarmentosum*. J. Ethnopharmacol. 128, 42-48.

[18] Felipe, F.C.B., Filho, J.T.S., Souza, L.E.O., Silveira, J. A., Uchoa, D.E.A., Silveira, E. R., Pessoa, O.D.L., and Viana, G.S.B. (2007). Piplartine, an Amide Alkaloid from *Piper tuberculatum*, presents Anxiolytic and Antidepressant Effects in Mice.*Phytomedicine* 14, 605-612.

[19] Hermoso, A., Jimenez, I.A., Mamani, Z.A, Bazzocchi, I.L., Pinero, J.E., Ravelo, A.G., and Valladares, B. (2003). Antileishmanial Activities of Dihydrochalcones from *Piper elongatum* and Synthetic Related Compounds; Structural Requirements for Activity. *Bioorg. Med. Chem.* 11, 3975-80.

[20] Bodiwala, H.S., Singh, G., Singh, R., Dey, C.S., Sharma, S.S., Bhutani, K.K., and Singh, I.P. (2007). Antileishmanial Amides and Lignans from *Piper cubeba* and *Piper retrofractum.J. Nat. Med.* 61, 418-421.

[21] Luis, O.R., Fernando, C., Janaina, R.S., Vanderlan, S.B., Dulce, H.S., Maysa, F., and Massuo, J.K. (2008). Radical Scavenging Capacity of *Piper arboreum* and *Piper tuberculatum* (Piperaceae). *Lat. Am. J. Pharm.* 27, 900-903.

[22] Turibio, K.T., Joseph, N., Jiawei, L., Anne, C.M.O., Sheraz, A.K.T., Shamsun, N.K., Muhammad, S.A., Bonaventure, T.N., Etienne, T., Marie, A.L.D., and Bang, L. (2008). Bioactive Aristolactams from *Piper umbellatum*. *Phytochemistry* 69, 1726-1731.

[23] Reddy, S.V, Srinivas, P.V, Praveen, B., Kishore, K.H, Raju, B.C, Murthy, U.S, and Rao, J.M. (2004). Antibacterial Constituents from the Berries of *Piper nigrum. Phytomedicine* 11, 697-700.

[24] Adams, R.P. (2001). *Identification of essential oil by gas chromatography-quadrupole mass spectroscopy*. Carol Stream IL, USA: Allured Publishing Corporation.

[25] Rajudin, E., Ahmad, F., Sirat, H.M., Arbain, D., and Aboul-Enein, H.Y. (2010) Chemical Constituents from Tiger's betel, *Piper porphyrophyllum* N. E. Br. (Fam. Piperaceae).*Nat. Prod. Res.* 24, 387-390.

[26] Jaya Chitra a/p Deagala Ramalu. (2001). *Kajian Sebatian Semulajadi daripada Empat Spesies Piper*. Thesis Master of Sciences (Chemistry), Universiti Teknologi Malaysia.

[27] Ahmad, F., Jamil, S., and Roger, W.R. (1995). Isobutylamides from *Piper ridleyi. Phytochemistry* 40, 1163-1165.

[28] Nor Akmalazura Bt. Jani. (2008). *Chemical Constituents and Bioactivty of Alpinia mutica and Two Varieties of Alpinia galanga*. Thesis Master of Sciense (Chemistry), Universiti Teknologi Malaysia.
[29] Chavi, Y., Sutthanut, K., Bungon, S., Nate, M., Chalayut, W., Sirriboon, M., and Teeraporn, K. (2007). Further Studies of Bioactive Flavonoids from *Kaemferia parviflora*. *KKU Sci. J.* 35, 37-41.
[30] Sutthanut, K., Sripanidkulchai, B., Yenjai, B., and Jay, M. (2007). Simulatneous Identification of 11 Flavonoids Constituents in *Kaempferia parviflora*. *J. Chromatogr. A* 1143, 227-233.
[31] Mackeen, M.M., Ali, A.M., El-Sharkawy, S.H., Manap, M.Y., Salleh, K.M., Lajis, N.H., and Kawazu, K. (1997). Antimicrobial and Cytotoxic Properties of Some Malaysian Traditional Vegetables. *Int. J. Pharmacog*. 35, 174-178.
[32] Zavala, M.A., Perez, G.S., and Perez, G.R.M. (1997). Antimicrobial Screening of Some Medicinal Plants. *Phytother. Res.*11, 368-371.
[33] Loo, A.Y., Jain, K., and Darah I. (2008). Antioxidant Activity of Compounds Isolated from the Pyroligneus Acid, *Rhizophora apiculata*. *Food Chem.* 107,1151-1160.
[34] Randrianarivelo, R., Sarter, S., Odoux, E., Brat, P., Lebrun, M., Romestand, B., Menut, C., Andrianoelisoa, H.S., Raherimandimby, M., and Danthu, P. (2009). Composition and Antimicrobial Activity of Essential Oils of *Cinnamosma fragrans*. *Food Chem.* 114, 680-684.
[35] Pessini, G.L., Filho, B.P.D., Nakamura, C.V., and Cortez, D.A.G. (2003). Antibacterial Activity of Extracts and Neolignans from *Piper regnellii* (Miq.) C.DC. var. *pallescens* (C.DC.) Yunck. *Mem. Inst. Oswaldo Cruz.* 98, 1115-1120.
[36] Jian-Qing, Y., Jia-Chuan, L., Xiu-Qiao, Z., Huai-Dong, Y., Dai-Zhi, T., Zhi-Xiong, L., and Guo-Lin, Z. (2011), Anticancer, Antioxidant and Antimicrobial Activities of the Essential Oil of *Lycopus lucidus* Turcz. var. hirtus Regel. *Food Chem.* 126, 1593-1598.
[37] Emrizal Rajuddin. (2008). *Phytochemicals and Biological Activities of Piper porphyrophllum N.E.Br., Piper magnibacum C.DC., Piper majusculum Ridl., and Piper sarmentosum Roxb.* Thesis Doctor of Philosophy (Chemistry). Universiti Teknologi Malaysia.
[38] Kong, J-M., Chia, L-S., Goh, N-G., Chai, T-F. And Brouillard, R. (2003). Analysis and biological of anthocyanins. *Phytochemistry*,64, 923-933.
[39] Delgado-Vargas, F. and Paredes, O (Eds) (2003). *Natural Colorants for Food and Nutraceutical Uses*. Washington, D.C.: CRC Press.
[40] Strack, D. (1997). *Plant Biochemistry*. London: Academic Press.
[41] Giusti, M.M., and Wrolstad, R.E. (2001). *Current Protocols in Food Analytical Chemistry*. New York: John Wiley & Sons.
[42] Stintzing, F.C., Stintzing, A.S., Carle, B.F and Wrolstad, R.E. (2002). Color and antioxidant properties of cyanidin-based anthocyanin pigments. *Journal of Agricultural and Food Chemistry*, 50, 6172-6181.
[43] Chaovanalikit, A., Dougherty, M.P., Camire, M.E., Darokar, and Briggs, J. (2003). Ascorbic acid fortification reduces anthocyanins in extruded blueberry-corn cereals. *Journal of Food Science*, 68, 2136-2140.
[44] Joniec, C.M., Mes, P. and Myers, J.R. (2003). Characterization and inheritance of the anthocyanin fruit (Aft) Tomato. *Journal of Heredity*, 94, 449-456.

[45] Kähkönen, M.P., Heinämäki, J., Olliainen, V. and Heinoneh, M. (2003). Berry anthocyanins: isolation, identification and antioxidant activities. *Journal of the Science of Food and Agriculture,* 83, 1403-1411.

[46] Kallithraka, S., Mohdaly, A.A.A., Makris, D.P. and Kefalas, P. (2005). Determination of major anthocyanin pigments in Hellenic native grape varieties (*Vitis vinifera* sp.): associated with antiradical activity, *Journal of Food Composition and Analysis*, 58, 375-386.

[47] Katsube, N., Iwashita, K., Tsushida, T., Yamaki, K. and Kobori, M. (2003). Induction of apoptosis in cancer cells by bilberry (*Vaccinium myrtillus*) and the anthocyanins .*Journal of Agricultural and Food Chemistry*,51, 68-75.

[48] Gómez-Plaza, E., Miñano, A. and Lóprz-Roca. (2006). Comparison of chromatic properties, stability and antioxidant capacity of anthocyanin-based aqueous extracts from grape pomace obtained from different vinification methods. *Food Chemistry*, 97, 87-94.

[49] Prata, R.B.A.E. and Oliveira L.S. (2007). Fresh coffee husks as potential sources of anthocyanins. *LWT-Food Science and Technology*, 40, 1555-1560.

[50] Muñoz, J., Garcia-Molina, F., Varen, R., Rodriguez-Lopez, J.N., García-Ruíz, P.A., Garcia-Cánovas and Tudela, J. (2007). Kinetic characterization of the oxidation of chlorogenic acid by polyphenol oxidase and peroxidase, characteristic of *o*-quinone. *Journal of Agricultural and Food Chemistry*,55, 920-928.

[51] Jiang, Y., Duan, X., Joyce, D., Zhang, Z. and Li, J. (2004). Advances in understanding of enzymatic browning in harvest litchi fruit. *Food Chemistry*, 88, 443-446.

[52] Jiménez-Atiénzar, M. Escribano, J., Cabanes, J., Gandía-Herrero, F. and García-Carmona, F. (2005). The flavonoid eriodityol as substrate of peach polyphenol oxidase. *Journal of Food Science*, 7, c540-c543.

[53] Zhang, Z., Pangi, X., Xuewu, D., Zuoliang, J. and Jiang, Y. (2005). Role of peroxidase in anthocyanin degradation in litchi fruit pericarp. *Food Chemistry*, 90, 47-52.

[54] Lee, C.Y. and Whitaker, J.R. (1995). *Enzymatic Browning and Its Prevention.* Washington, D.C.: American Chemical Society.

[55] Chisari, M., Barbagallo, R.N., and Spagna, G. (2007). Characterization of polyphenol oxidase and peroxidase and influence on browning of cold stored strawberry. *Journal of Agricultural and Food Chemistry*, 55, 3469-3479.

[56] Serrano-Martínez, A., Fortea, F.M., Del Amor, F.M. and Núñez-Delicado, E. (2008). Kinetic characterization and thermal inactivation study of partially purified red pepper (*Capsicum annuum* L.) peroxidase. *Food Chemistry*,107, 193-199.

[57] Richard-Forget, F.C. and Gauillard, F.A. (1997). Oxidation of chlorogenic acid catechins and 4-methylcatechol in model solution by combination of pear (*Pyrus communis* Cv. Williams) polyphenol oxidase and peroxidase: a possible involvement of peroxide in enzymatic browning. *Journal of Agricultural and Food Chemistry*, 45, 2475-2476.

[58] Muñoz, O., Sepúlveda, M. and Schwaerz, M. (2004). Effects of enzymatic treatment on anthocyanin pigments from grapes skin from Chilean wine. *Food Chemistry,*87, 487-490.

[59] Duan, X., Su, X., You, Y., Qu, H., Li, Y. and Jiang, Y. (2007). Effect of nitric oxide on pericarp browning of harvested longan fruit in relation to phenolic metabolism. *Food Chemistry,* 104, 571-576.

[60] Lee, M-K. and Park, I. (2007). Studies on inhibition of enzymatic browning in some foods by Du-Zhong (*Eucommia uimoides* Oliver) leaf extract. *Food Chemistry*, 1114, 154-163.

[61] Ye, S., Yo-Xin, Y., Heng, Z., Yuan-Peng, D., Feng, C. and Shu-Wei, W. (2007). Polyphenolic compound and the degree of browning in processing apple varieties. *Agricultural Sciences in China*, 6, 607-612.

[62] Gawlik-Dziki, U., Złotek, U. and Świeca,M. (2007). Characterization of polyphenol oxidase from butter lettuce (*Luctuca sativa var. capitata* L.). *Food Chemistry*, 107, 129-135.

[63] Markus, F., Daood, H.G., Kapitany, J. and Biass, P.A. (1999). Change in the carotenoid and antioxidant content of spice red pepper (paprika) as a function of ripening and some technological factors. *Journal of Agricultural and Food Chemistry*, 47, 100-107.

[64] Krinsky, N.I. (2001). Carotenoids as antioxidation. *Nutrition*, 17, 815-817.

[65] Perucka, I. and Masterska, M. (2001). Phenylalanine ammonia-lyase and antioxidant activities of lipophilic fraction of pepper fruits *Capsicum annuum* L. *Innovative Food Science and Emerging Technologies,* 2, 189-192.

[66] Navarro, J.M., Flores, P., Garrido, C. and Martinez, V. (2006). Changes in the contents of antioxidant compounds in pepper fruits at different ripening stages, as affected by salinity. *Food Chemistry*, 96, 66-73.

[67] Chatterjee, S., Niaz, Z., Gautan, S., Adhikari, S., Pasad, S.V. and Sharma, A. (2007). Antioxidant activity of some phenolic constituents from green pepper (*Piper nigum* L.) and fresh nutmeg mace (*Myristica fragrans*). *Food Chemistry*, 102, 515-523.

[68] Conforti, F., Statti, G.A. and Menichini, F. (2007). Chemical and biological variability of hot pepper fruits (*Capsicumannuum var. acuminatum* L.) in relation to maturity stage. *Food Chemistry*, 102, 1094-1104.

[69] Deepa, N, Kaur C., George, B., Singh, B. and Kapoor, H.C. (2007). Antioxidant constituents in some sweet pepper (*Capsicumannuum* L.). *LWT-Food Science and Technology,* 40, 212-219.

[70] Da Costa, C.T., Nelson, B.C., Margolis, S.A. and Horton, D. (1998). Separation of blackcurrent anthocyanins by capillary zone electrophoresis. *Journal of Chromatography A,* 799, 321-327.

[71] Longo, L., Scardino, A. and Vasapollo, G. (2007). Identification and quanitification of anthocyanins in the berries of *Pistacia lentiscus* L., *Phillyrea latifolia* L. and *Rubia peregrine* L. *Innovative Food Science and Emerging Technologies*, 8, 360-364.

[72] Lapornic, B., Prošek, M. and Wonder, A.G. (2005). Comparison of extract prepared from plant by-products using different solvents and extraction time. *Journal of Food Engineering,* 71, 214-222.

[73] Liyana-Pathirana, C., and Shahidi F. (2006). Optimization of extraction of phenolics from wheat using response surface methodology. *Food Chemistry,* 93, 45-56.

[74] Robards, K., Prenzler, P.D., Ticker, G., Swatsitang, P. and Glover, W. (1999). Phenolic compounds and their role in oxidative process in fruits. *Food Chemistry*, 66, 401-436.

[75] Revilla, E., Ryan, J.M. and Martín-Oetega, G. (1998). Comparison of several procedures used for the extraction of anthocyanins from red grapes. *Journal of Agricultural and Food Chemistry,* 46, 4592-4597.

[76] Mohapatra, D., Frias, J.M., Oloveira, Z.M., Bira, Z.M. and Kerry, J. (2008). Development and validation of a model to predict enzymatic activity during storage of

cultivated mushrooms (*Agaricus bisporus* spp.). *Journal of Food Engineering*,86, 39-48.

[77] Park, J-N., Fukumato, Y., Fujita, E., Tanaka, T., Washio, T., Otsuka, S., Shimazu, T., Watanabe, K. and Abe, H. (2001). Chemical composition of fish sauces produced in southeast and East Asian countries. *Journal of Food Composition and Analysis*, 14, 113-125.

[78] Boonsiri, K., Ketsa, S. and Van Doorn, W.G. (2007). Seed browning of hot peppers during low temperature storage. *Posthavest Biochemistry and Technology*, 45, 358-365.

[79] Göğüş, F., and Eren, S. (1998). Effect of temperature and pH on nonenzymic in minced dried pepper during storage. *Turkish Journal of Engineering Environmental Science*, 22, 33-38.

[80] Zhang, J., Wang, L., Shu, Q., Liu, Z., Li, C., Zhang, J., Wei, X. and Tian, D. (2007). Comparison of anthocyanins in non-blothes and blotches of the petals of Xibei tree peony. *Scientia Horticulturae*, 114, 104-111.

[81] Escribano-Bailón, M.T. and Santos-Buelga, C., (2003). *Method in Polyphenol Analysis*. United Kingdom: The Royal Society of Chemistry.

[82] Jiang, Y-M. (1999). Purification and some properties of polyphenol oxidase of longan fruit, *Food Chemistry*, 66, 75-79.

[83] Belcarz, A., Ginalska, G., Kowalewska, B. and Kulesza, P. (2008). Spring cabbage peroxidase – Potential tool in biocatalysis and bioelectrocatalysis. *Phytochemistry*, 69, 627-636.

[84] Doğan, S. and Doğan, M. (2004). Determination of kinetic properties of polyphenol oxidase from Thymus (*Thymus logicaulis* subsp. *chaubardii* var. *chaubardii*). *Food Chemistry,* 88, 69-77.

[85] Fatima, A. and Husain, Q. (2007). A role of glycosyl moieties in the stabilization of bitter gourd (*Momordica charantia*) peroxidase. *International Journal of Biological Macromolecules,* 41, 56-63.

[86] Saraiva, J.A., Nunes, C.S. and Coimbra, M.A. (2007). Purification ad characterization of olive (*Olea europaea* L.) peroxidase – Evidance for the occurrence of pectin binding peroxidase. *Food Chemistry*, 101, 1571-1579.

In: Natural Heterocycles
Editors: K. L. Ameta and Saksit Chanthai
ISBN: 978-1-63463-424-3

Chapter 4

MARINE BASED ANTICANCER AGENTS – APRATOXIN AND ITS ANALOGUES

N. C. Desai* and Amit Dodiya
Division of Medicinal Chemistry, Department of Chemistry,
(UGC NON-SAP & DST-FIST Sponsored),
Mahatma Gandhi Campus, Maharaja Krishnakumarsinhji Bhavnagar University,
Bhavnagar, Gujarat, India

ABSTRACT

Two thirds of the earth is covered by sea and till this day, we have not explored the rich unfathomable diversity of marine chemicals and biological diversity. Marine chemicals are obtained from the marine flora and fauna. Marine resources are thus very important for the new drug discovery program, the shining example is apratoxin. In this chapter our main focus is directed on the structures of various analogues of apratoxin and its total synthesis established by various chemists i.e. apratoxin (A-G), dehydroapratoxin A and oxaapratoxin. Reterosynthesis analysis for the synthesis of the apratoxin has been reported.

INTRODUCTION TO MARINE DRUGS

Despite the fact that oceans cover more than 70% of the earth‘s surface, the reconnaissance of marine ecosystems began only in the mid 1970’s with the rise of modern snorkeling, the commencement of scuba in 1970 and later, around 1990, with the use of remotely operated vehicles (ROVs) [1]. With the unceasing exploitation of marine environment, the attention of scientists diverted to microorganisms such as marine cyanobacteria, marine fungi, and several other groups of marine bacteria due to their biological and habitat diversity, which culminated in the ability to produce metabolites having unmatched structures [2]. Microorganisms present themselves as a prolific source of

*E-mail: dnisheeth@rediffmail.com; Telefax: +91 278 2439852

structurally diverse bioactive metabolites, generating some of the most important active ingredients known today [3]. It was discovered that many compounds previously isolated from marine microorganisms, such as sponges and tunicates, were in fact, metabolic products of associated microorganisms [4,5]. Due to an array of bioactivities such as anti-tumor, anti-microtubule, anti-proliferative, photo protective, antibiotic and anti-infective [6-11], marine natural products (MNP) are exceptionally intriguing high-value ingredients for utilization in the pharmaceutical industry and consequently, more and more companies are investing in this field.

The discovery and identification of two nucleosides, spongothymidine and spongouridine in the early 1950's from the Caribbean marine sponge *Cryptotethiacrypta* laid a concrete platform for MNP as promising new chemical entities of potential therapeutic value [12]. Since then, numerous other therapeutic agents are procured through isolation from natural sources, by chemical synthesis or by other methods. There are seven therapeutic agents (four anticancer, one antiviral, one pain control, and one for hypertriglyceridemia) that are derived from marine environment. [13] (Table 1). Of these, actual chemical structures of two agents are isolated, and five others are synthetic agents that capture their chemical idea from a marine product.

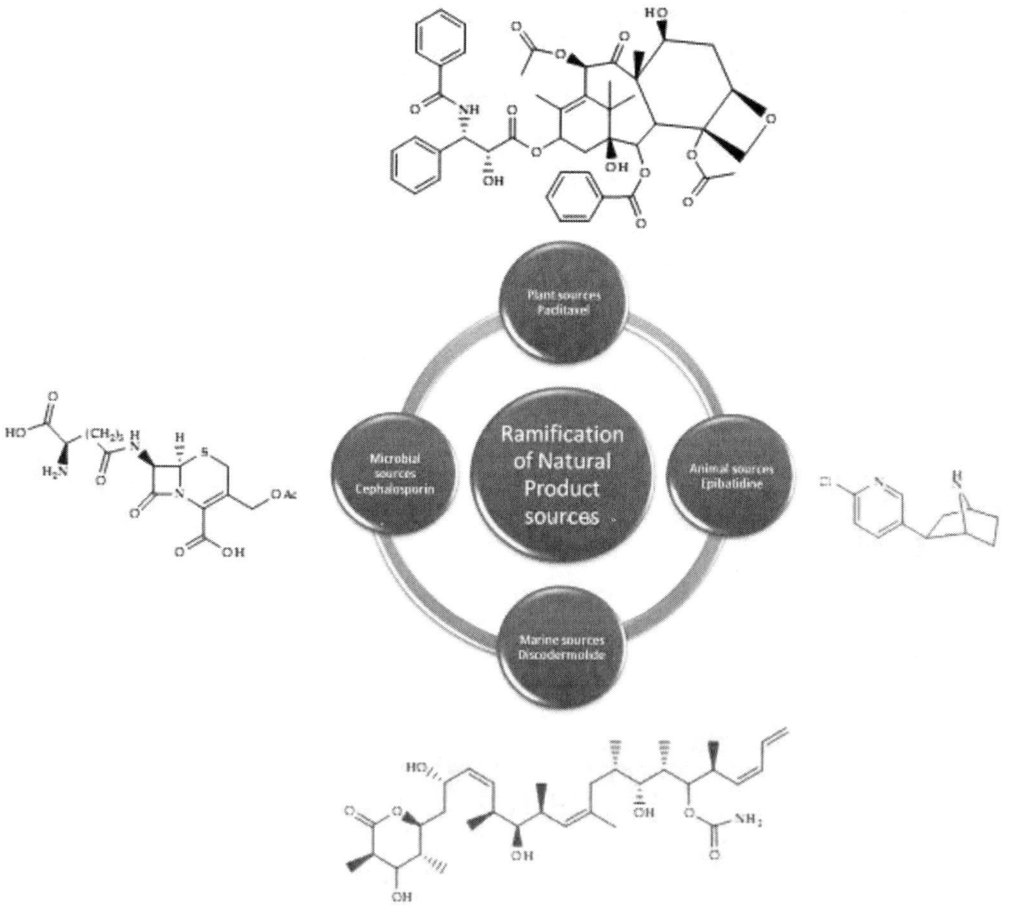

Figure A.various naturallyoccurring medicinal sources proliferate into four categories.

Table 1. Outstanding examples of Six Marine Natural Products inspired Compounds that are FDA-Approved Agents with details of their collected sources, predicted Biosynthetic Source, Molecular Target, and Disease Treated

Clinical Status	Compound Name	Natural Product or Derivative	Collected Source Organism	Predicted Biosynthetic Source	Biosynthetic Class of Agent	Molecular Target	Disease Area
FDA Approved	Cytarabine (Ara-C)	Derivative	Sponge	Bacterium	Nucleoside	DNA polymerase	Cancer
FDA Approved	Vidarabine (Ara-A)	Derivative	Sponge	Bacterium	Nucleoside	Viral DNA polymerase I	Antiviral
FDA Approved	Ziconotide	Natural Product	Cone Snail	Mollusk	Cysteine Knot Peptide	N-type Ca channel	Pain
FDA Approved	Eribulin Mesylate (E7389)	Derivative	Sponge	Bacterium	Complex Polyketide	Microtubule	Cancer
FDA Approved	Omega-3-acid ethyl esters	Derivative	Fish	Microalgae	Omega-3 fatty acids	Triglyceride synthesizing enzymes	Hypertriglyceridemia
FDA Approved	Trabectedin (ET-743) (EU registered only)	Natural Product	Tunicate	Bacterium	NRPS-derived Alkaloid	Minor groove of DNA	Cancer
FDA Approved	Brentuxi-mabvedotin (SGN-35)	Derivative	Mollusk	Cyano bacterium	Antibody drug conjugate (MM auristatin E) – Linear NRPS/PKS	CD30 and microtubule	Cancer

In the most generic technique for bioactive discovery, a natural product is first extracted from the source, screened against a specific target, isolated by a bioassay-guided isolation procedure, fractionated and purified, yielding typically a single biologically active compound. Several drug candidates derived from various naturally occurring medicinal sources proliferate into above four categories (Figure-A).

Looking at the emphasis given to MNPs in the drug discovery program notably for the treatment of cancer, in the present chapter, our focal point is marine cyanobacteria, which produce an ornate lineup of secondary metabolites embellished with a broad spectrum of bioactivities. These bacteria are known for their great biosynthetic potential and ability to produce a wide range of cytotoxic compounds, some of them with possible therapeutic applications such as curacins [14] and dolastatins [15-16]. Hendrik Luesch et al., described Apratoxin A (1) as the most potent cytotoxin produced by a variety of the cyanobacterium Lyngbya sp. from Finger's Reef, Apra Harbor, GU, USA [17]. Populations of the same cyanobacterium, recognized by their dark reddish color, strong attachment to the reef substrate, and predictable coexistence with the red alpheid shrimp Alpheus frontalis, were also chemically probed at other sites around the island of Guam and in Palauan waters. These organisms displayed complementary natural products chemistry, including the production of Apratoxin A (1).

Prominence is given to naturally occurring anticancer compound Apratoxin and its analogues which are isolated from marine cyanobacteria. Synthetic methods are discussed in the form of total synthesis and retrosynthesis analysis. Pharmacology of naturally occurring as well as synthetic apratoxin and its derivatives are discussed in detail. Finally we have chalked out a futuristic road map for the drug discovery program from MNPs.

BRIEF INTRODUCTION OF APRATOXIN

Apratoxin A (Figure-1) and its analogues are a recently discovered class of cyclodepsipeptides from marine origin, which possess high cytotoxicity [16]. Due to their potent antitumoral activity and uncommon mechanism of action, they have attracted much attention. Thus, total synthesis of apratoxins became such a competitive challenge for chemists and several routes for the total synthesis have been reported [18-19].

Apratoxin A (1), isolated from marine cyanobacterium Lyngbyamajuscula, exhibits potent cytotoxic activity against KB and LoVo cancer cells with IC_{50} values of 0.52 and 0.36 nm respectively [16]. This 25-membered macrocyclic structure depsipeptide is comprised of proline, three methylated amino acids (N-methylisoleucine, N-methylalanine, and o-methyltyrosine), α,β-unsaturated modified cysteine residue (moCys), a dihydroxylated fatty acid moiety, and 3,7-dihydroxy-2,5,8,8-tetramethylnonanoic acid (Dtena) (Figure 2-a). Apratoxin A has a thiazoline ring in its macrocycle and epimerization of thiazoline at C34 position as well as β-elimination of the hydroxy group at C35 position were observed under weak acidic conditions, such as in $CDCl_3$ and the resulting dehydroapratoxin A (2) (Figure-3), lost its cytotoxicity [20]. Apratoxin B–C-D (Figure-2-b) &Apratoxin F-G (Figure-2-c), which are structurally related to Apratoxin A, were also isolated and their biological activity was investigated [21]. For the first time Forsyth and Chen reported the total synthesis of Apratoxin A (1) [22-23].

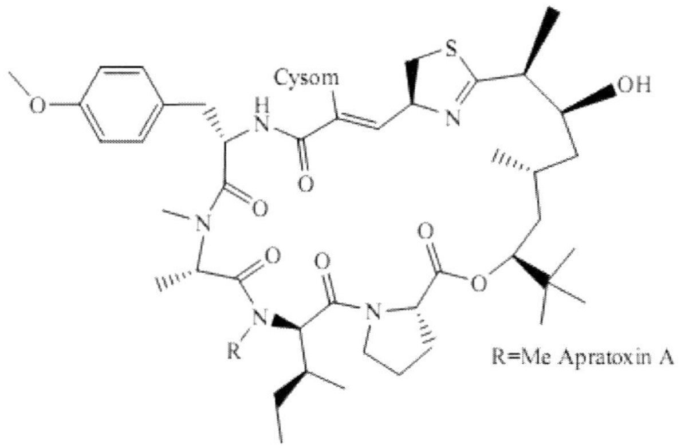

Figure 1. Apratoxin A.

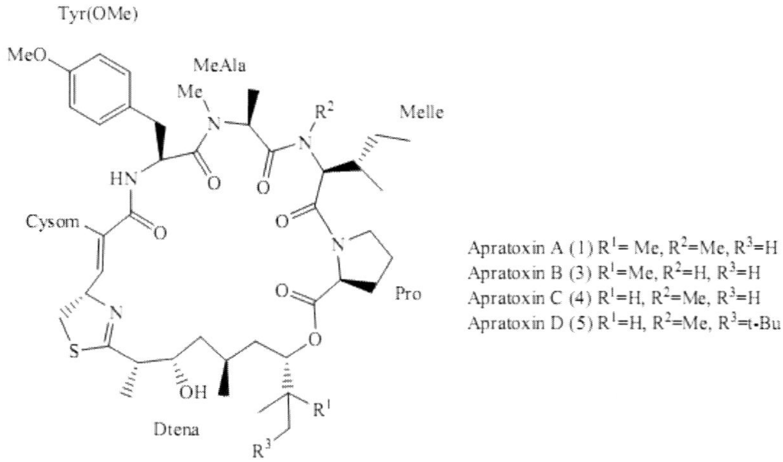

Figure 2a-2b. Apratoxin A, B, C and D.

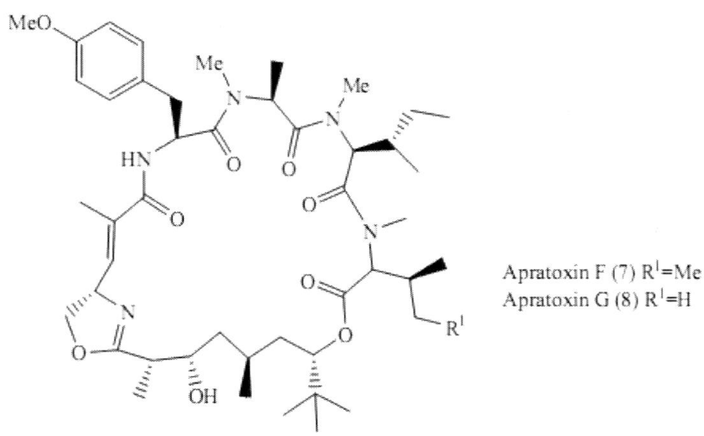

Figure 2c. Apratoxin F and G.

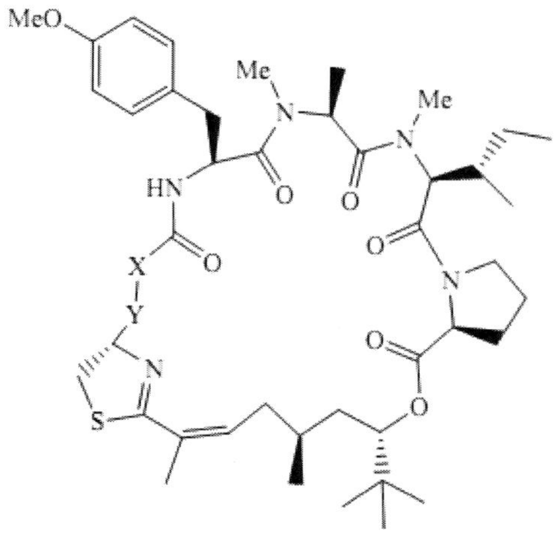

Dehydroapratoxin A (2) R^4 = Me, X-Y = CHMe=CH_2
apratoxin E (6) R^4 = H, X-Y = CH_2-CH_2

Figure 3. Dehroapratoxin A.

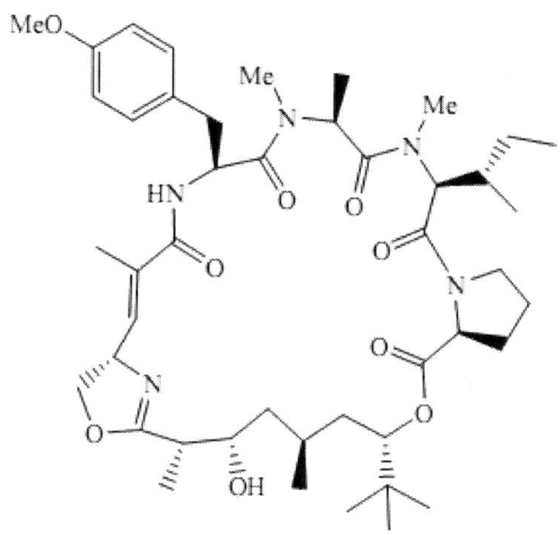

Figure 4. Oxoapratoxin (9).

Oxoapratoxin A (9) (Figure-4), an oxazoline analogue, is more accessible and has been synthesized by other groups. In particular, Cavelier et al. recently reported the solid-phase total synthesis of 9, which enabled the synthesis of various analogues of oxoapratoxin derivatives [24]. In this synthesis, two linear precursors on a polymer support, and macrolactamization were successfully performed between O-MeTyr and a modified serine, and also between the modified serine and Dtena. The oxazoline was finally obtained from the modified serine moiety using dehydrocyclization. They also mentioned that the formation of oxazoline prior to macrocyclization led to 9 without any adverse affect.

SYNTHETIC STRATEGY FOR SYNTHESIS OF APRATOXIN

The quantity of isolated bio-active is one of the major challenges faced in the drug discovery program. Continuous supply in sufficient quantity for screening of compounds, carrying out preclinical and clinical trials and developing a promising compound for market are the burning issues.

Many reliable and tested solutions are offered to address these problems taking aid of the advances in synthetic organic chemistry and retrosynthetic analysis. Apratoxin has grabbed attention of several organic chemists as a potential anticancer lead and extensive attempts are made to develop its synthetic routes. The synthetic analogues were screened for anticancer activity by the same groups. Some shining examples of the synthetic routes are discussed over here.

1. Thomas et al., [16] proposed Total Synthesis and Structure Determination of Apratoxin A, a potent novel cytotoxin from the marine Cyanobacterium Lyngbyamajuscula. The common structure of the compound is given below (Figure-5):

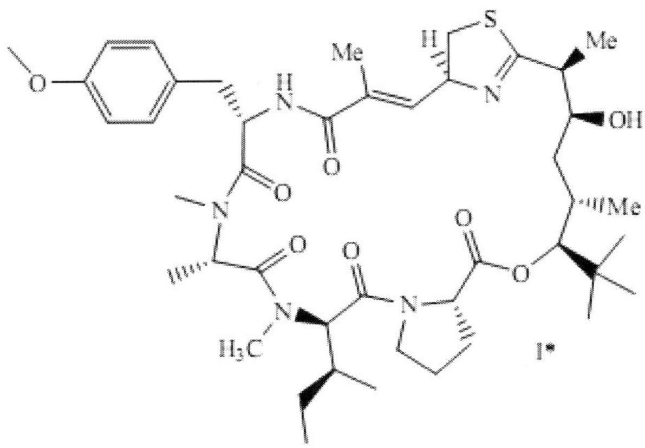

Figure 5. Common structure of marine cyanobacterium lyngbyamajuscula.

2. Paul et al., [25] have synthesized diverse structures of New Alkaloids from Palauan Collections of the Apratoxin-Producing Marine Cyanobacterium Lyngbya sp. The general structures of the compounds are given below:

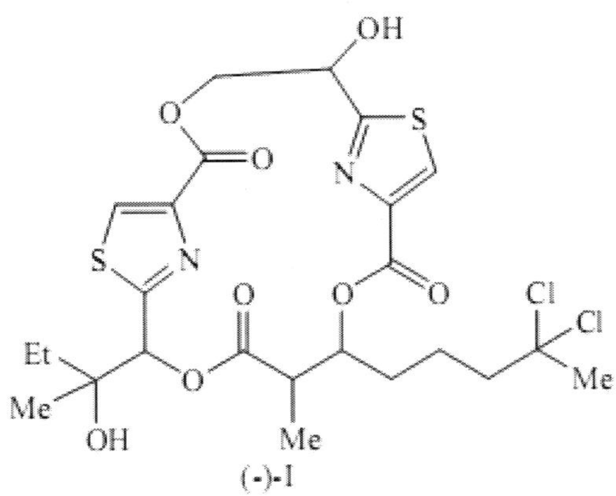

Figure 6. Lyngbyabellin C.

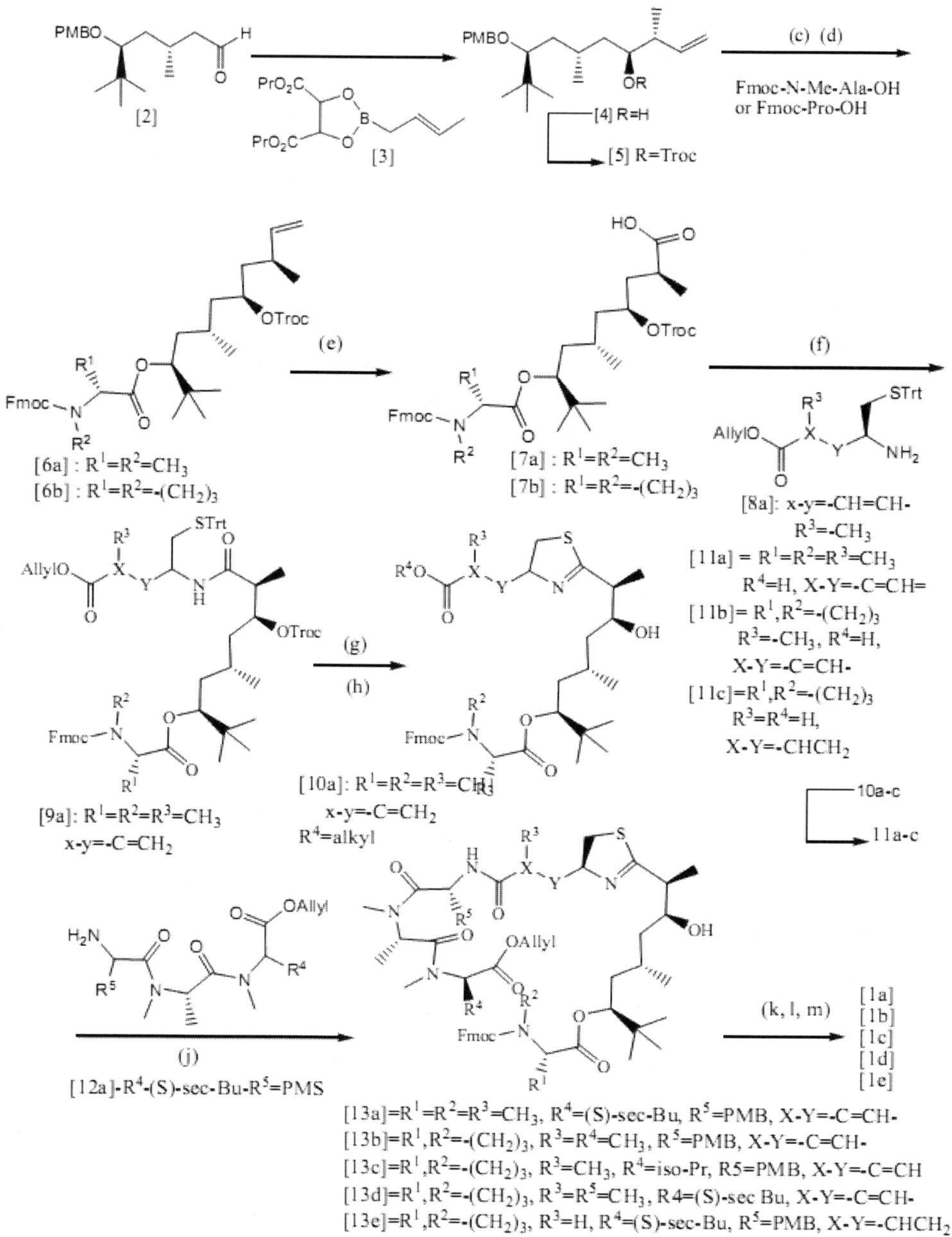

[a]Reagents and conditions: (a) MS 4 °A, toluene, -78°C, (b) TroCCl, DMAP, pyridine, CH_2Cl_2, (c) DDQ, CH_2Cl_2, $-H_2O$,(d) $Cl_3C_6H_2COCl$, DIEA, THF, DMAP, toluene (e) OsO_4, oxone, $NaIO_4$ (f) HATU, DIEA, CH_2Cl_2, (g) Ph_3P (O), Tf_2O, CH_2Cl_2, O°C, (h) Zn, NH_4OAc, THF, (i) $Pd(PPh_3)_4$, N-methylakniline, THF, (j) HATU, DIEA, CH_2Cl_2, (k) Pd-$(PPh_3)_4$, N-methylaniline, THF, (l)Et_2NH, MeCN, (m)HATU, DIEA, CH_2Cl_2, Troc, 2,2,2-trichloroethoxycaronyl; DMAP, 4-(dimethylamino)-pyridine; DDQ, 2,3-dichloro-4,5-dicynobenzoquionone; Fmo,9-fluorenyl-methoxycarbonyl; Trt, triphenylmethyl; HATU, o-(7-azabenzotriazol-1-yl)N,N,N',N'-tetramethyluroniumhexafluorophosphate; DIEA, diisopropylethylamine; and Tf, trifluoromethylsulfonyl.

Scheme 1. Synthesis of Analogues of Apratoxin A[a].

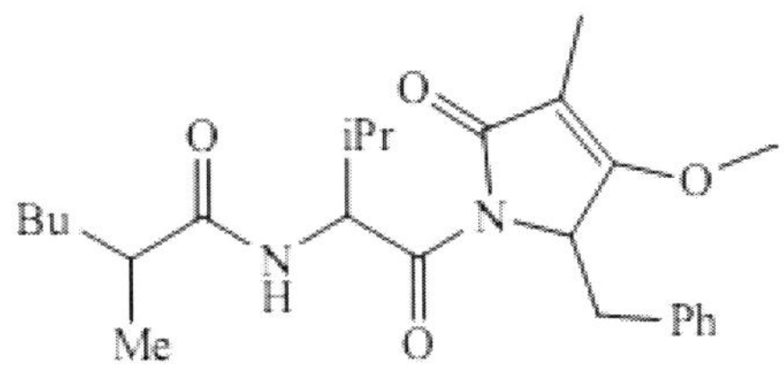

Figure 7. N-acylpyrrolinone derivative.

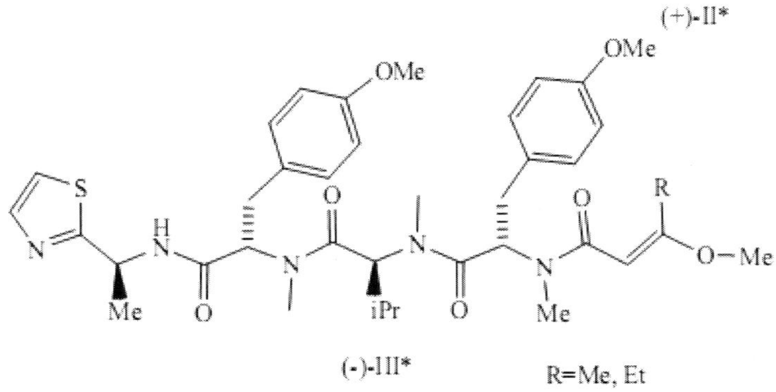

Figure 8. Lynggbyapeptins derivates.

Luesch et al., have also carried out an alanine scan (without changing the N-alkylation pattern) and synthesized a valine analogue to broadly probe the hydrophobicity requirements at the R3 position (Ile f Val f Ala; (1a)), which led to the first synthesis of the natural product apratoxin F (1a) and positional congeners (1bd), providing new insight into the SAR of Apratoxins.

Scheme 1 illustrates the main synthetic approaches to obtain the desired apratoxin analogues, closely paralleling reported synthetic strategies and retrosynthetic analysis for apratoxin A with minor modifications [26-29]. They have used the same strategy to synthesize Apratoxin A for direct comparison in their biological assays.

3. Forsyth et al., [30] reported the synthesis of apratoxin 2,4-disubstituted thiazoline via an intramolecular Aza-Wittig Reaction. In continuation with a work on the synthesis of thiazoline-containing domain of apratoxin natural products, the conversion of vicinal azido-thiolesters (14) into 2,4-disubstituted thiazolines (15) via a sequential one-pot Staudinger reduction and intramolecular Aza-Wittig reaction was investigated. The reaction scheme is given below (Reaction Scheme-2):

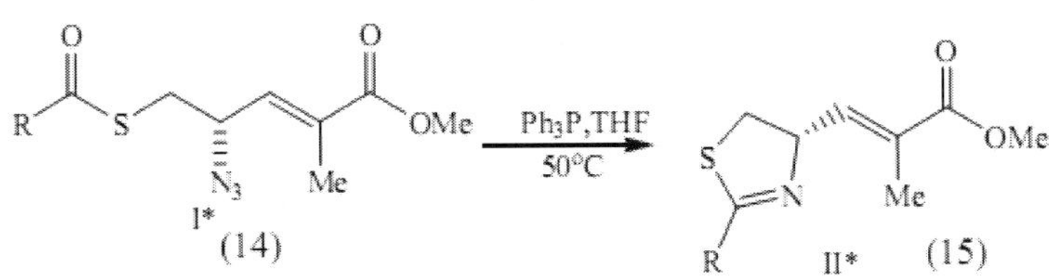

Scheme 2. (Continued).

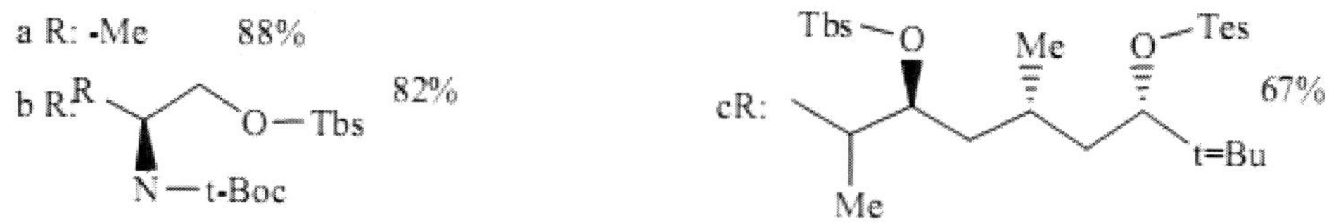

Scheme 2.synthesis of the Apratoxin 2,4-disubstituted thiazoline via an intramolecular Aza-Wittig Reaction.

4. Ma et al. [31] synthesized oxazoline analogue (a) of Apratoxin A (b). The structure is given below:

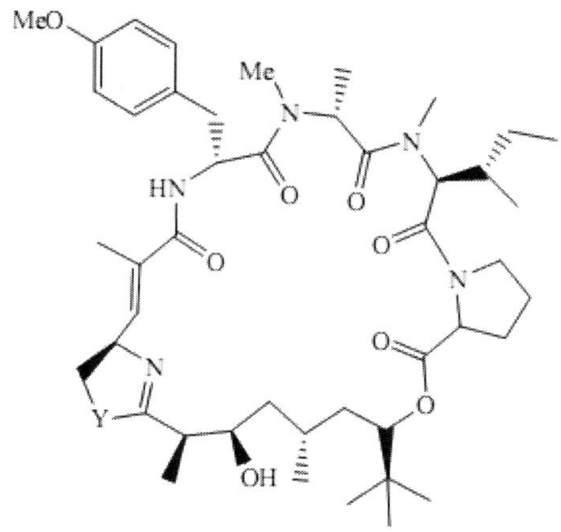

a : Y= O
b : Y= S

Figure 9. Oxazoline Analogue of Aprotoxin A.

5. Takahashi et al., [32] reported the total synthesis of Apratoxin A. The structure is as under:

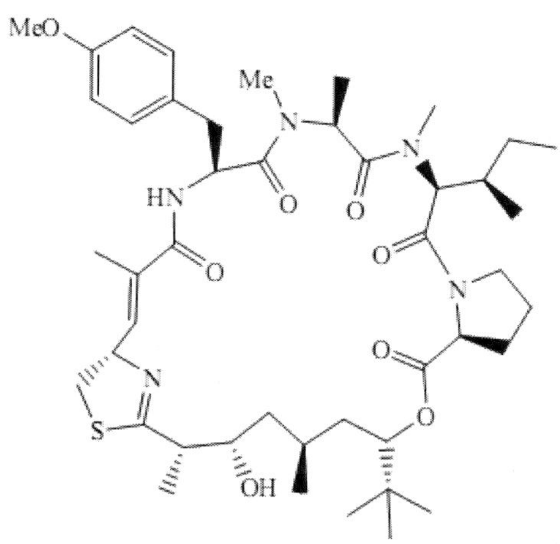

Figure 10. Thiazoline Analogue of Aprotoxin A.

6. Gilles et al., [33] synthesized a flexible synthetic sequence for the synthesis of polyketide region of apratoxin. The common step of the synthesis is a crotylation reaction. Stereospecific aldolisation, sulfate ring opening or Jacobsen HKR is also

highlighted. This synthetic scheme led to the synthesis of several analogues and these examples raised the possibility of synthesizing numerous analogues of Apratoxins. This route of synthesis has opened up the formation of oxoapratoxin analogues for future SAR studies of potent antitumoral compounds.

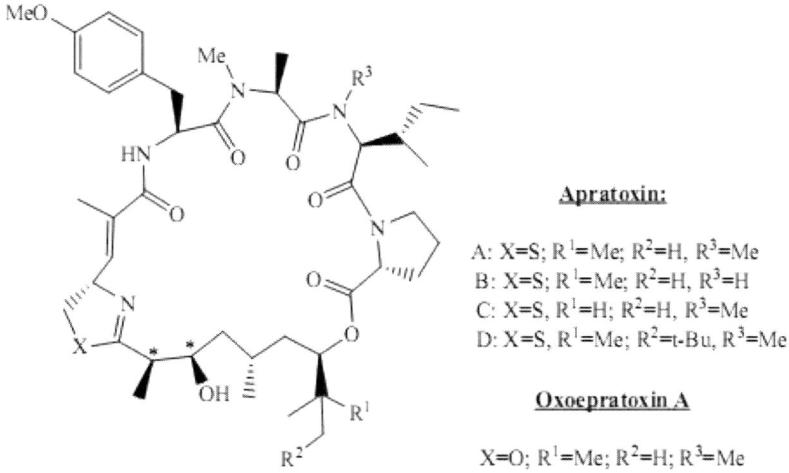

Figure 11. Apratoxin A and analogues.

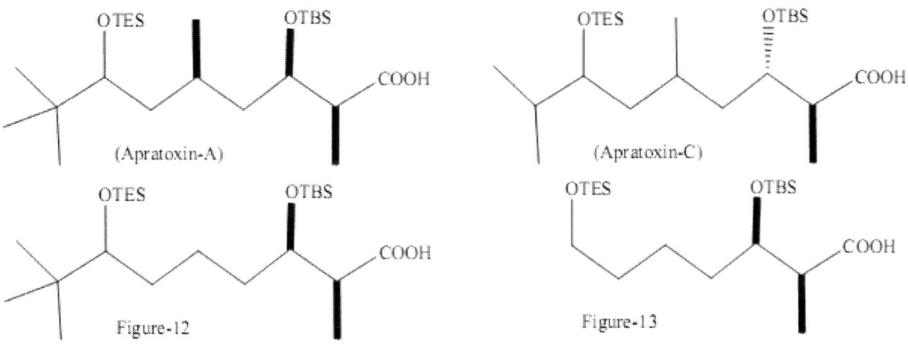

Figure 12 & 13. The four synthesized polyetides.

Apratoxin A and analogues arerecently discovered classes of cyclodepsipeptides from marine origin, which possess high cytotoxicity (Figure-11). We directed our efforts to synthesize versatile synthetic Apratoxin analogues, polyketide moiety of Apratoxin A being one among them. As a proof of efficiency, the polyketide fragment corresponding to Apratoxin A was synthesized on a gram scale. The four synthesized polyketides are displayed in (Figure-12 & 13).

7. Takahashi et al. [32] have described total synthesis of Apratoxin A in which thiazoline formation (16) was accomplished from the moCys containing amide by using $PPh_3(O)/Tf_2O$. Deprotection of the Troc and allyl ester coupled with tripeptide 3, and deprotection of the allyl ester and the Fmoc (17), followed by macrolactamization yielded ApratoxinA (1). The reaction is described below (Reaction scheme-3):

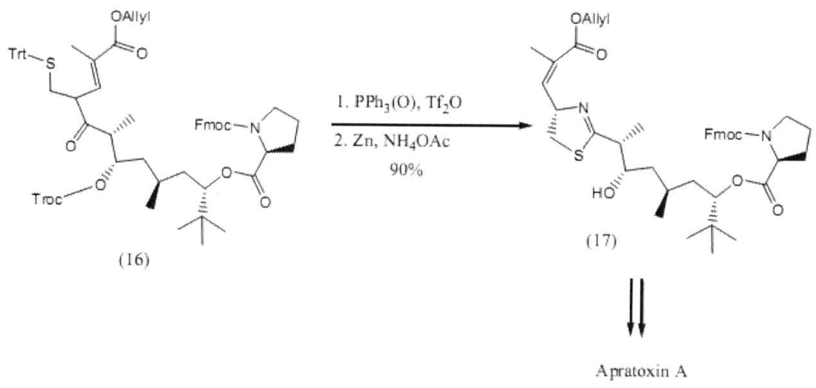

Scheme 3.Total synthesis of Apratoxin A.

RETROSYNTHESIS

Due to the technological advancements, retrosynthesis has become an impertinent tool for drug discovery from natural source. Corey received the Nobel Prize for Chemistry for his concepts of retrosynthetic analysis and brilliant contributions to organic synthesis [34]. The complicated scaffolds of natural products are simplified by the retrosynthetic analysis. In continuation to this, some leading workers have developed retrosynthetic pathways of Apratoxin and its analogues. The details of discussion are as under.

1. Liu et al., [35] described novel synthesis of Apratoxin A (Reaction scheme-4), with key steps including the assembly of its ketide segment through d-proline catalyzed by direct aldol reaction, Oppolzer's anti-aldol reaction and the preparation of its thiazoline unit in a biomimetic synthesis. An oxazoline analogue of apratoxin A is elaborated by a similar approach. This compound has potency against HeLa cell proliferation only slightly lower than that of Apratoxin A, whilst a C(40)-demethylated oxazoline analogue of Apratoxin A displays a much lower cytotoxicity and C(37)-epimer and C(37) demethylation product of these new analogues are inactive. These results suggest that the two methyl groups at C(37) and C(40) and the stereochemistry at C(37) are essential for the potent cellular activity of the oxazoline analogue of Apratoxin A. Further biological analysis revealed that both synthetic Apratoxin A and its oxazoline analogue inhibited cell proliferation by causing cell cycle arrest in the G1 phase.

As outlined in Scheme 4, synthetic strategy was based on the retrosynthetic degradation of apratoxin A into a tripeptide part (A) and a thiazoline unit (B). Further, bond disconnection of B gave a cysteine-derived α,β-unsaturated ester (D) and an L-proline-based carboxylic acid (19a). The installation of its β-hydroxy-2,4-disubstituted thiazoline unit is challenging for the assembly of intermediate B, as most existing methods for thiazoline ring elaboration are forbidden on this unstable moiety. Indeed, Forsyth and Chen's success in the total synthesis of Apratoxin A was highly dependent on their newly developed methodology for preparing thiazoline rings [30]. This problem prompted them to consider replacing the thiazoline ring with a more conveniently accessible oxazoline ring for its further SAR studies [36-37].

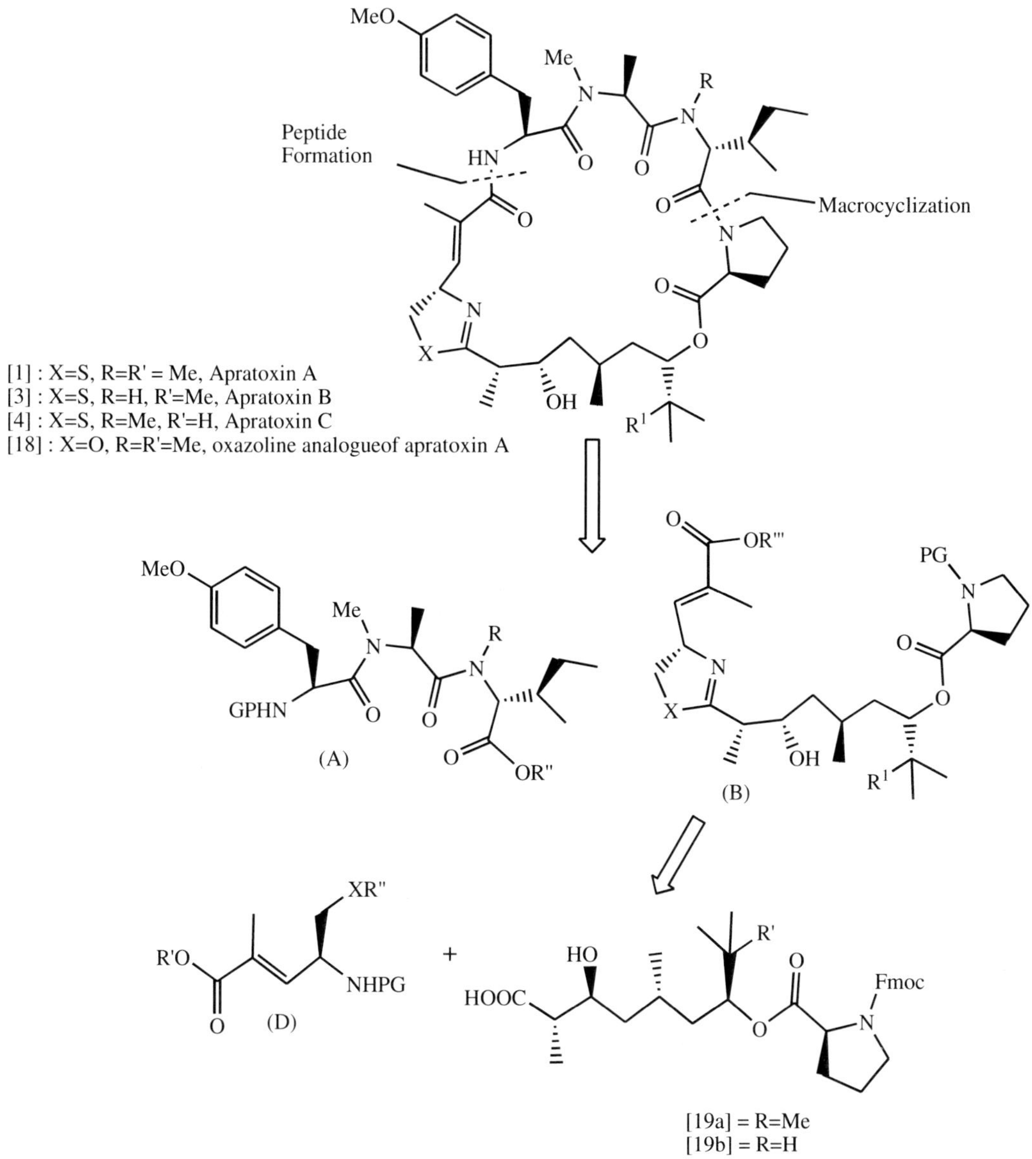

Scheme 4. Structures of Apratoxins and an oxazoline analogue with their reterosynthetic analysis.

2. Tao et al., [19] synthesized Apratoxin A 1 which is a potent cytotoxic agent is extracted from a marine cyanobacterium. Details of the reaction scheme are as follow:

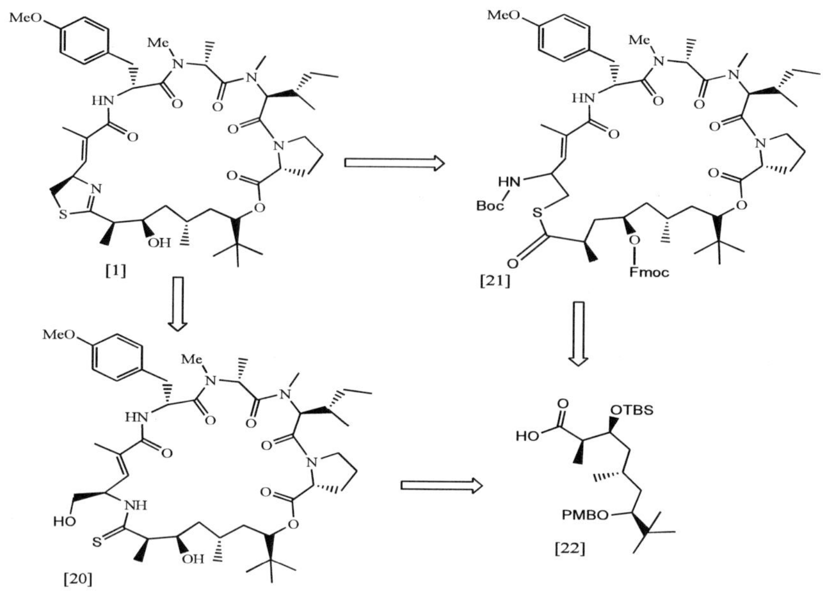

Scheme 5. Reterosynthesis analysis of Aprotoxin-A.

3. Doi et al., [38] described Solid-phase total synthesis of (-) Apratoxin A, its analogues and their biological evaluation. The reaction scheme is mentioned below:

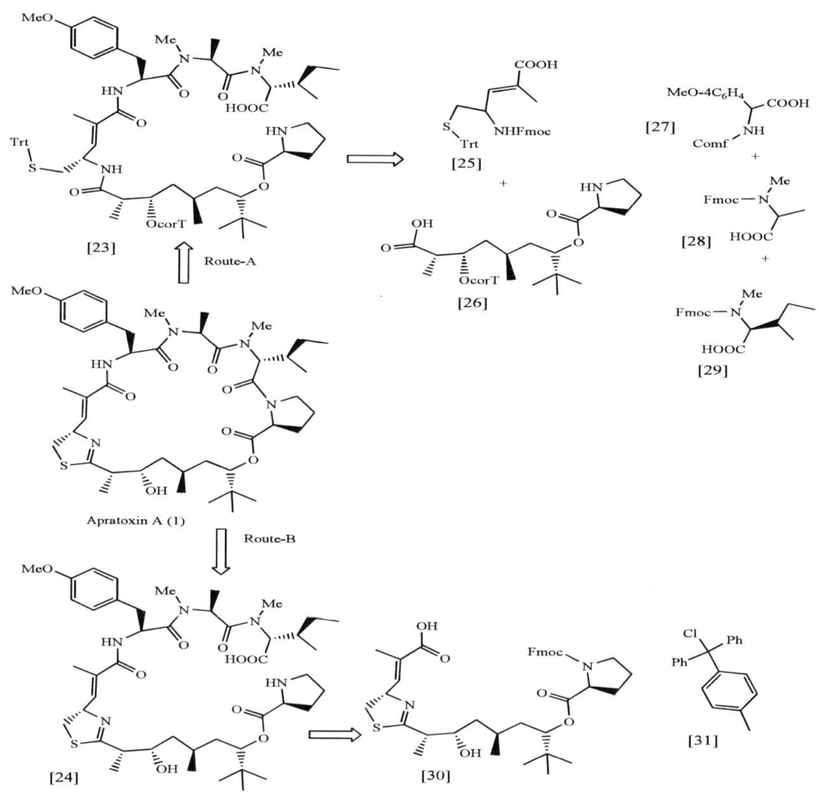

Scheme 6. Two synthetic routes toward solid phase synthesis of Apratoxin A (1).

4. Forsyth et al., [39] described total synthesis of 1, where the robust proline ester was installed early, while assembly of the sensitive thiazoline moiety was deferred until late stage. The details of the reaction are outlined in scheme 7.

Pharmacology: Apratoxins are cytotoxic marine natural products that prevent cotranslational translocation in the early stage of secretory pathway. Cancer cells are characterized by aberrant pro-growth signaling, usually induced by ligand binding to their corresponding receptor tyrosine kinases (RTKs), which are over expressed in many cancers [40]. Consequently, RTKs have become prime therapeutic targets resulting in the development of specific small molecules (e.g., erlotinib) and antibodies (e.g., cetuximab and panitumumab) targeting epidermal growth factor receptors [41]. However, drug resistance is a clinical problem, and combination therapy or the use of broader-spectrum RTK inhibitors (e.g., sorafenib and sunitinib), targeting vascular endothelial growth factor (VEGF) receptor (VEGFR) and platelet-derived growth factor receptor (PDGFR), may be more advantageous [42]. Alternatively, the ligands themselves may be targeted, whichled to the development and FDA approval of an anti-VEGF-A antibody (bevacizumab) for colorectal cancer treatment [43].

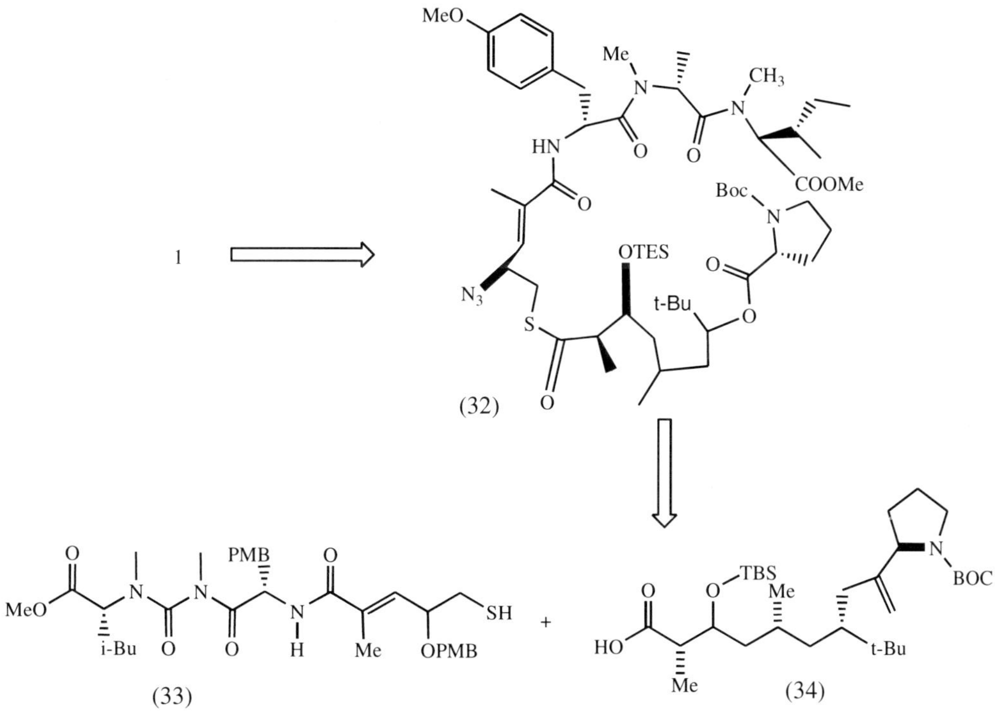

Scheme 7. Retrosynthetic disconnections of Apratoxin A.

Preliminary biological studies have shown that Apratoxin A affects neither microtubule polymerization dynamics nor topoisomerase I, but further investigation of its mode of action has been hampered by the scarcity of natural material. In addition, apratoxin A was poorly tolerated on mice, probably due to its intrinsic toxicity to normal cells and lack of selectivity for tumor cell lines [16]. One possible solution to this problem is to carry out structure-activity relationship (SAR) studies through synthesis and testing of new analogues of apratoxin A. In fact, different cytotoxicity patterns displayed by its new family members, apratoxins B (2) and C (3), suggest that the cytotoxicities of different members of this family of natural products are highly dependent on their structures [44]. It was reported that 3 exhibits *in vitro*IC_{50} cytotoxicity values almost identical to those of 1 whilst 2 is significantly less cytotoxic than 1. These observations together with the intriguing structure of apratoxinA, have attracted considerable synthetic attention, which culminated in the first total synthesis of apratoxin A by Forsyth and Chen [45]. Here we report a novel synthetic route forApratoxinA, which also enabled the synthesis of a number of analogues for preliminary SAR and mechanistic studies [31].

As Apratoxin A (1) exhibits potent cytotoxicity towards cancer cells, elucidation of its biological mode of action would be an interesting development and could lead to a new anticancer agent. In 2006, Luesch et al. proposed the mode of action of Apratoxin A by functional genomics, where Apratoxin A induces pronounced G1 cell cycle arrest and apoptosis by the inhibition of phosphorylation of the signal transducer and the activator of transcription (STAT) 3 [46]. Upstream events were recently reported in which Apratoxin A inhibits Janus kinase (JAK)/STAT signaling through rapid down-regulation of interleukin 6 signal transducer (gp130), and also depletes cancer cells of 1 cancer-associated receptor tyrosine kinase by preventing their N-glycosylation, which led to their rapid proteasomal degradation [47].

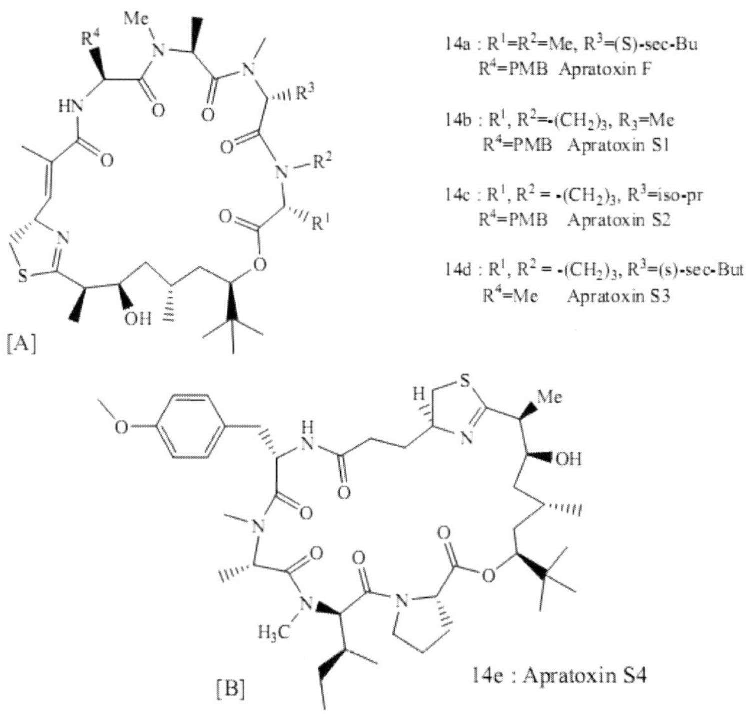

Figure 14. (A & B). Selected known apratoxins and synthetic targets 14a-e produced through alanine or valineincroporations within apratoxin A at various positions (apratoxins F and S1-S3) or in through "hybridization" of apratoxins A and E (apratoxin S4). Me, methyl; Bu, butyl; Pr, propyl; and PMB, p-methxybenzyl. Two epimers of 14e (2-epi-1e, apratoxin S5; and 34-epi-14e, apratoxin S6) and minor reactions products during 14e synthesis, were included in our SAR studies.

Luesch et al., studied the mechanism of action of the marine natural product Apratoxin A (Figure 1), belonging to a class of potent cytotoxins produced by marine cyanobacteria [48]. They have raised the possibility or inhibition of the secretory pathway at the level of cotranslational translocation by Apratoxins and, in general, may be exploited for anticancer therapy [49]. Inhibition of this process prevents export of receptors and secretory molecules from the cytoplasm, leading to receptor (including RTK) depletion and additionally prevents secretion of the corresponding ligands, growth factors, and cytokines [50]. This one-two punch is expected to have unusually potent antiproliferative activity, may be an effective alternative to combination therapy with multiple (or broad-spectrum) RTK and growth factor inhibitors and may be particularly useful to treat cancers where autocrine loops, that is, uncontrolled proliferation stimulated by secreted growth factors, play a major role (e.g., colorectal cancer). In this article, they tested this hypothesis through a combination of medicinal chemistry based on the [51] Apratoxin scaffold, molecular *in vitro* and *in vivo* pharmacology, and colorectal tumor xenograft efficacy studies. Since they found that ApratoxinA activity was reversible, it became evident that Apratoxins have to be administered on a chronic schedule at a low dose rather than given as a bolus injection; the latter was initially reported to cause toxicity without significant antitumor activity [16]. Preliminary data for chronic administration of Apratoxin A (Figure-14 (S3)), recently confirmed by others [52], indicate that this compound shows some antitumor efficacy in vivo but is not well tolerated. The therapeutic window is small; they found a 50% death rate (day 16) at the "therapeutic" concentration of 0.25 mg/kg (daily ip) with irreversible toxicity occurring after 2 weeks of treatment (Figure-14 (S3)). The main question remained whether this toxicity is mechanism-related or an off-target effect. In this article, they described the synthesis and biological evaluation of synthetic analogues of Apratoxin A, one which shows greater potency and efficacy than the parent compound and is much better tolerated *in vivo,* providing the groundwork for second-generation apratoxins with more promising anticancer potential and validating this novel mechanism of action for cancer therapy.

Apratoxin A/E hybrid (5e) was synthesized with a view to improving anticancer activity and also tumor selectivity. Furthermore, we interrogated modifications in the peptide portion, where it is known that lack of *N*-methylation of isoleucine reduces activity (Apratoxin B) [53] and replacement of proline by *N*-methyl alanine largely retains the activity (14a, apratoxin F)[54] (Figure-14 (14a)). (Note: Figure of Apratoxin-A, Apratoxin-B, Apratoxin-E and E-dehydroapratoxin A are described previously).

CONCLUSION

Cancer is considered as one of the most perilous diseases for mankind and, scientists are therefore trying to search new antitumor compounds from plant kingdom, marine resources, and through synthetic strategy. HendrikLuesch et al., in 2001 isolated apratoxin A (1) as the most potent cytotoxin produced by a variety of the cyanobacterium Lyngbya sp. from Finger's Reef, Apra Harbor, GU, USA. Since then, several workers have identified Apratoxin (A-G) and dehydroapratoxin A & oxaapratoxin. Despite the characteristic structural complexity associated with marine natural products, there is a remarkable development in their total chemical synthesis, which is often reported after their discovery. The success in developing a feasible approach to total chemical synthesis of a bioactive molecule has two major advantages: first, it ensures a large-scale supply of a rare bioactive compound; second, it allows further structure–activity relationship (SAR) studies and lead optimization. Due to these reasons, we have also discussed the synthetic routes of apratoxin and its analogues as reported by several workers. All these Apratoxins are screened for their anticancer activity by world class biological laboratories. On the basis of positive results reported in various research papers, we hope that modifications on the basic skeleton of apratoxin will lead to potent anticancer compounds. Several synthetic routes are established by synthetic strategy and by retrosynthetic analysis. Hence we conclude that Apratoxin and its analogues are a promising lead as anticancer agents.

FUTURISTIC ROADMAP

Despite the fact that natural products are of immense importance to mankind, with the advent of high- throughput screening technology, they have been deserted to some extent. Largely uncharted oceanic world that presumably harbors the largest biodiversity could be a vast resource to discover novel 'validated' structures with innovative modes of action that cover biologically relevant active chemical space.

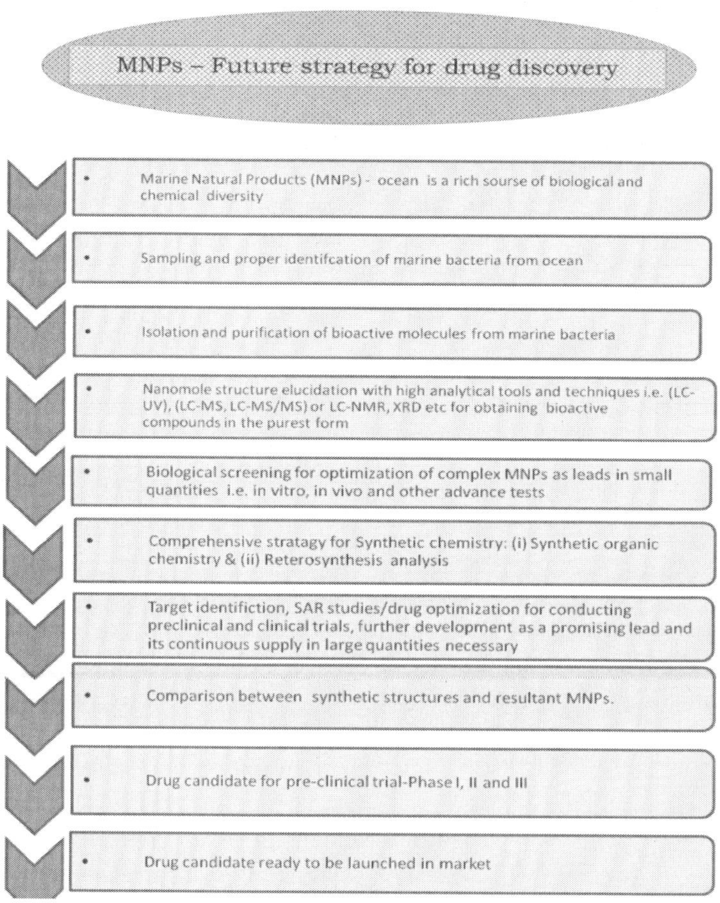

In the process of developing marine drugs, several challenges including the problem of quantity of bioactive supply and target identification, need to be met for successful drug development of the complex structures of MNPs. Are the ocean sources capable enough to provide a robust pipeline of marine drugs? In our view, high degree of innovation in the field of marine natural products will lead to successful marine drug discovery and development, and provide a ground for new wave of drugs that will flow into the market and pharmacies in future. In the present chapter; we have discussed brief introduction, synthetic strategies, retrosynthetic analysis and pharmacology of apratoxin and its analogues. On the basis of the screening results of biological activity and on the basis of SAR studies, we can predict that analogues of Apratoxin will play a very important role in cancer chemotherapy in the times to come. In our view, we should concentrate on the rich biodiversity of oceanic world for the next twenty five years. Following figure shows future drug discovery program of MNPs with combination of natural sources, total synthesis of complex structures and their pharmacology. It is given in the form of schematic diagram.

REFERENCES

[1] Dias, D.A.; Urban, S.; Roessner, U. (2012).A historical overview of natural products in drug discovery. *Metabolites*, *2*, 303–336.

[2] Bhatnagar, I.; Kim, S.K. (2010). Immense essence of excellence: Marine microbial bioactive compounds. *Mar. Drugs*, *8*, 2673–2701.

[3] Waters, A.L.; Hill, R.T.; Place, A.R.; Hamann, M.T. (2010). The expanding role of marine microbes in pharmaceutical development. *Curr. Opin. Biotechnol. 21*, 780–786.

[4] Penesyan, A.; Kjelleberg, S.; Egan, S. (2010). Development of novel drugs from marine surface associated microorganisms. *Mar. Drugs 8*, 438–459.

[5] Piel, J. Metabolites from symbiotic bacteria. (2009), *Nat. Prod. Rep. 26*, 338–362.

[6] Pettit, G.R.; Fujii, Y.; Hasler, J.A.; Schmidt, J.M. (1982). Isolation and characterization of palystatins A–D. *J. Nat. Prod. 45*, 272–276.

[7] Berdy J. (2005). Bioactive microbial metabolites. *J. Antibiot.*, *58*, 1–26.

[8] Sudek, S.; Lopanik, N.B.; Waggoner, L.E.; Hildebrand, M.; Anderson, C.; Liu, H.; Patel, A.; Sherman, D.H.; Haygood, M.G. (2007). Identification of the putative bryostatinpolyketide synthase gene cluster from —Candidatus Endobugulasertula, the uncultivated microbial symbiont of the marine bryozoan Bugulaneritina. *J. Nat. Prod. 70*, 67–74.

[9] Molinski, T.F.; Dalisay, D.S.; Lievens, S.L.; Saludes, J.P.(2009). Drug development from marine natural products. *Nat. Rev. Drug Discov.*, *8*, 69–85.

[10] Schumacher, M.; Kelkel, M.; Dicato, M.; Diederich, M. (2011). Gold from the sea: Marine compounds as inhibitors of the hallmarks of cancer. *Biotechnol. Adv.29*, 531–547.

[11] Mishra, B.B.; Tiwari, V.K. (2011). Natural products: An evolving role in future drug discovery. *Eur. J. Med. Chem. 46*, 4769–4807.

[12] Mayer, A.M.S.; Rodriguez, A.D.; Berlinck, R.G.S.; Hamann, M.T. (2009). Marine pharmacology in: Marine compounds with anthelmintic, antibacterial, anticoagulant, antifungal, anti-inflammatory, antimalarial, antiprotozoal, antituberculosis, and antiviral

activities; affecting the cardiovascular, immune and nervous systems, and other miscellaneous mechanisms of action. *Biochim. Biophys. Acta 1790*, 283–308
[13] Mayer, A.M.S., Glaser, K.B., Cuevas, C., Jacobs, R.S., Kem, W., Little, R.D., McIntosh, J.M., Newman, D.J., Potts, B.C., and Shuster, D.E. (2010). The odyssey of marine pharmaceuticals: a current pipeline perspective. *Trends Pharmacol. Sci.* 31, 255–265.
[14] Gerwick, W. H.; Proteau, P. J.; Nagle, D. G.; Hamel, E.; Blokhin, A.; Slate, D. L. (1994). J. Org. Chem. 59, 1243 Structure of Curacin A, a Novel Antimitotic, Antiproliferative and Brine Shrimp Toxic Natural Product from the Marine Cyanobacterium Lyngbyamajuscula.
[15] Luesch, H.; Moore, R. E.; Paul, V. J.; Mooberry, S. L.; Corbett, T. H. (2001). Isolation of Dolastatin 10 from the Marine Cyanobacterium Symploca Species VP642 and Total Stereochemistry and Biological Evaluation of Its Analogue Symplostatin. *J. Nat. Prod.* 64, 907-910.
[16] Luesch H. Corbett T. (2001). Total Structure Determination of Apratoxin A, a Potent Novel Cytotoxin from the Marine Cyanobacterium Lyngbyamajuscule. *J. Am. Chem. Soc.* 123 23, 5418-5423.
[17] Luesch, H., Wesley A., Yoshida Y., Richard a Moorea, E. and. Paulb V. J. (2002). 1973–1978 New apratoxins of marine cyanobacterial origin from guam and palau. *Bioorganic & Medicinal Chemistry.* 10 1973-1978.
[18] Chen J. and Forsyth C.J. (2003). Total synthesis of apratoxin A. 125, 8734-5.
[19] Tao Yea,; Zhengshuang Xu, Zhiyong Chena. (2004). Synthesis of the polyketide segment of apratoxin A Tetrahedron: Asymmetry 15, 355–363.
[20] Luesch, H., Yoshida, W. Y., Moore, R. E. and Paul V. J. (2002). *Bioorg. Med.Chem.* 10, 1973 –1978. New apratoxins of marine cyanobacterial origin from Guam and Palau.
[21] Matthew, S., Schupp, P. J., Luesch H. (2008). Apratoxin E, a cytotoxic peptolide from a guamanian collection of the marine cyanobacterium Lyngbyabouillonii. *J. Nat. Prod.* 71, 1113 –1116.
[22] a) Jiehao Chen and Craig J. Forsyth. (2004). Total synthesis of the marine cyanobacterial cyclodepsipeptide apratoxin A. *Proc. Natl. Acad. Sci.USA,* 101, 12067 – 12072; b) Chen J., Forsyth, C. J., (2003). Synthesis of the apratoxin 2,4-disubstituted thiazoline via an intramolecularaza-Wittig reaction. *Org. Lett.* 5, 1281 –1283.
[23] Ma, D., Zou, B., Cai, G., Hu X., Liu, J. O. (2006). Total synthesis of the cyclodepsipeptide apratoxin A and its analogues and assessment of their biological activities. *Chem. Eur. J.* 12, 7615 – 7626.
[24] Gilles, A., Martinez, J., Cavelie, F. (2009). Supported synthesis of oxoapratoxin A. *J. Org. Chem.*, 74, 4298 –4304.
[25] Paul, V. J.; Luesch, H.; Yoshida, W. Y.; Moore*, R. E. (2002). Structurally Diverse New Alkaloids from Palauan Collections of the Apratoxin-Producing Marine CyanobacteriumLyngbya sp. Tetrahedron 39, 7959-7966;
[26] Doi, T.; Numajiri, Y.; Munakata, A.; Takahashi, T. (2006).Total synthesis of apratoxin A. *Org. Lett.*, 8, 531–534.
[27] Inanaga, J.; Hirata, K.; Saeki, H.; Katsuki, T.; Yamaguchi, M. (1979).A rapid esterification by means of mixed anhydride and its application to large-ring lactonization. *Bull. Chem. Soc. Jpn.,* 52, 1989–1993.

[28] Travis, B. R.; Narayan, R. S.; Borhan, B. (2002). Osmium tetroxidepromoted catalytic oxidative cleavage of olefins: an organometallic ozonolysis. *J. Am. Chem. Soc.*, 124, 3824–3825.

[29] You, S.-L.; Razavi, H.; Kelly, J. W. (2003). A biomimetic synthesis of thiazolines using hexaphenyl oxodiphosphonium trifluoromethane sulfonate. *Angew. Chem., Int. Ed.*, 42, 83–85.

[30] Forsyth, Chen, J. (2003). Synthesis of the Apratoxin 2,4-Disubstituted Thiazoline via an Intramolecular Aza-Wittig Reactions. J.; *Org. Lett.* 8, 1281-1283;

[31] Ma, D, Zou, B.; Wei, J.; Cai, G. (2003). Synthesis of an Oxazoline Analogue (Ia) of Apratoxin A (Ib). *Org. Lett.* 19, 3503-3506.

[32] Takayuki Doi,; Yoshitaka Numajiri,; Asami Munakata,; Takashi Takahashi (2006).Total Synthesis of Apratoxin A. — (DOI10.1021/ol052907d, *Org. Lett.* 8 3, 531-534.

[33] Arnaud Gilles *, Jean Martinez, Florine Cavelier. (2011). A flexible synthesis of C33-C39 polyketide region of apratoxin: Synthesis of natural and unnatural analogues. C. R. *Chimie* 14 437–440.

[34] Corey, E. J. *Angew. Chem., Int. Ed. Engl.* (1991). *30*, 455. Corey, E. J. In *Nobel Lectures: Chemistry 1981–1990*; Malström, B., Ed.; World Scientific: River Edge, NJ, 1992; p 677. (Sir's ref).

[35] Dawei Ma, Bin Zou, GuorongCai, Xiaoyi Hu, and Jun O. Liu. (2006). Total Synthesis of the Cyclodepsipeptide Apratoxin A and Its Analogues and Assessment of Their Biological Activities. *Chem. Eur. J.*, 12, 7615 – 7626.

[36] Wipf, P. (1995). *Chem. Rev.*, 95, 2115 –2134; Synthetic Studies of Biologically Active Marine Cyclopeptides Peter. Wipf.

[37] Wipf, P. Venkatraman, S. (1997). *Synlett*, 1–10. "From Aziridines to Oxazolines and Thiazolines: The Heterocyclic Route to Thiangazole." Wipf, P.; Venkatraman, S. Synlett 1997, 1-10.

[38] Doi Takayuki, Numajiri Yoshitaka,Takahashi Takashi,Takagi Motoki,(2011). Solid-phase Total Synthesis of (-)-Apratoxin A and Its Analogues and Their Biological Evaluation. *Chem. Asian J.*, 6, 180 – 188.

[39] Chen Jiehao and Forsyth Craig J. (2003). Total Synthesis of Apratoxin A. *J. AM. CHEM. SOC.*, 125, 8734-8735.

[40] Baselga, J. (2006). Targeting tyrosine kinases in cancer: the second wave.*Science*, 312, 1175–1178.

[41] Lurje, G.; Lenz, H.-J. (2009). EGFR signaling and drug discovery.*Oncology*, 77, 400–410.

[42] Martinelli, E.; Troiani, T.; Morgillo, F.; Rodolico, G.; Vitagliano, D.; Morelli, M. P.; Tuccillo, C.; Vecchione, L.; Capasso, A.; Orditura, M.; De Vita, F.; Eckhardt, S. G.; Santoro, M.; Berrino, L.; Cardiello, F.(2010). Synergistic antitumor activity of sorafenib in combination with epidermal growth factor receptor inhibitors in colorectal and lung cancer cells. Clin. Cancer Res., 16, 4990–5001.

[43] Hurwitz, H.; Fehrenbacher, L.; Novotny, W.; Cartwright, T.; Hainsworth, J.; Heim, W.; Berlin, J.; Baron, A.; Griffing, S.; Holmgren, E.; Ferrara, N.; Fyfe, G.; Rogers, B.; Ross, R.; Kabbinavar, F. (2004). Evacizumab plus irinotecan, fluorouracil, and leucovorin for metastatic colorectal cancer. *N. Engl. J. Med.* 350, 2335–2342.

[44] Leusch H., Yoshida, W. Y. Moore, R. E. Paul, V. (2002). New apratoxins of marine cyanobacterial origin from Guam and Palau. *J. Bioorg. Med. Chem.* 10, 1973 –1978.
[45] Chen, J. Forsyth, C. J. (2003). Total synthesis of apratoxin A. *J Am Chem Soc.* 125(29):8734-5.
[46] Luesch, H., Chanda,S. K., Raya, R. M., Dejesus, P. D., Orth, A. P., Walker, J. R., Belmonte, J. C. I., Schultz, P. G. (2006). A functional genomics approach to the mode of action of apratoxin A. *Nat. Chem. Biol.*2, 158 – 167.
[47] Liu, Y. Law, B. K. Luesch, H. (2009). "Apratoxin A Reversibly Inhibits the Secretory Pathway by Preventing Cotranslational Translocation".
[48] Ruan, W.-J.; Lai, M.-D. (2004). Autocrine stimulation in colorectal carcinoma (CRC): Positive autocrine loops in human colorectal carcinoma and applicable significance of blocking the loops. *Med. Oncol.* 21, 1–7.
[49] Liu, Y.; Law, B. K.; Luesch, H. (2009). Apratoxin A reversibly inhibits the secretory pathway by preventing cotranslational translocation. *Mol. Pharmacol.,* 76, 91–104.
[50] Shen, S.; Zhang, P.; Lovchik, M. A.; Li, Y.; Tang, L.; Chen, Z.; Zeng, R.; Ma, D.; Yuan, J.; Yu, Q. (2009). Cyclodepsipeptide toxin promotes the degradation of Hsp90 client proteins through chaperone-mediated autophagy. *J. Cell. Biol.* 185, 629–639.
[51] Ruan, W.-J.; Lai, M.-D. (2004). Autocrine stimulation in colorectal carcinoma (CRC): Positive autocrine loops in human colorectal carcinoma and applicable significance of blocking the loops. *Med. Oncol.*, 21, 1–7.
[52] Tidgewell, K.; Engene, N.; Byrum, T.; Media, J.; Doi, T.; Valeriote, F. A.; Gerwick, W. H. (2010). Evolved diversification of a modular natural product pathway: Apratoxins F and G, two cytotoxic cyclic depsipeptides from a Palmyra collection of Lyngbyabouillonii. *Chem- BioChem*, 11, 1458–1466.
[53] Luesch, H.; Yoshida, W. Y.; Moore, R. E.; Paul, V. J.; Corbett, T. H. (2002). New apratoxins of marine cyanobacterial origin from Guam and Palau. *Bioorg. Med. Chem.* 10, 1973–1978.
[54] Tidgewell, K.; Engene, N.; Byrum, T.; Media, J.; Doi, T.; Valeriote, F. A.; Gerwick, W. H. (2010). Evolved diversification of a modular natural product pathway: Apratoxins F and G, two cytotoxic cyclic depsipeptides from a Palmyra collection of Lyngbyabouillonii. *Chem-BioChem,* 11, 1458–1466.

In: Natural Heterocycles
Editors: K. L. Ameta and Saksit Chanthai
ISBN: 978-1-63463-424-3

Chapter 5

AURONE: THE VERSATILE NATURAL PRODUCT HAVING BIOLOGICAL POTENTIAL

Chetna Ameta[1], Arpit Kumar Pathak[1] and K. L. Ameta[2*]
[1]Department of Chemistry, University College of Science,
M L S University, Udaipur
[2]Department of Chemistry, FASC, Mody Institute of Technology and Science,
Lakshmangarh, Rajasthan, India

ABSTRACT

Aurones [2-benzylidenebenzofuran-3(2H)-ones] are natural as well as synthetic compounds, belonging to the flavonoid family. They are isomeric to flavones and provide a bright yellow colour to the plants in which they occur. Flavonoids have been studied not only for their physiological properties and effects on nature, but also for their therapeutic potential. Aurones are recently attracting the interest of an increasing number of research groups and some interesting advances have been made in understanding the chemistry of aurones. Naturally occurring aurones have emerging biological potential and can extracted from many plants species like *Withania somnifera, Gomphrena agrestis, Bidens parviflora Asarum longerhizomatosum* etc. Aurones and even its crude extract isolated from different parts of plants exhibits many biological activities like antimicrobial, anticancer, antiviral, anti-malarial, antioxidant, anti-inflammatory, anti-carcinogenic etc.

1. INTRODUCTION

Aurone is a flavonoid type of heterocyclic compound having two isomeric forms with (*E*)- and (*Z*)-configurations. The molecule contains a benzofuran skeleton associated with a benzylidene linked in position 2. In aurone, a chalcone-like group is closed into a 5-membered ring instead of the 6-membered ring more typical of flavonoids. (Figure 1)

* Email: klameta@hotmail.com

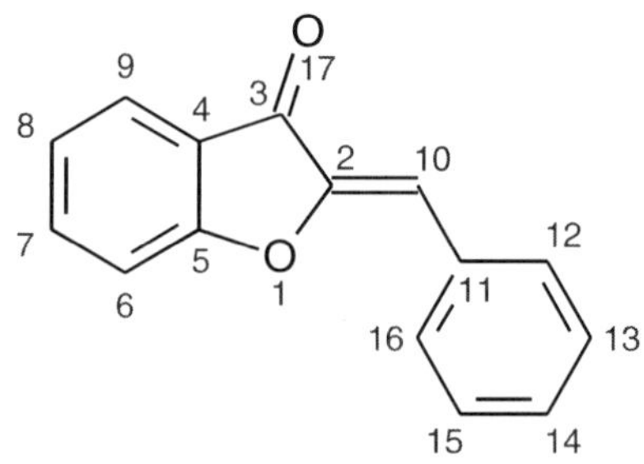

Figure 1. Skeletal structure of an (Z)-aurone.

Aurone forms the core for a family of derivatives which are known collectively as aurones. Aurones are plant flavonoids that provide yellow colour to the flowers of some popular ornamental plants, such as snapdragon and cosmos. Aurones including 4'-chloro-2-hydroxyaurone ($C_{15}H_{11}O_3Cl$) and 4'-chloroaurone ($C_{15}H_9O_2Cl$) can also be found in the brown alga *Spatoglossum variable*.

Most aurones are in a (Z)-configuration, which is the more stable configuration but there are also some in the (E)-configurations such as (E)-3'-O-β-d-glucopyranosyl-4,5,6,4'-tetrahydroxy-7,2'-dimethoxyaurone, found in Gomphrena agrestis. Analogy with flavonoids suggests that aurones could have interesting biological properties [1]. Aureusidin (Figure 2), a common aurone, is an inhibitor of iodothyronine- deiodinase, an enzyme involved in hormone synthesis and regulation. Synthetic aurones bind to the nucleotide-binding domain of P-glycoprotein, to inhibit cyclin-dependent kinases in connection with antiproliferative properties, and to act as anticancer agents.

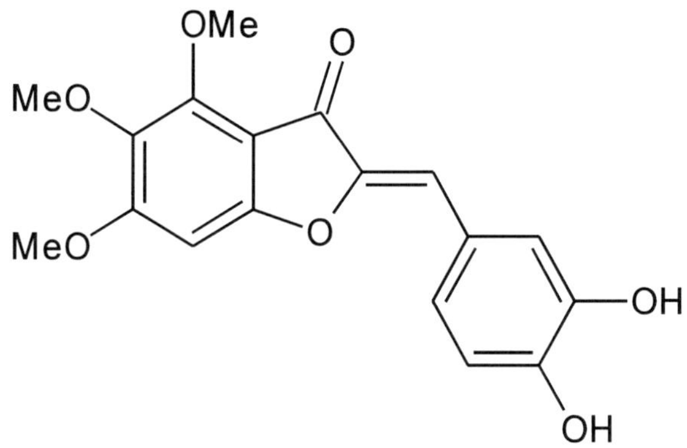

Figure 2. Aureusidin: an example of a common aurone.

2. Extraction and Isolation of Aurones

Isolation from *Withania Somnifera* [2]

Withania Somnifera, known as ashwagandha, Indian ginseng and winter cherry has been an important herb in the Ayurvedic and indigenous medical systems for over 3000 years. Historically, the plant has been used as an aphrodisiac, liver tonic, anti-inflammatory agent, stringent and to treat bronchitis, asthma, ulcers, emaciation, insomnia and senile dementia [3]. It is grows about 30- 150 cm in height in India, Pakistan, Bangla Desh, Sri Lanka and parts of northern Africa. It is also found in Congo, South Africa, Egypt, Morocco, Jordan and Afganistan.

The flavonoids belong to one of the most bioactive compounds which naturally exist in the plant kingdom. Different naturally occurring flavonoids have been described and subcategorized into flavones, flavans, flavanones, isoflavonoids, chalcones, aurones and anthocyanidines [4, 5]. These flavonoids have remarkable biological activities, including inhibitory effects on enzymes, modulatory effect on some cell types, protection against allergies, antiviral, anti-malarial, antioxidant, anti-inflammatory and anti-carcinogenic properties. A number of flavones, flavonols, flavanones, and isoflavones, as well as some of their methoxy, isoprenyl, and acylated derivatives, show antibacterial activity [6].

Flavonoids can be considered as derivatives of the 2-phenylchromone parent compound composed of two phenyl rings, and a heterocyclic ring referred to as the A-, B and C- rings [7]. Flavonoids are secondary metabolites characterized by flavan nucleus [8] and a C6-C3-C6 carbon skeleton [9]. These are group of structurally related compounds with a chromane-type skeleton having phenyl substituent in C2-C3 position [10].

General Experimental Procedure for Extraction

Collection of the Plant Materials: Extraction of *Withania somnifera* Leaves constituents powdered air-dried leaves (1.0 Kg) of *Withania somnifera* were ground in a Warring Blender to a fine powder and extracted with chloroform to remove chlorophyll, resinous and waxy material. The ground material was extracted with 85% aqueous methanol and the slurry was allowed to stand at room temperature for 24 hours with occasional stirring. The solvent containing extract was then decanted and filtered by vacuum filtration. The extraction of the ground leaves were further repeated twice with the same solvent and twice with 50% aqueous methanol [11].The four filtrates from each extraction was combined and the excess solvent was evaporated under reduced pressure at 40°C using a rotary evaporator to give crude extracts, (53g) of a solid mass and then kept into a glass container to use.

Preliminary Phytochemical Screening of the *Withania Somnifera*

Crude Methanolic Extract*:* The methanolic extracts were tested for the presence of active constituents such as flavonoids, steroids, terpenoids, alkaloids, saponins and tannins. The following standard procedures were used:

- ***Flavonoids***

i. To about (3 ml) of the methanolic extract few drops of methanolic aluminum chloride were added. Formation of a yellowish colour indicates the presence of flavonoids [12].

ii. To about 3ml of the methanolic few drops of aqueous potassium hydroxide solution were added. A dark yellow colour is taken as a positive test for flavonoid compounds.
iii. Shinoda Test: Small pieces of Magnesium ribbon followed by few drops of concentrated hydrochloric acid were added to a small amount of methanolic extract of the plant material. Formation of a crimson red colour indicates the presence of flavones in glycosidic linkage [13].

- ***Steroids and Terpenoids***

(40 ml) aliquot of the methanolic extract was evaporated to dryness on a water bath and the cooled residue was stirred with petroleum ether to remove most of the colouring matter. The residue was extracted with (20 ml) chloroform. The chloroform solution was dehydrated over anhydrous sodium sulphate. (5 ml) portion of the solution was mixed with (0.5 ml) acetic anhydride, followed by two drops of concentrated sulphuric acid. A green colour was observed indicating the presence of steroids and no pink-violet colour was observed indicating the absence of terpenoids [14].

- ***Alkaloids***

About (30 ml) aliquot of the methanolic extract was evaporated to dryness on a water bath. (5 ml) of 2N hydrochloric acid were added and the solution was heated with stirring in a water bath for 10 minutes. The mixture was cooled and filtered. (2 ml) of the filtrate, were added to few drops of Mayer reagent (Potassium mercuric iodide solution). A precipitate was observed indicating the presence of alkaloids.

- ***Saponins***

20 ml of water was added to 0.25 g of the extract in 100 ml beaker and boiled, filtered and then the filtrates used for the tests:

Froth Test**:** 5 ml of the filtrate was diluted with 20 ml of water and shaken vigorously. A stable froth (foam) up on standing indicates the presence of saponins [15].

- ***Tannins***

To about (25 ml) of the methanolic extract was evaporated to dryness on a water bath and the residue was extracted with n-hexane and filtered. The hexane- insoluble portion was stirred with (10 ml) of hot saline solution (0.9% w/v of sodium chloride and freshly prepared distilled water). The mixture was cooled and filtered and the volume adjusted to (10 ml) with more saline solution. (5 ml) of this solution was treated with few drops of ferric chloride solution, a blue colour was formed indicating the presence of tannins [16].

- ***Isolation of Flavonoids***

The dry extract was loaded on a polyamide 6S column chromatography (80 x 3 cm). The column was eluted with water, and then water -Methanol mixtures of decreasing polarity and 7 fractions were collected. The major fractions obtained were combined into three fractions after chromatographic analysis. Fraction A (5 g) was fractionated by PC, using Whatman No 3 sheets and (n-butanol: acetic acid: water) upper layer as solvent system followed by Sephadex LH- 20 column eluted with methanol (100%) and methanol/ water (9: 1), to afford two pure compounds W_1 (29 mg) and compound W_2 (36mg). Fraction B (3 g) was subjected to column chromatography on cellulose and n-BuOH saturated with water as an eluted solvent to give two major sub fractions, then each of them was separately fractionated on a Sephadex LH-20 to yield pure compound W_3 (27 mg) and compound W_4 (21 mg). The phytochemical screening of *Withania somnifera* constituents in methanolic extracts revealed the presence of alkaloids, Steroids, saponins and phenolic compounds (flavonoids and tannins). The weight and percentage yield of pure compounds extracted from *Withania somnifera* are shown in Table 1.

Table 1. Weight and percentage yield of some compounds extracted from *W. somnifera* leaves

Components	Weight	Percentage yield
Compound W1 (mg)	29	0.58
Compound W2(mg)	36	0.72
Compound W3 (mg)	27	0.77

A detailed phytochemical study of *Withania somnifera* leaves showed that from the methanolic extract of *Withania somnifera* three flavonoids were isolated and purified using different chromatographic techniques. These compounds (W_1, W_2 and W_3) were identified via spectroscopic tools: IR, UV, ^{1}H NMR and MS spectroscopy.

- ***Antimicrobial Activity***

Micro-organisms collection and maintenance: *Bacillus subtilis, Escherichia coli, Neisseria gonorrhoeae, Pseudomonas aeruginosa* and *Staphylococcus aureus* were obtained from stock culture in the Department of Botany, University of Cairo, and Cairo City, Egypt. The organisms were stored on agar slant in McCartney bottles and kept in the refrigerator, prior to subculture.

Preparation of standard test organisms: One ml aliquots of 24 hours broth cultures of test organisms (*Bacillus subtilis, Staphylococcus aureus, Escherichia coli, Pseudomonas aeruginosa* and *Neisseria gonorrhoeae*) were aseptically distributed onto nutrient agar slopes and incubated at 37 C for 24 hours. The bacterial growth was harvested and washed off with sterile normal saline, to produce a suspension containing about 108- 109 colony forming units per ml. The suspension was stored in the refrigerator at 4 C till used. Serial dilution of stock suspension, were made in sterile normal saline, and 0.02 ml volumes of the appropriate dilution were transferred onto the surface of dried nutrient agar and the drop was left to dry, and then incubated at 37°C for 24 hours [17].

Testing for antibacterial activity: To determine the activity of methanolic extract and pure component of *Withania somnifera* against the five standard organisms. The cup-plate agar was adopted with some minor modification. (2ml) Of the standard bacteria stock suspension were mixed with 200 ml of sterile molten nutrient agar which was maintained at 45°C. (20 ml) Aliquots of incubated agar were distributed into sterile Petri dishes. The agar was left to settle and each plate was cut and agar discs were removed.

Alternated cups were filled with 0.1 ml sample of each of the extract (pure component) and allowed to diffuse at room temperature for two hours. The Petri dishes were then incubated in the upright position at 37°C for 18 hours. After incubation, the diameter of the resultant growth inhibition zones were measured and averaged.

Table 2. The antibacterial activity of crude and pure compounds extracted from *Withania somnifera*

Sample	Inhibition zone diameter (mm / mg sample)				
	Bacillus subtilis (G+)	*Escherichia coli* (G-)	*Neisseria gonorrhoeae* (G-)	*Pseudomonas aeruginosa* (G-)	*Staphylococcus aureus* (G+)
Control: DMSO	0.0	0.0	0.0	0.0	0.0
Standard: Tetracycline	27	31	33	29	26
Crude extract	9	13	9	13	13
W_1	12	11	12	12	11
W_2	9	9	8	9	9
W_3	13	14	15	10	13

The antibacterial activity results of crude extract and pure compounds of *Withania somnifera* against five human pathogens was carried out in Table 2. The cup-plate agar diffusion methods were adopted with some minor modifications. The test organisms were: *Bacillus subtilis, Escherichia coli, Neisseria gonorrhoeae, Pseudomonas aeruginosa* and *Staphylococcus aureus.* The crude extract of *Withania somnifera* showed moderate inhibition against *Escherichia coli, Pseudomonas aeruginosa* and *Staphylococcus aureus* and weak inhibition against *Neisseria gonorrhoeae* and *Bacillus subtilis.* Compound W_1 exhibited a moderately inhibition against all five organisms but less than their crude extract. Compound W_2 showed weak inhibition against all organisms. Compound W_3 exhibited high inhibition against all organisms accepts *Pseudomonas aeruginosa* where it is low activity.

The crude extract and pure compounds exhibited moderate activity against the human pathogens: *Bacillus subtilis, Escherichia coli, Neisseria gonorrhoeae, Pseudomonas aeruginosa* and *Staphylococcus aureus.*

- ***Isolation from Gomphrena Agrestis [18]***

In laboratory, several plant extracts from Amaranthaceae were screened *in vitro* against amastigote forms of *Leishmania amazonensis* and for antimicrobial activity. In an introductory assessment, the ethanolic crude extract (dry, whole plant) of *G. agrestis* has shown good antimicrobial and leishmanicidal activities. Therefore, in the course of chemical and biological investigations of Amaranthaceae plants, the isolation and identification of a new aurone 1 and two known substances, aurantiamide acetate 2 and tiliroside 3 isolated from the ethanolic extract of *G. agrestis.*The genus Gomphrena belongs to the tribe Gomphreneae of the Amaranthaceae family. *Gomphrena agrestis* is herbaceous plant commonly found in all Brazil.

The ethanolic crude extract and compounds 1, 2 and 3 were screened for antimicrobial activity *in vitro* as well as their toxicity against *L. amazonensis* amastigotes from axenic culture was evaluated.

Collection of plant material: Gomphrena agrestis was collected from Brazil. The whole plants of *G. agrestis* were dried in an open stove at 40°C and powdered affording 1300 g. Exhaustive extraction with ethanol after hexane extraction gave 37 g of the crude extract (dry weight). To obtain the aurone, 36.0 g of the crude ethanolic extract of *G. agrestis* (whole plant) was partitioned. After subsequent extractions with hexane, followed by CH_2Cl_2, the hydroalcoholic phase was column filtered with amberlite XAD-2. The ethanolic phase (4.5 g) obtained from the filtration was column fractionated with Sephadex LH-20 eluted with MeOH and yielded 14 samples. After being subjected to a HPLC separation, the 11th sample yielded 20 mg of compound 1 and the 14th sample yielded 28 mg of compound 3. Aurantiamide acetate (2) was obtained from the dichloromethane phase (4.6 g).

Biological Activities

Antimicrobial activity: 22 strains, including gram positive and gram negative bacteria and yeast strains, were used for the susceptibility assays. The following microorganisms were used: *Escherichia coli* D ATTC 10538; *E. coli* D ec 26.1 (field strain); *Pseudomonas aeruginosa* D ATCC 27853; *P. aeruginosa* D 290D (field strain); *Micrococcus luteus* DATTC 9341; *Staphylococcus aureus* D ATTC 25923, 6538 and 25213; *S. aureus* 7+ penicilinase producer; *S. aureus* 8D penicilinase non-producer; *Staphylococcus epidermidis* 6epi, epiC (field strains), cultivated for 24 h at 37 C in Mueller Hinton broth (MHb) (Difco); *Enterococcus faecalis* D ATCC 10541; *Streptococcus mutans* D ATCC 25175; *S. mutans* (strains Fab 3; 9.1; 11.1; 11.22.1) and *Streptococcus sobrinus* 180.3 (field strains), incubated for 24 h at 37 C in Brain Heart Infusion (BHI) (Difco); *Candida albicans* D ATTC 1023; *C. albicans* cas (field strain) and *Candida tropicalis* ct (field strain), incubated for 24 h at 37°C in Sabouraud glucose agar slants. Sensitivity tests were performed using a modified agar-well diffusion method.

Leishmanicidal activity: The effect of crude extract and isolated compounds on viability of *L. amazonensis* (strain designation MPRO/BR/72/M 1841) amastigotes was performed using parasites cultured and incubated at 33°C for 18 h with the ethanolic crude extract (1 mg/ml) or isolated compounds (0.02, 0.1 and 0.5 mg/ml) dissolved previously in dimethyl sulfoxide.

Result: Initial observations indicated the activity of the crude ethanolic extract of *G. agrestis* (whole plant) against bacteria and *L. amazonensis*. This extract inhibited the growth of the gram positive bacteria as *S. aureus* ATTC 25923 and *S. epidermidis* (strains 6epi and epiC) with inhibition zones of 8, 6 and 6 mm, respectively, and of the gram negative *P. aeruginosa* strains ATTC 27853 and 290D with inhibition zones of 7 and 6 mm, respectively. The compounds 1, 2 and 3 were obtained from this extract, which results are showed in Table

II. These isolated compounds were active mainly against strains of *Staphylococcus aureus*, *S. epidermidis* and *Pseudomonas aeruginosa* with MIC values between 0.1 and 0.5 mg/ml, while the crude ethanolic extract markedly reduced the viability of *L. amazonensis* amastigotes to 1.7% (at 1 mg/ml), the isolated compounds did not appreciably interfere with parasite viability at 0.5 mg/ml.

This latter result is in agreement with previous findings with this class of compounds (aurones), indicating that the introduction of bulky substituents (benzylation or glycosylation), oxygenation pattern of the aromatic ring and structural rigidity correlate with reduction of leishmanicidal activity [19, 20]. Therefore, subsequent fractionation and purification resulted in substances displaying low activity against *L. amazonensis* amastigotes compared to the ethanolic extract.

- ***Isolation from Bidens Parviflora Willd [21]***

Five chalcones, two flavanones and two aurones were isolated from *Bidens parviflora Willd* and identified as 2′-hydroxy-3, 4, 4′–trimethoxy chalcone (I), 4,2′,4′,6′-tetramethoxy dihydrochalcone (II), 3, 4, 2′, 4′, 6′-pentamethoxy dihydro chalcone (III), 7, 3′, 4′-trihydroxy flavanones (IV), 3′, 4′-dihydroxy flavanones 7-*O*-*β*-D-glucoside (V), 3, 4, 2′-trihydroxy chalcone 4′-*O*-*β*-D-glucoside (VI), 3-Methoxy-4, 2′, 3′-trihydroxy chalcone 4′-*O*-*β*- D-glucoside (VII), sulfuretin 6-*O*-*β*-D-glucoside (VIII), and maritimetin 7-*O*-*β*-D-glucoside (IV), by their spectral data (UV, FABMS, 1H-NMR, 13C-NMR, HMQC, HMBC and NOE). Anti-allergic testing was carried out based on nitric oxide (NO) production in LPS and IFN-γ activated murine macrophages (RAW264.7) and histamine release from rat mast cells stimulated by an antigen-antibody reaction.

General Experimental Procedures

Collection of Plant materials: The whole plant of *Bidens parviflora Willd* was collected in the Dahishan region of Liaoning province in China, in July, 1999 and was identified by Prof. Weichun Wu (Department of Medical Plants, Shenyang Pharmaceutical University, China).

Acid hydrolysis: A solution of the compounds (10 mg) in 0.2 N HCl-dioxane (1:1, 6 mL) was refluxed for 40 min. The reaction mixture was then extracted with AcOEt, and the AcOEt layer was evaporated to dryness *in vacuo* after being washed with H_2O. The aqueous layer was neutralized with Amberlite IRA-45 (OH-form) and evaporated to dryness *in vacuo*. The residue showed the presence of D-glucose ($Rf = 0.25$) on TLC (solvent: $CHCl_3$ - MeOH - H_2O, 65:30:5).

Extraction and Isolation: The air-dried whole plant (5.5 kg) was extracted twice with 60% ethanol (10 each) under reflux. The extract was dissolved and partitioned with hexane, ethyl acetate and *n*-butanol, respectively, then the butanol phase was evaporated under reduced pressure below 40ºC, to give a butanol extract (176 g). This *n*-butanol extract was subjected to silica- gel column chromatography (SiO_2, 500 g, eluted with $CHCl_3$ and MeOH in increasing order of polarity) to give 12 fractions. Fractions 6-8 (10.5 g) were applied to a Sephadex LH-20 column and eluted with 50% MeOH to give fractions 4-9 then purified by

HPLC (INW 125 Fluofix, 10 mm i.d. × 250 mm, UV detector, 254 nm) and eluted with 18% CH_3CN in water to give chalcones and aurones (I - IX).

Wang and co-workers [21] also examined the inhibitory effects of these compounds on histamine release from rat peritoneal exudates cells induced by an antigen-antibody reaction, and also compared the inhibitory effects of these compound (I-IX) on NO production by the murine macrophage–like cell line RAW264.7 activated by lipopolysaccharide (LPS) and recombinant mouse interferon-γ (IFN-γ). Based on the data in Table 3, it appears that the presence of the double bond between C-2 and C-3 and the carbonyl group at C-4 are essential for inhibitory activity, and free hydroxyl groups are also necessary. The presence of two *o*-hydroxyl groups at C-3′ and C-4′ produced a greater inhibition of histamine release. They also found that the glycosides exhibit greater activity than the glycones. These findings suggest that the Flavonoids have great prospects for further development as anti-allergic or anti-inflammatory drugs.

Table 3. Anti-allergic activities of compounds I – IX

Compounds	A IC50 (μg / mL)	B IC50 (μg / mL)
Indomethacin (positive control)	52.6	52.5
2′-hydroxy-3, 4, 4′–trimethoxy chalconee (I)	-	-
4, 2′,4′,6′-tetramethoxy dihydrochalcone (II)	-	-
3, 4, 2′, 4′, 6′-pentamethoxy dihydrochalcone (III)	-	-
7, 3′, 4′-trihydroxy flavanone (IV)	--	-
3′, 4′-dihydroxy flavanone 7-*O*-*β*-D-glucoside (V)	57.7	89.2
3, 4, 2′-trihydroxy chalcone 4′-*O*-*β*-D-glucoside (VI)	55.8	70.2
3-methoxy-4, 2′, 3′-trihydroxy chalcone	-	-
4′-*O*-*β*-D-glucoside (VII)	89.7	64.9
Sulfuretin 6-*O*-*β*-D-glucoside (VIII)	83.3	72.1
Sulfuretin (VIIIa)	-	-
Maritimetin 6-O-β-D-glucoside (IX)	44.5	46.8

A Inhibitory effects on histamine release.
B Inhibitory on NO production by macrophages.

- ***Isolation from Asarum longer hizomatosum [22]***

Two new glycosyl flavonoids including a glycosyl aurone, together with six known flavonoids were isolated from the roots and rhizomes of *Asarum longerhizomatosum*. The structures of the two new compounds were elucidated as 4,6,40-trihydroxy-aurone-4,6-

di-O-b-D-glucopyranoside (7, caulesauroneside) and naringenin-7,40-di-O-b-Dglucopyranoside (8, caulesnarinside). The six known flavonoids were identified as naringenin (1), naringenin-5-O-b- D-glucopyranoside (2), naringenin-7-O-b-D-glucopyranoside (3), chalcononaringenin-20-O-b-D-glucopyranoside (4), naringenin-5,7-di-O-b-D-glucopyranoside (5), chalcononaringenin-20,40-di-O-b-D-glucopyranoside (6), respectively. This is the first report of the isolation of aurones in the family Aristolochiaceae.

Collection of Plant Material: Roots and rhizomes of *A. longerhizomatosum* was collected in the Guangxi province of China by Professor Shou-Yang Liu and was taxonomically identified by Professor Shao-Qing Cai. The voucher specimen (No. 1255) was deposited in the specimen room of the School of Pharmaceutical Sciences, Peking University.

Extraction and isolation: A powder of the roots and rhizomes of *A. longerhizomatosum* was extracted with EtOH. After concentration under reduced pressure, the aqueous residue was partitioned with petroleum ether, EtOAc and n-BuOH, respectively. The EtOAc-soluble portion was chromatographed over silica gel and eluted with petroleum ether–EtOAc and $CHCl_3$– MeOH gradient solvent. Combination of similar fractions on the basis of TLC analysis afforded 8 fractions. Fraction 4 was chromatographed over silica gel with $CHC1_3$–Me_2CO to give narigenin (1, 14 mg). Fraction 7 was chromatographed over silica gel with $CHCl_3$– MeOH, then purified by ODS column chromatography to afford naringenin-5-O-b-Dglucopyranoside (2, 22 mg) and naringenin-7-O-b-D-glucopyranoside (3, 8 mg). Fraction 8 was isolated by the same methods above to afford chalcononaringenin-20-O-b-Dglucopyranoside (4, 4 mg). The n-BuOH-soluble part was subjected to D101 resin column chromatography and eluted with H_2O, 20% MeOH, 50% MeOH and MeOH, respectively.

The 20% MeOH eluted part was subjected to silica gel column chromatography eluted with $CHCl_3$–MeOH–H_2O in gradient manner, then purified by Sephadex LH-20 and RP-18 column chromatography to obtain naringenin-5,7-di-O-b-D-glucopyranoside, naringenin- 7,40-di-O-b-D-glucopyranoside (5, 2.3 g), chalcononaringenin-20,40-di-O-b-D-glucopyranoside (6, 18mg), 4,6,40 -trihydroxyaurone-4,6-di-O-b-D-glucopyranoside (7, 12mg), naringenin-7,40-di-O-b-D-glucopyranoside (8, 23 mg) were obtained respectively.

- ***Aurone in Bidens ferulifolia [23]***

The formation of 4-deoxyaurones, which serve as UV nectar guides in *Bidens ferulifolia* (Jacq.) DC., was established by combination of UV photography, mass spectrometry, and biochemical assays and the key step in aurone formation was studied. The yellow flowering ornamental plant accumulates deoxy type anthochlor pigments (69-deoxychalcones and the corresponding 4-deoxyaurones) in the basal part of the flower surface whilst the apex contains only yellow carotenoids. For UV sensitive pollinating insects, this appears as a bicoloured floral pattern which can be visualized in situ by specific ammonia staining of the anthochlor pigments. The pet al. back side, in contrast, shows a faintly UV absorbing centre and UV absorbing rays along the otherwise UV reflecting pet al. apex. Matrix-free UV laser desorption/ionisation mass spectrometric imaging (LDI-MSI) indicated the presence of 9 anthochlors in the UV absorbing areas. The prevalent pigments were derivatives of okanin and maritimetin. Enzyme preparations from flowers, leaves, stems and roots of *B. ferulifolia* and from plants, which do not accumulate aurones e.g. Arabidopsis thaliana, were able to convert chalcones to aurones.

Thus, aurone formation could be catalyzed by a widespread enzyme and seems to depend mainly on a specific biochemical background, which favours the formation of aurones at the expense of flavonoids. In contrast to 4-hydroxyaurone formation, hydroxylation and oxidative cyclization to the 4-deoxyaurones does not occur in one single step but is catalyzed by two separate enzymes, chalcone 3-hydroxylase and aurone synthase (catechol oxidase reaction). Aurone formation shows an optimum at pH 7.5 or above, which is another striking contrast to 4-hydroxyaurone formation in Antirrhinum majus L. This is the first example of a plant catechol oxidase type enzyme being involved in the flavonoid pathway and in an anabolic reaction in general.

- ***Isolation from Marine Organism***

In recent years several new bioactive compounds have been isolated from marine organisms. However, despite recent progress in the area of marine chemistry, organisms of the Arabian Sea are little explored, which stimulated us to investigate the chemical constituents of such marine organisms. *Spatoglossum variabile* is a brown alga belonging to the family Dictyoaceae. *S. variabile* grows on the mid and littoral rocks along the coastline of Pakistan near Karachi city. Rahman et al. describes the isolation of two aurone derivatives (Figure 3, 4) from the methanolic extracts of this seaweed [24]. They also reported the isolation of two new aurones derivative from ethanolic extract of brown algae *Spatoglossum variabile* marine organism.

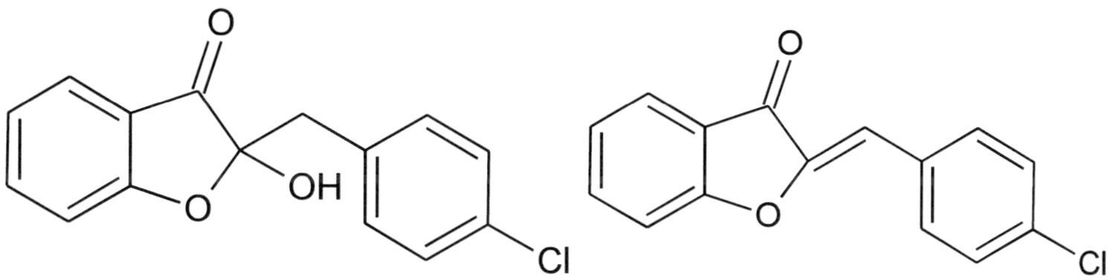

Figure 3 & 4. Aurone derivatives extracted from algae Spatoglossum variabile.

General Experimental Procedures

Collection of plant material: S. variabile (35 kg) was collected from the Buleji coastline of Karachi in December 1996. A voucher specimen has been deposited in the Department of Botany, University of Karachi.

Extraction and Isolation: The alga was washed with water, air-dried for five days and then soaked in methanol for one week. The methanolic extract of the alga was filtered and then evaporated under vacuum. The residue (512.0 g) was triturated with distilled water and successively fractionated with *n*-hexane, $CHCl_3$, EtOAc and *n*-BuOH. The concentrated EtOAc extract (49.2 g) was loaded onto a silica gel (804 g) column and then subjected to gradient elution with mixtures of *n*-hexane–$CHCl_3$ and $CHCl_3$–MeOH. The fraction obtained on elution with $CHCl_3$–MeOH (9: 1) was subjected to preparative TLC using $CHCl_3$–MeOH–NH_4OH (9.5: 0.5:3drops) as the solvent system to afford the new compounds.

3. Biological Roles, Targets and Activities

Flavonoids represent a large class of natural products in the plant kingdom, exhibiting multiple biological activities. Aurones play an important role in the pigmentation of some flowers and fruits and contribute especially to the bright yellow color of flowers (25). Aurones and others subclasses of flavonoids, such as flavones and chalcones, have been studied for their numerous biological activities. Aurones, however, are only studied sparingly compared with the others sub-classes. Nevertheless, recent studies have revealed promising biological properties of this group of natural molecules.

3.1. Aurones As Antiparasitic Agents

- ***Antimalarial Activity***

Malaria is a vector-borne infectious disease that is widespread in tropical and subtropical regions. Every year approximately 450 million clinical cases are caused by malaria parasites and 3.2 billion people in the world are living under the risk of malaria. Malaria is caused by a protozoal parasite of the genus *Plasmodium* and remnants worldwide problem and there is an urgent need for identifying new class of antimalarials. In view of above and as a part of our effort to develop newer anti-malarial agents, molecular modeling analysis was performed to develop QSAR models that show substantial predictive promise for aurones. The major problem in malaria treatment is the spread of resistance to natural or semi-synthetic antiplasmodial drugs.

Aurones are structural isomers of flavones, and compared to other flavonoids subclasses, somehow, aurones still by large less studied. The natural origin of aurones, theirs inhibitory effects on erythrocytic stages of *Plasmodium falciparum* strains in vitro [26]

Souard et al., [27] report the synthesis and in vitro antiplasmodial activity of 35 compounds, designed as analogues of the naturally occurring aurones. Several of these analogues showed submicromolar antimalarial activity against a chloroquine-resistant strain of Plasmodium falciparum (FcB1-Columbia strain) cultured on human erythrocytes.

Substitution of the intracyclic oxygen in aurones by a nitrogen atom and systematic variation of the substituent at the B-ring revealed promising leads showing good activity on the CQ-resistant strain. The most active compound was tested *in vivo* on mice and was not toxic to the mouse itself, but the anti plasmodial effect appeared to be less efficient compared with the *in vitro* studies due to its low solubility. The structure–activity relationship analysis revealed that dimethoxylation at positions C4 and C6, a halogen atom at position 4’, and the substitution of the intracyclic oxygen atom with an N-H group increased the activity of the compounds. However, a long chain had an adverse effect. Concerning the azaaurones, an ethyl moiety at C4’ rather than C2’, substitution of the ethyl group by an acetylenyl group, methoxylation at the 4’-position, and a dimethylamino moiety improve the efficiency of the molecules. (Figure 5)

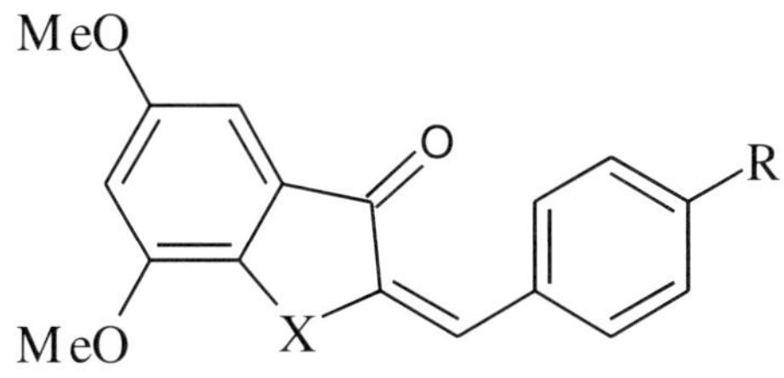

X= N, H, O

R= hydrophobic substituant

Figure 5. Chemical Structure of aurone scaffold.

3.2. Anticancer Activity

Cancer is a disease which is caused either directly or indirectly by physical, environmental, metabolic, chemical and genetic factors. Nearly 90% of all cancers are suggested to be caused by potentially controllable factors. A series of aurones have been synthesized and analyzed for their ability to target ABCG2 (ATP – binding cassette sub – family G member 2), an ABC protein transporter responsible for the breast cancer multidrug resistance mechanism. Results have shown that aurones are able to inhibit the ABCG2 efflux transporter in a dose-dependent manner and have a low cytotoxic effect against several cancer and healthy cell lines. Functionalized 4,6-dimethoxyaurones are promising ABCG2 inhibitors that combine good activity at submicromolar concentrations with limited antiproliferative activity [28].

In another study, Cheng and Zhang synthesized a series of 4'–substituted 5–hydroxyaurone derivatives. These compounds have been tested for their cytotoxic effects on both cancer and noncancer cell lines. Results show that some aurones are cytotoxic toward cancer cell lines, but exert weaker activities against noncancer cell lines. These compounds have also been tested *in vitro* regarding their ability to inhibit cell motility and angiogenesis, two processes implicated in cancer cell attack and metastasis development. The authors identified two synthetic compounds that are able to beneficially modulate these two mechanisms involved in cancer growth [29].

Human umbilical vein endothelial cells are involved with endothelial cell function in human beings. Aurones exhibit good anticancer activities against human umbilical vein endothelial cell lines and hence provide for a demanding and potential area of development of aurone derivatives as anticancer agents. Parate and his co-workers [30] are correlated the anticancer activity of the aurones to the physicochemical descriptors through QSAR studies. KNN- MFA and other advanced methods of analysis were opted for a set of seventeen aurone derivatives to figure out the correlationship between steric and electrostatic descriptors. (Figure 6)

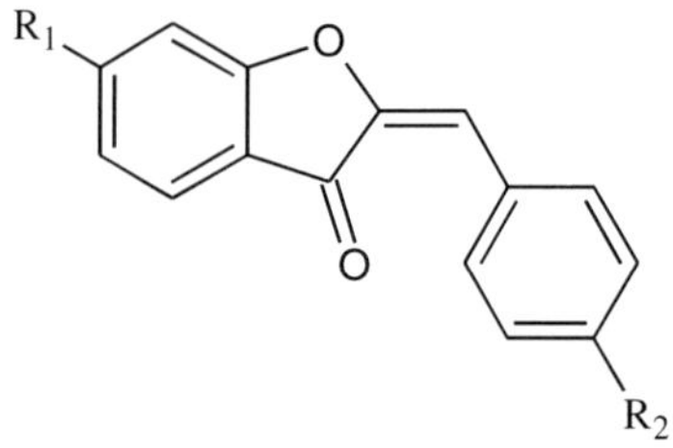

R_1= H, Ac
R_2= OH, OMe, F, Cl, Br, CF_3, CN

Figure 6. Chemical Structure of aurone scaffold.

Nicholas et al. [31] reported the naturally occurring aurone (Figure7), isolated from Uvaria hamiltonii, and a series of aurones analogues based structurally on known tubulin binding agents were prepared and evaluated for anticancer activity. Aurone (Figure 8) was the most active (IC_{50} K562 50 nM) and caused significant G_2/M cell-cycle arrest.

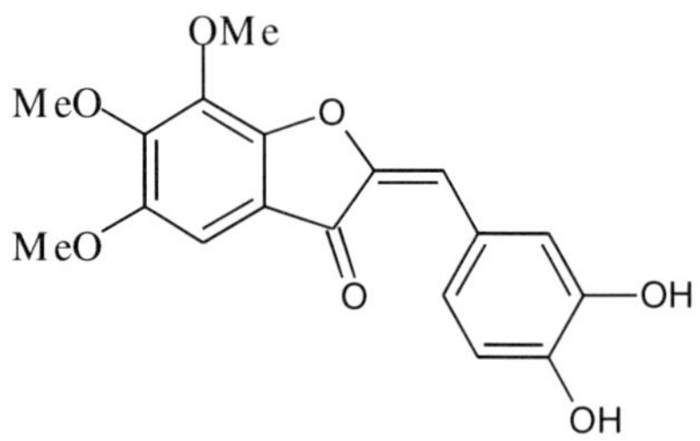

Figure 7. (IC_{50} 12μM).

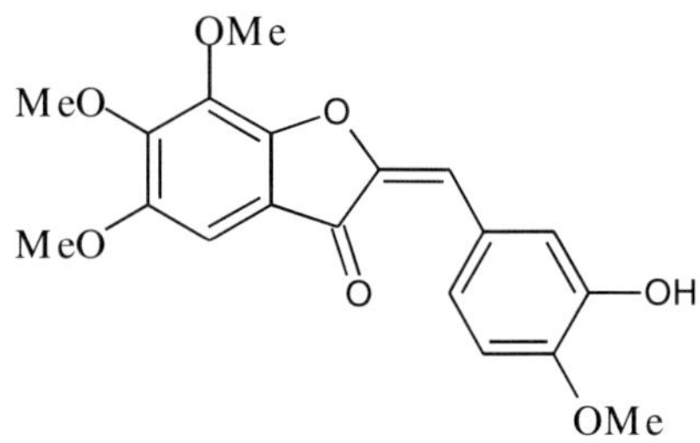

Figure 8. (IC_{50} 50nM).

Figure 7 & 8. Aurone derivatives as anticancer agents. IC50 Cytotoxicity against K562 cell line.

Benzofuranone part of a classical aurone with either a chromone or a coumarin scaffold which proved to feature interesting biological activities including antimicrobial, antiviral, anticancer, anti-inflammatory and antioxidant properties. Herein Zwergel et al. [32] presented series of 26 novel benzofuran derivatives with the first biological results in K562 human leukemia cells showing that compounds are able to induce around 24% apoptosis.

3.3. Anti-VSMC (Anti-Vascular Smooth Muscle Cells)

Rui et al. [34] reported the series of new aurone derivatives was prepared by means of a practical route and their anti-vascular smooth muscle cells(VSMC) vegetation activities were evaluated by the 3-(4,5-dimethylthlazol-2-yl)-2,5-diphenyltetrazolium bromide(MTT) method with tetrandrine as a positive contrast drug. Anti-VSMC vegetation activities of compounds 1a—1i, 2a—2i and 3a—3i were evaluated *in vitro via* the 3-(4,5- dimethylthlazol-2-yl)-2,5-diphenyltetrazolium bromide(MTT) method with tetrandrine as the positive control. (Figure 9)

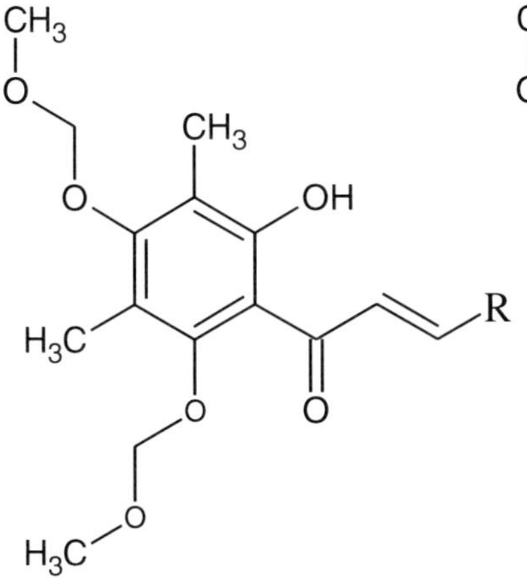 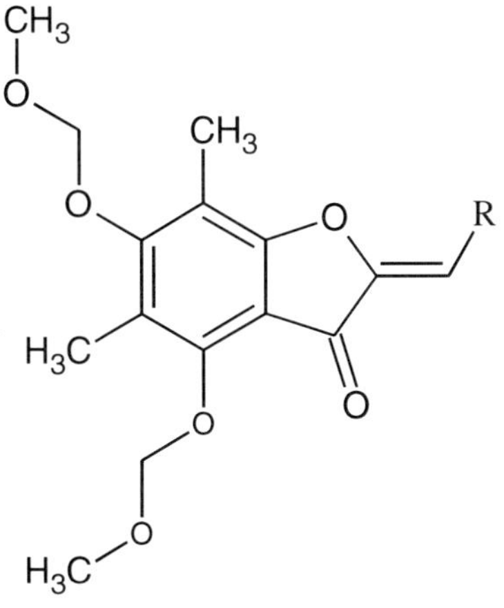

5a-5i 6a-6i

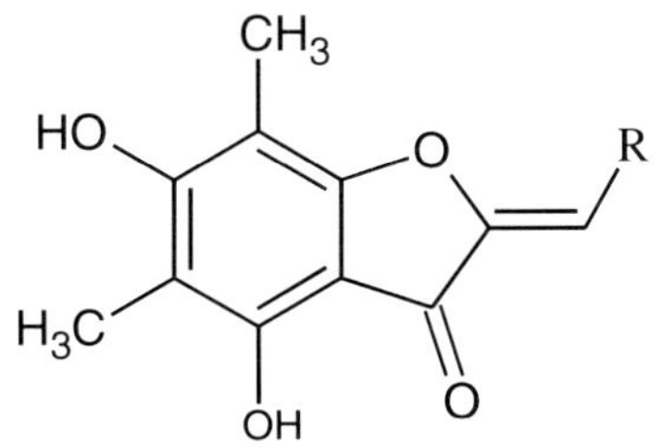

7a-7i

a=toluene, b=p-chlorotoluene, c=o-chlorotoluene d= 2,4-dichlorotoluene, e=p- fluorotoluene, , f = p-bromotoluene, g = m-nitrotoluene, h = 3-methylpuridine, i = p-xylene

Figure 9. Different aurones derivative synthesis by Rui et al. having anti-vascular smooth muscle cells (VSMC) vegetation activities.

The results show that substituents at different positions could exhibit important effects on the bioactivity of the hydroxyl protected compounds. For example, the compounds with groups on B ring at *para*- and *meta*-substituted positions had a good anti-VSMC vegetation activity, while the compounds with groups on B ring at *ortho*-substituted position showed no activity. Meanwhile, the activity of the hydroxyl protected compounds with a *para*-halogen group on B ring was in the order, Br>Cl>F, such as compounds 1b, 1e, 1f, 2b, 2e and 2f.

3.4. Antimicrobial Activity

Patel et al. [34] synthesized oxine containing aurone derivatives (Figure 10) by the condensation of substituted iodoacetophenone and 5-formyl- 8-hydroxy quinoline. All this aurone derivative show good activity towards antibacterial and anti-fungal.

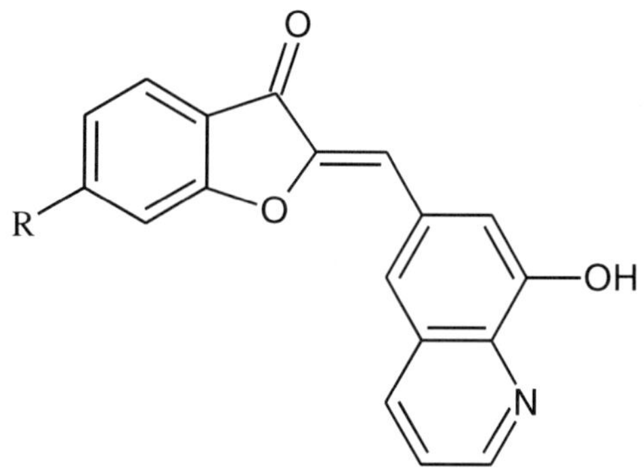

R= H, CH_3, OCH_3, Cl, Br

Figure 10. Oxine containing aurones derivatives.

All the aurones metal chelates were also screened for antifungal activity. Antimicrobial activity of all the compounds and metal complexes (Table 4) were screened for their antibacterial activities in "in vitro" at doses of 100 mg in 0.1 ml of DMF against gram negative bacterium *Escherichia coli* and the gram positive bacterium *Bacilas ciroflagallous* and for their antifungal activity in vitro agents the fungi *Aspergills niger* and *Candida ablicans*. DMF was used as a solvent control.

PDA (potato-dextrose-agar) medium was used as a culture medium and method employed was cup method. The zones of inhibition was measured in mm and are represented by (+), (++), (+++) depending upon the diameter and clarity.

Table 4. Antibacterial and Antifungal activities of oxine containing aurones and metal complexes of 2d

Microbes	2a	2b	2c	2d	2e	2d-Cu^{2+}	2d-Ni^{2+}	2d-Co^{2+}	2d-Mn^{2+}	2d-Zn^{2+}
E.Coil	++	++	+++	+++	+++	+++	+++	+++	++	++
B. Cirrofla Gallosus	++	++	+++	+++	+++	+++	+++	+++	+++	++
A. Niger	++	+++	+++	+++	+++	+++	+++	+++	++	++
C. Albicans	++	++	+++	+++	+++	+++	+++	+++	+++	++

(+) Poor active, (++) Active, (+++) More active.

3.5. Antileishmanial Agents

Leishmaniasis, an infectious disease caused by protozoan parasites belonging to the genus *Leishmania* (*L.*), is transmitted to humans through the bite of female phlebotomine sand flies infected with the parasite. The antileishmanial activity of a series of aurones was first reported by Kayser and Kiderlen in 1999 [35]. Certain aurones exhibit a remarkably strong capacity to kill intracellularly persisting amastigotes of *Leishmania donovani* and *L. major* with no significant cytotoxicity for host cells.

In this study, five aurones and one aurone glucoside were tested against extracellular promastigotes of *L. donovani*, *L. infantum*, *L. enriettii*, and *L. major*, and against intracellular amastigote *L. donovani,* and the results revealed that the compounds were active against the parasites in concentrations ranging between 0.4 and 5.0 μg/mL while they showed moderate cytotoxicity.

Kayser et al. [36] reported data indicate that the antileishmanial activity of certain aurones may also be explained by their interference with respiratory mitochondrial enzymes of the pathogen. A series of aurones was analyzed for the ability to inhibit respiratory functions of mitochondria of *Leishmania* parasites. The presented data show strong inhibitory activity of certain aurones and auronols on *L. major* mitochondrial fumarate reductase. The efficacy of the tested aurones and auronols appeared to be both dose- and structure-dependent, albeit the limited number of compounds tested allows only a preliminary evaluation of structure-activity relationships.

Roussaki et al. [37] reported a series of (*Z*)-2-benzylidenebenzofuran-3-(2H)-ones (aurones) bearing a variety of substituents on rings A and B were synthesized and evaluated for their antiparasitic activity against the intracellular amastigote form of *Leishmania infantum* and their cytotoxicity against human THP1-differentiated macrophages. The position of the substituents as well as their electronic nature seem to play an important role in the antiparasitic activity. Among the twelve tested compounds, the aurones 1 and 2, possessing methoxy groups at positions 4 and 6 of ring A and electron-donating groups on position 2 of ring B exhibited antiparasitic activity comparable to that of Amphotericin B. (Figure 11)

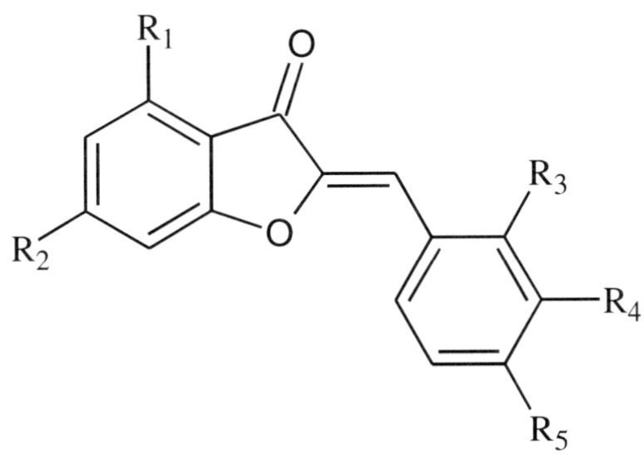

1 R_1=R_2=R_3=OCH_3, R_4=R_5= H (Structure 1)
2 R_1=R_2=OCH_3, R_3=CH_3, R_4=R_5=H (structure 2)

Figure 11. (*Z*)-2-benzylidenebenzofuran-3-(2H)-ones (aurones) derivatives with antiparasitic activity.

3.6. Aurones As Anti-Viral Agents

A series of different flavonoids has been analyzed for neuraminidase inhibition potency, a glycoprotein involved in the infection process of the influenza virus. Liu et al. [38] described the SAR studies of flavonoids, most of the aurones tested were described as good neuraminidase inhibitors. (3) The structure–activity relationship revealed that a glycosyl group at any position, and a hetero-function at C3 or C4 decrease this effect, and an OH group at C6 or C4', a double bond between C-2 and the phenylidene, and a heterofunction at C3 are essential for the activity. Most of the aurones tested were described as good neuraminidase inhibitors. (Figure 12)

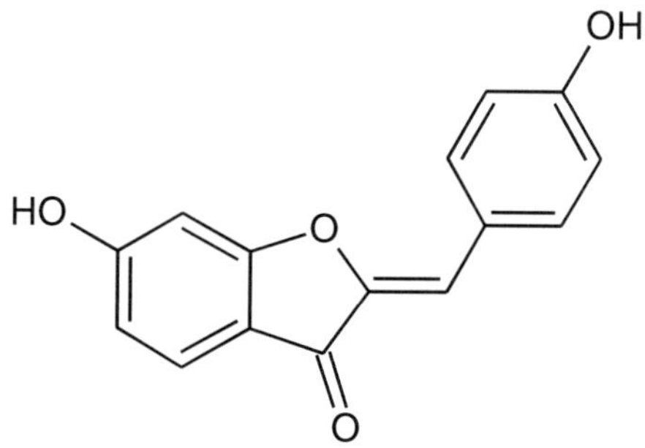

Figure 12. (2Z)-6-hydroxy-2-(4-hydroxybenzylidene)-1-benzofuran-3(2H)-one.

3.7. Aurones As Potent Allosteric Inhibitors of Hepatitis C Virus RNA-Dependent RNA Polymerase

Haudecoeur et al. [39] identified naturally occurring 2-benzylidenebenzofuran-3-ones (aurones) as new templates for non-nucleoside hepatitis C virus (HCV) RNA-dependent RNA polymerase (RdRp) inhibitors. The aurone target site, identified by site-directed mutagenesis, is ocated in Thumb Pocket I of HCV RdRp. The RdRp inhibitory activity of 42 aurones was rationally explored in an enzyme assay. Molecular docking studies were used to determine how aurones bind to HCV RdRp and to predict their range of inhibitory activity.

Seven aurone derivatives were found to have potent inhibitory effects on HCV RdRp, with IC s below 5 M and excellent selectivity. 50 μ The most active aurone analogue was (*Z*)-2-((1-butyl-1*H* -indol-3-yl)methylene)-4,6-dihydroxybenzofuran-3(2*H*)-one (Figure 13), with an IC of 2.2 M. Their potent RdRp inhibitory activity, together with their low toxicity, makes these molecules attractive 50 μ candidate direct-acting anti-HCV agents.

They investigated the capacity of aurones, non-toxic natural compounds; to inhibit HCV RdRp. Aureusidin potently inhibited the activity of this enzyme. By using classical medicinal chemistry, they identified positions and substituents that are crucial for the inhibitory activity of aurones on RdRp. A combination of site-directed mutagenesis and molecular modeling was successfully used to understand and optimize aurones anti-HCV activity. Our findings constitute a milestone in the development of new non-nucleoside inhibitors of HCV RdRp. Studies are underway to synthesize more active compounds warranting preclinical evaluation.

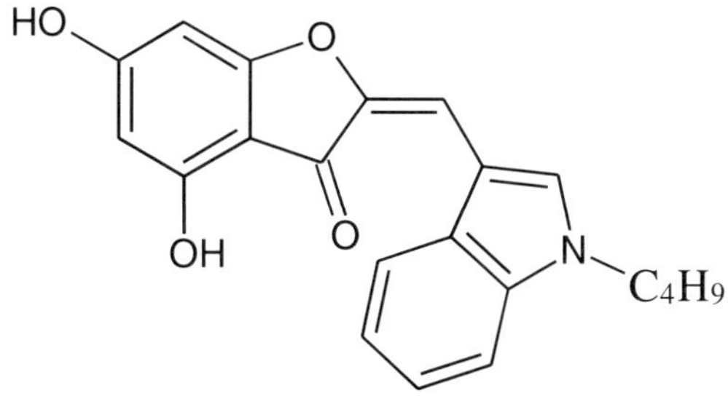

Figure 13. (Z)-2-((1-butyl-1H -indol-3-yl)methylene)-4,6-dihydroxybenzofuran-3(2H)-one.

3.8. Aurones Use for Skin Diseases

Some aurones are able to inhibit human tyrosinase, an enzyme which plays a role in the melanin synthesis pathway. Melanin is the natural pigment of human skin and has been implicated in several dermatological diseases.

Okombi and Rival [40] reported the SAR study of aurones for dermatological diseases. Analysis of the structure–activity relationship indicates that hydroxyl groups on the B ring are necessary for the activity and hydroxyl groups at the 4,6 and 4' positions strongly increase efficiency. (Figure 14)

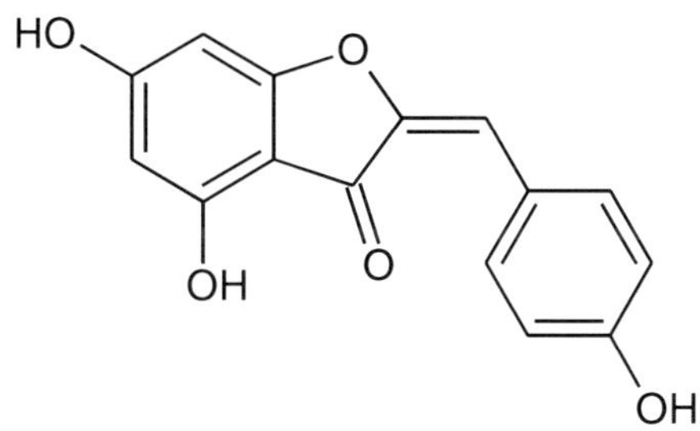

Figure 14. (2E)-4, 6-dihydroxy-2-(4-hydroxybenzylidene)-1-benzofuran-3(2H)-one.

3.9. Alzheimer's Disease

Alzheimer's disease, the most common form of dementia among the elderly, is a progressive, degenerative disorder of the brain with a loss of memory and cognition. AChE inhibitors are still the major and most developed class of drugs approved for Alzheimer's disease therapy, such as donezepil, rivastigmine and galanthamine have been approved by FDA and EMEA for the symptomatic treatment of Alzheimer's disease. A defining characteristic of Alzheimer's disease is the deposition of amyloid fibrils and neurofibrillary tangles in the brain of afflicted individuals. Biochemically, they are mainly composed of amyloid protein and phosphorylated tau proteins, respectively. A novel series of aurone derivatives has been synthesized as radiolabelled probes to detect β-amyloid plaques in Alzheimer's disease. *In vitro* results show that these compounds have only one high affinity binding site for β- amyloid peptides. Biodistribution performed on mice revealed that these molecules have a good brain uptake and a fast clearance from the brain, which are two important properties for *in vivo* amyloid probes [41].

Maya et al. [42] report a novel series of radioiodinated aurone derivatives as probes for imaging Abeta plaques in the brains of patients with Alzheimer's disease (AD) using single photon emission computed tomography (SPECT). In binding experiments in vitro, aurone derivatives showed very good affinity for Abeta aggregates.

REFERENCES

[1] Villemin, D., Martin, B. and Bar, N. (1998). Application of Microwave in Organic Synthesis. Dry Synthesis of 2-Arylmethylene-3(2)-naphthofuranones. *Molecules* 3 (3), 88-93.

[2] Bashir, H. S., Mohammed, A. M., Magsoud, A. S. and Shaoub, A. M. (2013). Isolation of Three Flavonoids from *Withania Somnifera* Leaves (*Solanaceae*) and their Antimicrobial Activities. *J forest products & industries*, 2(5), 39-45.

[3] Pattipati, S., Amanpreet, S. and Shrinivas, K. (2003). Effect of *Withania somnifera* Root Extract on Haloperidol-Induced Orofacial Dyskinesia: Possible Mechanisms of Action. *J. Med. Food,* 6(2), 107-114.

[4] Amanlou, M., Ataie, S. and Farsam, H. (2005). Xanthohumol from Humulus lupulus cultivated in iran. *Journal of Medicinal and Aromatic Plant Sciences,* 27, 469 – 475.

[5] Veitch, N. C. and Grayer, R. J. (2008). Flavonoids and their glycosides, including anthocyanins. *Natural Product Reports,* 25, 555 –611.

[6] Harborne, J. B. and Williams, C. A. (2000). Advances in flavonoid research since 1992. *Phytochemistry*, 55, 481-504.

[7] Yao, L. H., Jiang, Y. M., Shi, J., Tomas-Barberan, F. A, Datta, N., Singanusong, R. and Chen, S. S. (2004). Flavonoids in food and their health benefits. *Plant Food Human Nutr.*, 59, 113-122.

[8] Heim, K. E., Tagliaferro, A. R. and Bobliya, D. J. (2002). Flavonoids antioxidants: Chemistry, metabolism and structure-activity relationships. *The Journal of Nutritional Biochemistry*, 13, 572-584.

[9] Tsuchiya, H. (2010). Structure-dependent membrane interaction of flavonoids associated with their bioactivity. *Food Chemistry*, 120, 1089-1096.

[10] Rijke, E. D., Out, P., Niessen, W. M. A., Ariese, F., Goojer, C. and Brinkman, U. A. T. (2006). Analytical separation and detection methods for flavonoids. *Journal of Chromatography*, 1112 A, 31-63.

[11] Warashina, T. and Noro, T. (1994). Cardenolide glycosides from Asclepias fruticosa. *Phytochemistry*, 37(3), 801-806.

[12] Harbone, J. B. (2001). Phytochernical methods. Chapman and Hall Ltd: London, pp111-113.

[13] Shinoda J. (1928). A new biologically active flavone glycoside from the roots of *Cassia fistula* Linn. *J. of pharmaceutical society Japan,* 1928, 48, 214.

[14] Schmidt J. (1964)."Text book of organic chemistry" Oliver and Poyed Ed. London, 673.

[15] Sofowora A., (2005). Medicinal plants and traditional medicine in Africa. Spectrum Books Ltd: Ibadan. Nigeria, p. 289.

[16] Gonzalez, E. E. and Delgando J. N. (1962). Phytochemical investigation of *Amphipterygium adstringens. J. Pharm. Sci.*, 51,786-790.

[17] Kavanagh, F. (1972): "Analytical Microbiology" Vol. II,Academic Press (Pub) New York and London, P. 11.

[18] Eliane, O., Ferreiraa, M. J. S., Elizabeth M. F. P., Silvia C. A., Izabel Y. I. and Diones A. D. (2004). A New Heptasubstituted (*E*)-Aurone Glucoside and Other Aromatic Compounds of *Gomphrena agrestis* with Biological Activity. *Z. Naturforsch.* 59c, 499-505.

[19] Kayser, O., Kiderlen A. F., Folkens U. and Kolodziej, H. (1999). *In vitro* leishmanicidal activity of aurones. *Planta Med.*, 65, 316-319.

[20] Kayser, O., Chen, M., Kharazmi, A. and Kiderlen, A. F. (2002). Aurones interfere with *Leishmania major* mitochondrial fumarate reductase. *Z. Naturforsch.* 57c, 717-720.

[21] Wang, J., Wang, N., Yao, X. and Kitanaka, S., (2007). Structures and Anti-histamine Activity of Chalcones & Aurones Compounds from *Bidens parviflora* Willd. *Journal of Traditional Medicines*, 2 (1), 23-29.

[22] Zhanga, S.-X., Tanib, T., Yamajib, S., Maa, C.-M., Wanga, M.-C., Caia, S.-Q. and Zhao.Y. –Y. (2003). Glycosyl flavonoids from the roots and rhizomes of *Asarum longerhizomatosum*. *Journal of Asian Natural Products Research*, 5 (1), 25–30.

[23] Miosic, S., Knop, K., lscher, D., Greiner, J., Gosch, C., Thill1, J., Kai, M., Shrestha. B. K., Schneider, B., Crecelius, A. C., Schubert, U. S., Svatos, A., Stich, K. and Halbwirth, H. (2013). 4-Deoxyaurone Formation in Bidens ferulifolia (Jacq.) DC. *PLOS ONE*, 8 (5), e61766.

[24] Rahman, A., Choudhary, M. I., Hayat, S., Khan, A. M. and Ahmed, A. (2001). Two New Aurones from Marine Brown Alga Spatoglossum variabile. *Chem. Pharm. Bull.*, 49(1), 105-107.

[25] Eiichiro, O., Masako, F. M., Noriko, N., Yuko, F., Keiko, Y. S., Masaatsu, Y., Toru, N., Takaharu, T., Takaaki, K. and Yoshikazu, T. (2006). Yellow flowers generated by expression of the aurone biosynthetic pathway. *PNAS*, 103, 29, 11075–11080.

[26] Boumendjel, A. (2003). Aurones: a subclass of flavones with promising biological potential. *Curr. Med. Chem.*, 10, 2621-2630.

[27] Florence, S., Sabrina, O., Chantal Beney, Severine, C., Alexis V. and Ahcene B., Bioorganic & Medicinal ChemistryVolume 18, Issue 15, 2010, 5724–5731.

[28] Sim H.M. and Lee, C. Y. (2008) Dimethoxyaurones: Potent inhibitors of ABCG2 (breast cancer resistance protein). *European Journal of Pharmaceutical Sciences: official Journal of the European Federation for Pharmaceutical Sciences*, *35*, 293-306.

[29] Cheng, H. and Zhang, L. (2010). Design, synthesis and discovery of 5-hydroxyaurone derivatives as growth inhibitors against HUVEC and some cancer cell lines. *European Journal of Medicinal Chemistry*, *45*, 5950-5957.

[30] Parate, A. Sharma, R. and Maheshwari, I. (2013) Structural indents for 5-hydroxyaurone derivatives as potent anticancer agents against HUVEC cancer cell lines-kNN MFA approach. *Der Pharmacia Sinica*, 4(3), 121-129.

[31] Nicholas, J., Lawrence, D. R., Alan, T. and McGown, J. A. (2003). The total synthesis of an aurone isolated from *Uvaria hamiltonii*: aurones and flavones as anticancer agents. *Hadfield, Bioorganic & Medicinal Chemistry Letters*. 13, (21), 3759–3763.

[32] Zwergel, C., Valente, Sergio, Salvato, A., Xu, Z., Talhi, O., Mai, A., Silva, A., Altucci, L. and Kirsch G. (2013) Novel benzofuran–chromone and –coumarin derivatives: synthesis and biological activity in K562 human leukemia cells. *Med. Chem. Commun.* 4, 1571-1579.

[33] Rui F., Ban S., Feng X. E., Zhao C. X. and Li Q. S. (2012). Synthesis and Anti-VSMC Vegetation Activities of New Aurone Derivatives. *chem. res.* chinese universities 28(3), 438—442.

[34] Patel, V. D., Thakor, S. B., Patel, V. G. and Patel, K. R. (2013). Synthesis and biological activities of oxine containing aurones. *indian journal of applied research.*3, (7), 573-575.

[35] Kayser, O. and Kiderlen, A. F. (1998). "Leishmanicidal activity of aurones," *Tokai Journal of Experimental and Clinical Medicine*, 23, (6), 423–426.

[36] Kaysera, O., Chenb, M., Kharazmib, A. and Kiderlen, A. F. (2002). Aurones Interfere with *Leishmania major* Mitochondrial Fumarate Reductase.17-20.

[37] Roussaki, M., Lima S. C., Kypreou A. M., Kefalas P., Silva, A. C. and Detsi, A. (2012). Aurones: A Promising Heterocyclic Scaffold for the Development of Potent Antileishmanial Agents. *International Journal of Medicinal Chemistry*, 1-8.

[38] Liu, A. L. and Wang, H.D. (2008) Structure-activity relationship of flavonoids as influenza virus neuraminidase inhibitors and their *in vitro* anti-viral activities. *Bioorganic & Medicinal Chemistry*, *16*, 7141-7147.

[39] Haudecoeur, R. and Ahmed, B. A. (2011). Discovery of naturally occurring aurones that are potent allosteric inhibitors of hepatitis C virus RNA-dependent RNA polymerase. *Journal of Medicinal Chemistry*, *54*, 5395–5402.

[40] Okombi, S. and Rival, D. (2006). Discovery of benzylidenebenzofuran-3(2H)-one (aurones) as inhibitors of tyrosinase derived from human melanocytes. *Journal of Medicinal Chemistry*, *49*, 329-333.

[41] Ono, M. and Maya, Y. (2007). Aurones serve as probes of beta-amyloid plaques in Alzheimer's disease. *Biochemical and Biophysical Research Communications*, *361*, 116-121.

[42] Maya, Y., Ono, M., Watanabe, H., Haratake, M., Saji, H. and Nakayama, M. (2009). Novel radioiodinated aurones as probes for SPECT imaging of beta-amyloid plaques in the brain. *Bioconjug Chem.* 20(1), 95-101.

In: Natural Heterocycles
Editors: K. L. Ameta and Saksit Chanthai
ISBN: 978-1-63463-424-3

Chapter 6

ASCORBIC ACID AND FURFURALS IN CHILLI PEPPER, WINE AND FRUIT JUICE: THE NON-ENZYMATIC BROWNING STUDY

Jatesadaporn Yaisoongnern and Saksit Chanthai*
Department of Chemistry, Faculty of Science, Khon Kaen University,
Khon Kaen, Thailand

ABSTRACT

This study presented HPLC determination of ascorbic acid, dehydroascorbic acid (DHA), 5-hydroxymethylfurfural (5-HMF) and furfural in chilli fruits, wine and fruit juice. The contents of ascorbic acid and DHA in chilli peppers were ranged from 1.31-99.58 and 0.13-2.75 mg/100 g, respectively. Furfural was rarely found while 5-HMF was not detectable in the samples tested. Ascorbic acid, DHA, 5-HMF and furfural found in wine and fruit juice samples were ranged from 0.38-46.47, 0.91-7.39, 0.20-12.41 and 0.01-0.70 mg/100 mL, respectively. The DHA content in these samples was determined using the proposed method compared with that of DHA as the difference in total ascorbic acid after DHA reduction with dithiothreitol and that ascorbic acid of the original sample. This method could also be used to evaluate their quality of chilli pepper, wine and fruit juice during processing and storage. In addition, the relationship between the degradation of ascorbic acid and the formation of browning of fresh chilli pepper was investigated. The browning index was determined using an absorbance measured at 420 nm. From the results, the ascorbic acid content in the chilli samples decreased while their browning index increased a long with storage time.

1. INTRODUCTION

Browning reactions in foods are widespread phenomena which take place during processing and storage. These reactions occur during the manufacture of meat, fish, fruit and

* Tel. +66-4300-9700 ext. 12371, Fax. +66-4320-2373. E-mail: sakcha2@kku.ac.th.

vegetable products, as well as when fresh fruits and vegetables are subject to mechanical injury [1]. Browning usually impairs the sensory properties of products due to associated changes in the color, flavor and softening (due probably to the action of peptic enzymes) besides nutritional properties [1-2]. Non-enzymatic browning has been considered one of the major causes of quality and color loss during the processing and storage of food products [3]. This includes Maillard reaction, caramelization, pigment destruction and ascorbic acid oxidation [4]. Theories have been proposed which implicate ascorbic acid, based on the observations that browning formation follows the ascorbic acid loss [5]. Degradation of ascorbic acid proceeds both aerobic and anaerobic pathways and depends upon many factors such as pH, oxygen, heat, light, storage temperature and storage time [6-8].

Pathways of ascorbic acid degradation are shown in Figure 1. The oxidation of ascorbic acid leads to the formation of the intermediates, dehydroascorbic acid (DHA) and diketogulonic acid and subsequently to L-xylosone, the further degradation to various five carbon compounds [7-11].

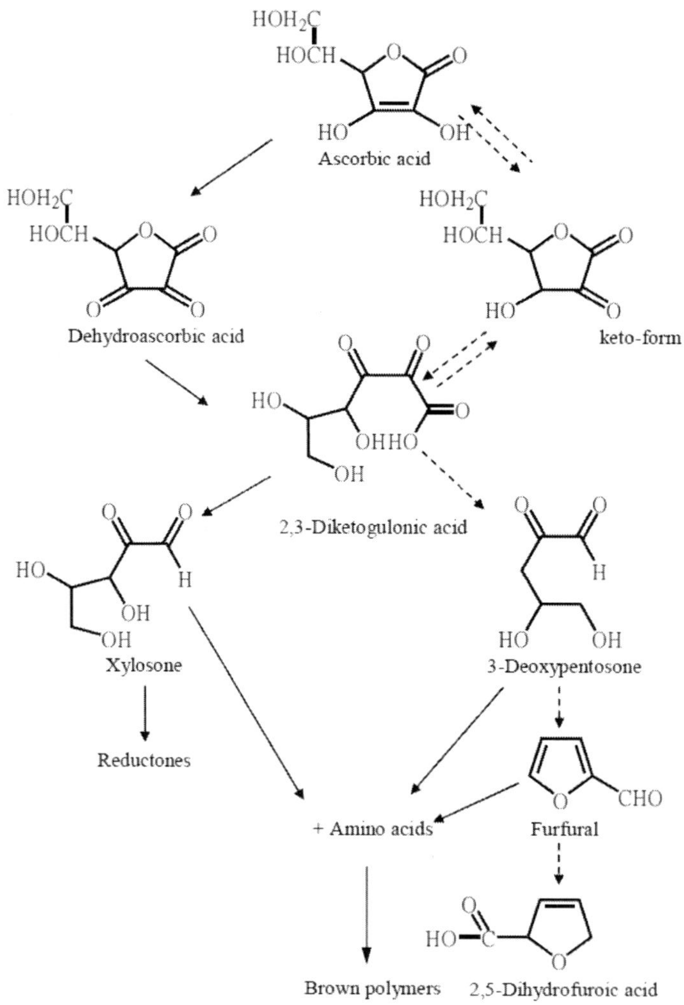

Figure 1. Pathways of ascorbic acid degradation (solid line, aerobic route; dashed line, anaerobic route) [13].

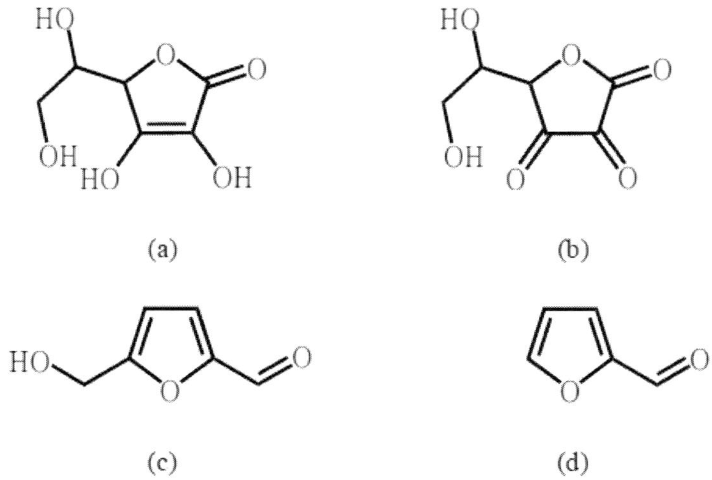

Figure 2. The heterocyclic structures of (a) ascorbic acid, (b) dehydroascorbic acid, (c) 5-hydroxymethylfurfural and (d) furfural.

Under anaerobic conditions, ascorbic acid degrades via several steps to furfural instead of DHA [9]. 5-Hydroxymethylfurfural (5-HMF) is one of the decomposition products of ascorbic acid and is suggested to be a precursor of brown pigments [1, 6, 12]. It was reported that several decomposition reactive products occur via the degradation of ascorbic acid and these compounds may combine with amino acids, thus result in the formation of the brown pigment [1,6]. Thus, furfural and 5-HMF are considered to be important intermediates and indicators of the browning [7-8]. The structures of ascorbic acid, DHA, furfural and 5-HMF are shown in Figure 2.

Peppers (*Capsicum annuum* L.) are popular spices in many parts of the world, valued for their sensory attributes of color, pungency and flavor. Peppers are economically important because of the vast consumption of the diverse varieties. Food industry is the largest user of capsicums, where they are used as coloring and flavoring agents in sauces, soups, processed meats, snacks, candies, soft drinks and alcoholic beverages [14]. Peppers are a good source of vitamin C and E as well as provitamin A and carotenoids with well known antioxidant properties [15]. The nutritional quality of pepper is related primarily to its content of ascorbic acid and DHA, the decomposition of ascorbic acid, together with non-enzymatic browning is the main deteriorative reaction that occurs during storage [16]. It has long been known that pepper is darkening during harvesting, transporting, processing and storage. Color is one of the most important qualities of pepper, which affects the consumers' preferences [17]. Therefore, non-enzymatic browning is the most common quality problem in pepper.

Wines and fruit juices are highly consumed in many countries and nowadays, consumer desires high quality foods, with freshly flavor, texture and color, with minimal or no chemical preservatives, and above all safe. During storage, wine and fruit juice suffer an important number of deterioration reactions (ascorbic acid degradation, cloud loss, microbial spoilage, development of off-flavor, changes in color, texture, appearance), that produce an important quality loss. The presence of furfural and 5-HMF in stored products is an indicator of their quality loss; furfural and 5-HMF are related with the browning of products and they are also good indicators of the excess of temperature and storage time in some food. Consequently, the analysis of these compounds has special importance in the food industry [18-19].

Since ascorbic acid degradation causes browning which is the problem of quality loss in pepper, wine, and fruit juice during processing and storage, it is necessary to describe the ascorbic acid degradation resulted in both 5-HMF and furfural formation. Thus, the aim of the present work is to develop a reversed phase-high performance liquid chromatographic (RP-HPLC) method for the simultaneous determination of ascorbic acid, DHA, 5-HMF and furfural and to investigate the optimum conditions for the extraction of these compounds in hot pepper. The ascorbic acid, DHA, furfural and 5-HMF contents in hot pepper, wine and fruit juice and the browning index in hot pepper were determined. Moreover, the relationship between ascorbic acid degradation and browning formation was investigated.

1.1. Determination of Ascorbic Acid and Furfurals

Ascorbic acid is a good indicator of retention of nutritional quality for fruits and vegetables. It can participate in browning effect through the ascorbic acid browning reactions, leading to the formation of a wide variety of end products [10]. Ascorbic acid is reversibly oxidized to form dehydroascorbic acid (DHA). Since DHA can be easily converted into ascorbic acid, it is important to measure both ascorbic acid and DHA [20]. Several methods are available for the determination of ascorbic acid in food samples. These analytical methods include titrimetry [21-23], spectrophotometry [15, 24-26], derivative spectrophotometry [27-28], flow-injection spectrophotometry [29], fluorometry [30-32] and high performance liquid chromatography (HPLC) [33-39]. Most of these methods, other than HPLC, are time-consuming and may give overestimates due to the presence of oxidizable species other than ascorbic acid, and/or do not measure DHA. HPLC methods have some advantages regarding specificity, sensitivity or easy operation [40]. Numerous papers for the determination of ascorbic acid have described the HPLC technique, based on reversed-phase C_{18} [33-35, 37-40], bonded-phase NH_2 [36, 40] or cation-exchange [3] columns and UV detection [33, 36-38]. On the other hand, a photodiode array gives spectral information. HPLC combined with the diode array system proved to be able to detect a wide range of molecules and ensure identification of target compounds. The retention time and spectrum provide strong evidence of their identity [38]. However, sensitivity is still an important problem for the direct measurement of DHA by HPLC using a common detection system like a UV-Visible detector. Usually, DHA is determined as the difference between the total ascorbic acid after DHA reduction and ascorbic acid content of the original sample. Various reducing agents, such as dithiothreitol (DTT) [20, 34] and L-cysteine [16] have been previously studied for the reduction of DHA to ascorbic acid. Moreover, DHA can be determined after HPLC separation and detection by fluorimetry after a post-column derivatisation with *o*-phenylene diamine (OPDA) [42].

Extraction methods may also differ between various fruits or maturity stages because of their different matrices. It is essential to inactivate degradative enzymes which can destroy ascorbic acid during the extraction and to fix the ascorbic acid/DHA redox equilibrium. Ascorbic acid is readily oxidized under alkaline conditions, so the use of a high ionic strength, acidic extraction solvent is required to suppress metabolic activity upon disruption of the cell and to precipitate proteins [20]. Various types of extracting acids have been recommended, such as oxalic acid [20], *meta*-phosphoric acid (MPA) [21-22,35,39], trichloroacetic acid (TCA) [26], hydrochloric acid (HCl) [27], acetic acid (HOAc) [27], the mixtures of MPA with HOAc [30,38] and various other mixtures. The content of ascorbic acid during paprika processing was determined [33]. Capsicum pericarp and paprika were extracted with 40 g/L MPA contained 250 mg polyvinyl-pyrrolidone (PVP). The extract was injected into a HPLC equipped with an UV detector. Ascorbic acid was separated on a C_{18} column with isocratic mobile phase (20 g/L *di*-ammonium hydrogen phosphate (w/v), pH 2.8) and flow rate of 0.4 mL/min. The chromatograms were monitored at 210 nm. Total vitamin C, ascorbic acid and DHA in fruits and vegetables were determined [34]. De-ionized water was used as an extractant in this case. The complete separation of ascorbic acid and DHA could be achieved on a C_{18} column using 0.2 M KH_2PO_4 as the mobile phase at a flow rate of 0.5 mL/min. Since the detection sensitivity was poor for DHA even at 210 nm, it was estimated as the difference between the total ascorbic acid after DHA reduction and the ascorbic acid content of the original sample, using DTT as the pre-column reductant.

The determination of ascorbic acid in tomato (*Lycopersicum esculentum* Mill) and bell pepper (*Capsicum annuum*) during development, maturation and senescence was reported [35]. Fresh tissue was extracted with 6% (w/v) MPA before injection into the HPLC. The HPLC equipped with C_{18} column and photodiode array detector was used. The mobile phase consisted of 5.0×10^{-4} M tridecylammonium-formate in water/methanol (40:60, v/v), pH 3.5. Quantification of ascorbic acid was made by external standard method. HPLC was used for evaluation of ascorbic acid content in fresh-cut bell peppers at different intervals of storage [36]. For extraction, de-ionized water was also used as an extractant. The ascorbic acid was detected using bonded-phase NH_2 column. The mobile phase was acetonitrile : KH_2PO_4 (75:25, v/v) at a flow rate of 1.5 mL/min. The detector wavelength was at 268 nm. Navarro et al. [37] studied the change in the content of ascorbic acid in pepper fruits (*Capsicum annuum* L.) at different ripening stages, as affected by salinity. Sample was liquefied and determined by HPLC coupled with an UV-Visible wavelength detector (245 nm). The mobile phase was 20 g/L $(NH_4)_2HPO_4$, pH 2.8 at a flow rate of 0.9 mL/min.

HPLC technique was used for quantitation of ascorbic acid in ripe fruits of five *Capsicum annuum* cultivars, grown in Turkey [38]. Samples were extracted with 3% (w/v) MPA containing 10^{-6} M ethylenediaminetetraacetic acid and 10^{-7} M diethyldithiocarbamic acid. Then the extract was flushed through C_{18} cartridge, preconditioned with 3% *meta*-phosphoric acid. Chromatography was carried out on a reversed phase C_{18} column with UV detection (254 nm). For elution, 0.2 M KH_2PO_4 pH 2.2 was used at a flow rate of 0.7 mL/min. The ascorbic acid contents were found in various fruits such as fresh *Emblica* fruits (*Emblica officinalis*), quava fruits (*Psidium guajava*), amaranth leaves (*Amaranthus gangeticus*) and green chilli (*Capsicum annuum*) [39]. Four methods for the determination of ascorbic acid were compared including 2,4-dinitrophenylhydrazine method, indophenol-xylene method, enzymatic method and HPLC method. 2% (w/v) MPA was used for the extraction of ascorbic acid in samples. For HPLC method, ascorbic acid was oxidized to DHA in the presence of Norit. The oxidized form was then reacted with OPDA to form a fluorophore. Quantitation of the DHA-OPDA derivative in the extract was made by HPLC consisting of a fluorescence detector at the excitation wavelength 350 nm and emission wavelength 430 nm. Chromatographic separation was accomplished using C_{18} column and elution with a mobile phase containing methanol:water (55:45) at a flow rate of 1 mL/min.

5-Hydroxymethylfurfural (5-HMF) and furfural are the intermediate products in the formation of brown pigments, occurring during processing and storage [43]. Thus, 5-HMF and furfural are recognized as an indicator of quality deterioration in a wide range of foods, such as grape juice [4], orange juice [12,46], apple juice [43-45], honey [46], milk [47], jams [46,48], baby foods [49] and alcoholic beverages [50]. Classical methods for the quantitative determination of 5-HMF and furfural in food products are based on colorimetric measurements [51,52], though they exhibit several drawbacks. These methods are unspecific, time consuming and make use of hazardous or toxic chemicals. Gas chromatographic techniques have also been employed [46]. In recent years, HPLC methods have been proposed, mostly using reversed-phase C_{18} with UV detection [12,47,48,50,53,54]. HPLC techniques produce more accurate results and allow analyzing various furfural compounds in the same run. Various methods have been used to extract 5-HMF and furfural in foods. In complex matrices like cookies, spaghetti, milk and fruit products, water extraction followed by clarification by Carrez (I and II), TCA or mixtures of TCA with oxalic acid is commonly used, prior to quantification by HPLC [47,48,53,54]. The formation of 5-HMF and furfural in orange juice stored at 8°C for 52 days was investigated [12]. Orange juice was diluted with water before injected into HPLC. 5-HMF and furfural were separated using C_{18} column coupled with an UV detector (280 nm). The mobile phase was methanol : water (15:85, v/v). The flow rate was 1 mL/min. The separation and identification of furanic compounds in fruit juices and drinks by HPLC with a photodiode array detector was reported [55]. Cation-exchange column was chosen for the separation of 5-HMF, furfural, 2,5-dimethyl-4-hydroxy-3(2*H*)-furanone, 2-furoic acid, 2-acetylfuran and furfuryl alcohol. The mobile phase consisted of acetonitrile and 0.005 M sulfuric acid aqueous solution with a flow rate of 0.6 mL/min.

RP-HPLC method was used for the determination of 5-HMF, furfural, furylmethylcetone and methylfurfural in store-brand and name-brand ultra-high-temperature-treated cows' milk [47]. Mixtures of 40% (w/v) TCA with 0.15 M oxalic acid were used as an extractant. Separations were carried out isocratically using the mixture of acetonitrile:water (5:95, v/v) as mobile phase with a flow rate of 1 mL/min and detected at 284 nm for 5-HMF and furfural, at 274 nm for furylmethylcetone, and at 293 nm for methylfurfural. Mendoza et al. [48] determined the 5-HMF content in commercial jams and fruit-based infant foods. Sample extraction in water followed by clarification by Carrez (I and II) was used. The extract was separated and quantified by RP-HPLC with UV detection at 283 nm. Analytical separation was performed using C_{18} column and gradient elution with methanol and water at a flow rate of 1 mL/min. Quantitation was carried out by the external standard method. The method for determination of furfural and 5-HMF in alcoholic beverages by HPLC was described [50]. Sample was injected into HPLC without extraction and detected at 280 nm. A C-18 column and an acetonitrile-water mobile phase with isocratic elution were used. The content of 5-HMF in cookies during the backing process was reported [50]. Sample was extracted with 40% (w/v) TCA with stirring and quantified by RP-HPLC with UV detection at 284 nm. The mobile phase was composed of sodium acetate (0.04 M) and methanol (80:20), pH 3.6 at a flow rate of 1 mL/min. Doxastakis et al. [54] studied the non-enzymatic browning of white lupin protein enriched spaghetti. The mixture of 0.3 N oxalic acid with 40% (w/v) TCA for the extraction of 5-HMF was used. The HPLC was carried out on reverse phase C_{18} column and UV detection at 284 nm. For elution, 0.08 mM sodium acetate buffer pH 3.6 was used at a flow rate of 1 mL/min.

Moreover, studies concerning the simultaneous determination of ascorbic acid, DHA, 5-HMF and furfural have been reported [3,5]. The HPLC method for simultaneous separation and determination of sugars, ascorbic acid, 5-HMF, furfural and four other furanic compounds as the possible degradation products of sugars and ascorbic acid on heating and by Maillard reaction was developed [3]. Sucrose, glucose, fructose, ascorbic acid, DHA, 5-HMF, furfural, 2,5-dimethyl-4-hydroxy-3(2H)-furanone), 2-furoic acid, 2-acetylfuran and furfuryl alcohol were separated on an Aminex HPX-87H column. The mobile phase consisted of acetonitrile and 0.005 M sulfuric acid aqueous solution (16:84, v/v) with a flow rate of 0.5 mL/min. Two detectors; a refractive index and a photodiode array, were used to detect these compounds. Several fruit juice and drink samples were analyzed using the HPLC method. The occurrence of non-enzymatic browning during storage of citrus juice was studied. Separation of ascorbic acid, some of its precursors and oxidation products was straightforward without the need for pre-column or post-column derivatization reactions. The HPLC-UV was performed on two macroreticular reverse phase PLRP-S columns (100 Å and 5 μm), connected in series. The eluent used was an aqueous solution of 0.2 M sodium dihydrogenphosphate, pH 2.14 and pumped at a flow rate of 0.5 mL/min.

1.2. Determination of Browning Index

Various processed food products are susceptible to browning deterioration occurring during processing and storage. Due to browning reactions miscellaneous chemical changes develop, resulting in undesirable flavors, nutrient losses, and formation of brown color pigments [43, 56]. Several methods were applied for the measurement of visual browning changes in food products [12, 18, 43, 57-62]. These methods are based on three main steps: (1) extraction of soluble color materials in the case of solid foods; (2) clarification of the resulting extract, or direct clarification in liquid foods; (3) colorimetric measurement of the clarified extract. The effects of oxygen and fluorescent light on the quality of orange juice during storage at 8°C were studied [12]. Browning was measured as light absorbance at 420 nm with a spectrophotometer. Cohen et al. [43] reported the quality of apple juice during thermal processing. Ethyl alcohol was added to an apple juice sample, centrifuged (10 min; 7800×*g*), and the absorbency of the supernatant was read at 420 nm. The value obtained was considered as the non-enzymatic browning index. The effects of temperature and pH on non-enzymatic browning in minced dried pepper during storage were investigated [54]. Before measurement of brown pigments in minced dried pepper, the samples were diluted with distilled water and stirred continuously with a magnetic stirrer for 15 min in order to extract the color. Then they were filtered with filter paper. Color was determined as absorbance in the samples. The absorbance of the filtrate was read at 420 nm.

The effect of storage on non-enzymatic browning of apple juice concentrates was investigated [58]. Browning development was determined by using two color measurement methods. The first method, samples were diluted to 11.2°Bx with distilled water. The absorbance was determined on a spectrophotometer in 10 mm cell against water at 420 nm. The second method, the CIE L*, a* and b* values were measured using colorimeter. Vercet [59] reported the browning of white chocolate during storage. Brown pigments were determined by reading absorbance at 420 nm, using a spectrophotometer. The effect of extrusion cooking of soy–sweet potato mixtures on available lysine content and browning index of extrudates was studied [60]. Browning index was determined with 1 g sample. The optical density of centrifuged, clear filtrate was read on a spectrophotometer at 420 and 550 nm. Water used as blank and browning index was calculated as: browning index = $A_{420\ nm} - A_{550\ nm}$. Bozkurt and Erkmen [61] described the effects of salt, starter culture and production techniques on the quality of hot pepper paste. Brown pigment formation was followed by measuring the color as optical density (OD). After the proper dilution of 15 g of sample with distilled water, the absorbance was read at 420 nm against blank by a spectrophotometer.

The color of orange juice treated by high intensity pulsed electric fields during refrigerated storage and comparison with pasteurized juice was studied [18]. Non-enzymatic browning index was determined with an absorbance measured at 420 nm. Three milliliters of orange juice were centrifuged at 2000 rpm for 20 min at 18°C, and the supernatant is mixed equally distributed with ethanol 95%. After filtering the mixture, the absorbance was measured at 420 nm. The mechanism of browning of pomegranate *(Punica granafum* L. *cv.* Ganesh) peel in different storage conditions was investigated [60]. The browning index of peel was calculated by the following formula: $\sum$ (Browning scale x Percentage of corresponding fruit within each class)/100. The appearance scale was as follows: 0 (0 score), no browning and glabrous surface; 1 (25 score), a slight browning and smooth surface; 2 (50 score), obvious browning and rough surface; 3 (75 score), serious browning and hollow surface depression; and 4 (100 score), black and encrusting surface.

2. Extraction and Analytical Methods

Chemicals: All chemicals and reagents used were of analytical or HPLC grade. Ascorbic acid, furfural, glacial acetic acid (HOAc), hydrochloric acid, sodium hydroxide, trichloroacetic acid (TCA) and zinc acetate were obtained from Carlo Erba (Italy). 5-Hydroxymethylfurfural (5-HMF) was purchased from Fluka (Germany). Dithiothreitol (DTT) was obtained from Fluka (Canada). Sodium thiosulphate pentahydrate was purchased from Fluka (Switzerland). Iodine was purchased from Riedel-de Haën (Germany). Potassium dihydrogen phosphate was obtained from Scharlau Chemie S.A. (Spain). Potassium hexacyanoferrate (II) trihydrate was obtained from BDH (England). *meta*-Phosphoric acid (MPA) and potassium iodide were purchased from Merck (Germany). Methanol and acetonitrile (HPLC grade) were obtained from Lab Scan (Thailand). De-ionized water of 18.2 MΩ/cm resistivity was purified with a Milli-Q system (Millipore, USA).

Instruments: Chromatographic analysis was carried out using a Waters liquid chromatograph (Waters, USA) equipped with a Waters 600E multi-solvent delivery system, a Waters in-line degasser AF and a Waters 2996 photodiode array detector. Sample injection was made through a Rheodyne injector (model 7725i) with a sample loop size of 20 μL. Data collection and processing were performed using Empower software. The separation was performed throughout the experiments on Vertisep AQS C18 column (5 μm particle size, 250×4.6 mm i.d.) and Vertisep AQS C18 guard column (5 μm particle size, 10×4.6 mm i.d.) at ambient temperature. Hot plate stirrer, model HTS 1003 (Laboratory & Medical Supplies, Japan), shaker, model VS-202D (Scientific Co. Ltd., Korea) and vortex mixer, model G-560E (Scientific Industries Inc., USA) were used for sample extraction. Centrifuge, model EBA20 (Hettich Zentrifugen, Germany) was used. Browning index was obtained with a spectrophotometer, model JENWAY 6400 (Jenway Ltd., U.K.) equipped with a 1 cm quartz cell. A pipetman P20, P200 and P1000 (Pipetman, France) micropipette and membrane filter, 0.45 μm, (Whatman International Ltd., U.K.) were used.

Table 1. Description of wine and fruit juice samples

Sample type	Code	Description
Red wine	RW1	Roselle (Krajeab) wine, OTOP, Thailand
	RW2	Somkoi wine, OTOP, Thailand
	RW3	Mao wine, OTOP, Thailand
White wine	WW1	Bael fruit (Matoom) wine, OTOP, Thailand
	WW2	Apple wine, Boone's, USA
	WW3	White fruit flavored wine, Fresco, Thailand
Fruit juice	FJ1	Grape mixed, Tipco, Thailand
	FJ2	Passion fruit juice, Doi Kham, Thailand
	FJ3	Tomato juice 100%, Doi Kham, Thailand
	FJ4	Apple juice 100%, Tipco, Thailand

Chilli samples: Sixteen samples of hot chilli pepper (six red chilli; RP/RF, one orange chilli; OP and nine green chilli; GP/GF) were brought to be representative pepper fruits. The maturity stages of these samples were also recorded. These peppers were purchased from a local market in Khon Kaen province, Thailand and stored in plastic bag at 4°C before use.

Wine and fruit juice samples: The drink samples were choice of this study including red wine, white wine and fruit juice. Table 1 lists the description and the codes of the sample studied. The roselle (*Abelmoschus esculentus* L. Moench), somkoi (*Tetrastigma matabite* (Blume) Planch.), mao (*Antidesma thwaitesianum* Müll. Arg.), bael fruit (*Aegle marmelos* L.), apple (*Malus domestica*) and white fruit flavored wine were purchased from OTOP centre in Khon Kaen, Thailand. The grape (*Vitis vinifera* L.) mixed, passion fruit (*Passiflora edulis* v. flavicarpa), tomato (*Lycopersicon esculentum* Mill.) and apple (*Malus domestica*) juice were purchased from local supermarkets in Khon Kaen, Thailand. All samples were stored in glass bottles and kept in a refrigerator (4°C) until analysis. All drink samples were filtered through 0.45 μm membrane filter before analysis by HPLC under the same optimized conditions as used for the chilli extract.

2.1. Analysis of Ascorbic Acid, DHA, 5-HMF and Furfural by RP-HPLC-PDA

The stock standard solution (1000 μg/mL) of ascorbic acid was prepared by dissolving 0.0100 g of ascorbic acid in 10.0 mL of mixture of 3% (w/v) MPA with 8% (v/v) HOAc, transferred to amber vial and kept in cool at 4°C prior to use. The stock standard solution (500 μg/mL) of DHA was prepared from ascorbic acid stock standard solution by adding small amounts of a 0.1 M iodine solution stepwise to 5.0 mL of ascorbic acid standard solution (1000 μg/mL) until the solution had a constant light yellow color. To reduce the surplus iodine two crystals of sodium thiosulphate were added and the standard DHA solution was filled up to 10.0 mL with mixture of 3% (w/v) MPA with 8% (v/v) HOAc, transferred to amber vial and kept cool at 4°C prior to use [42]. The stock standard solution (1000 μg/mL) of 5-HMF and F were prepared by dissolving 0.0100 g of each standard compound in 10.0 mL of deionised water, transferred to vial and kept in cool at 4°C prior to use. Working standard solutions were prepared daily by an appropriate dilution of the stock solution.

Separation optimization by HPLC: The optimum mobile phase for the high performance liquid chromatographic analysis of ascorbic acid, DHA, 5-HMF and furfural was investigated by varying the composition of solvent systems (methanol : water, acetonitrile : water, methanol : 1% (v/v) HOAc and acetonitrile : 1% (v/v) HOAc) whereas the other conditions used throughout were as follow: Flow rate of 0.7 mL/min at ambient temperature, photodiode array detector at 225 nm for DHA, 245 nm for ascorbic acid, and 280 nm for 5-HMF and furfural. The mobile phases were filtered through a 0.45 μm nylon membrane and degassed before use. The column was equilibrated with the mobile phase for 30 min or until a steady detector baseline was achieved.

Method validation: The analytical characteristics of the proposed method were investigated using the optimum conditions. The studied method was validated through their linearity, sensitivity, precision and accuracy.

The linearity was evaluated through the relationship between peak area (y) and the corresponding concentration (x) on the calibration curves of ascorbic acid, DHA, 5-HMF and furfural. The calibration curves were studied by analysis of the standard mixtures at various concentrations ranging from 1 to 100 μg/mL for these compounds. All standard mixtures were prepared by dilution of the stock standard solutions. The multiple determination coefficients (r^2) were calculated by means of the least-squares analysis.

The limit of detection (LOD) and limit of quantitation (LOQ) under the present chromatographic conditions were calculated as the analyte concentration giving a signal to noise ratios (S/N) of 3 and 10, respectively. Mixed standard solutions of ascorbic acid, DHA, 5-HMF and furfural were diluted from their stock standard solutions for LOD and LOQ determination.

The precision of the proposed method was presented as the repeatability and reproducibility of retention time and peak area. The relative standard deviation (RSD) is a target value for quantitation. The repeatability (intra-day precision) was deduced from ten times per day (n = 10) and reproducibility (inter-day precision) was calculated from five times a day for three consecutive days (n = 3x5). Mixed standard solutions of ascorbic acid, DHA, 5-HMF and furfural were used at concentration of 5.0 μg/mL.

Recovery test was used to evaluate the accuracy of the proposed method. For the percentage recovery experiment, two concentration ranges of the standard compounds (10, 50 μg/mL for ascorbic acid, 5, 50 μg/mL for DHA and 15, 50 μg/mL for 5-HMF and furfural) were spiked into the pepper samples. The spiked samples were extracted with the suitable solvent system and analyzed by RP-HPLC.

The reversed-phase chromatographic conditions used were optimized with the aim of obtaining the separation with a good resolution of adjacent peaks within a short analysis time. The stationary phase was Vertisep AQS C18 which permitted excellent stability, efficiency, column-to-column reproducibility and long lifetime at high and low pH levels and ideal for LC application. Several mobile phases have been described in the literatures for the analysis of ascorbic acid, dehydroascorbic acid, 5-hydroxymethylfurfural and furfural using a reversed phase column [5, 12, 33-35, 37-39, 47, 48, 50, 53-55]. In this study, Two organic modifiers of mobile phase (methanol : water and acetonitrile : water) were attempted to elute the investigated four components. In order to enhance resolution and eliminate tailing of the peak of the target compounds, HOAc was added into the mobile phase. Considering total resolution of the chromatographic separation, running time and solvent/reagent consumption, methanol and 1% (v/v) HOAc (25:75, v/v) was chosen. The flow rate of 0.7 mL/min was used to maintain satisfactory separation and reasonable analytical time.

Using the optimal isocratic conditions, ascorbic acid, DHA, furfural and 5-HMF are separated within 12 minutes, as the chromatogram is shown in Figure 3. All compounds are well resolved with resolution higher than 2.4. Table 2 lists the retention time and resolution data of the studied compounds.

HPLC combined with the diode array system proved to be able to detect a wide range of molecules and ensure identification of target compounds. The retention time and spectrum provide strong evidence of their identity [41]. On the basis of the absorption maxima of these compounds in UV spectra acquired by use of photodiode array detector, the monitoring wavelengths were set at 225 nm for DHA, 245 nm for ascorbic acid and 280 nm for 5-HMF and furfural. The HPLC chromatogram for the mixture of the standard compounds is shown in Figure 3. Their absorption spectra are demonstrated in Figure 4. The detections at three wavelengths were carried out to improve their sensitivity and selectivity for quantitative analysis and provided good response with minimum interferences.

Since the mobile phase containing methanol and 1% (v/v) HOAc (25:75, v/v), the studied compounds were in their molecular form. Therefore, the separation was expected to be the reversed phase mechanism. These compounds were eluted according to their polarity as listed in Table 2.

Table 2. Absorption maxima, retention times and resolution of standard compounds

Standard	Absorption maxima (nm)	Retention time (min)	Resolution
Ascorbic acid	244.1	4.57	-
DHA	225.2	5.20	2.4
5-HMF	283.1	8.02	6.7
Furfural	277.2	10.87	5.1

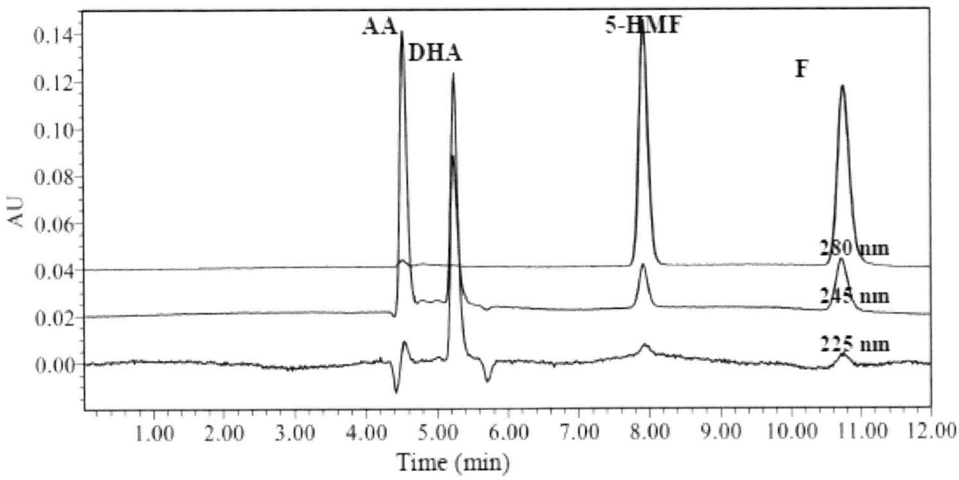

Figure 3. The superimposed typical HPLC chromatograms of the mixed standard solution (ascorbic acid, AA; dehydroascorbic acid, DHA; 5-hydroxymethylfurfural, 5-HMF; furfural, F) at each wavelength.

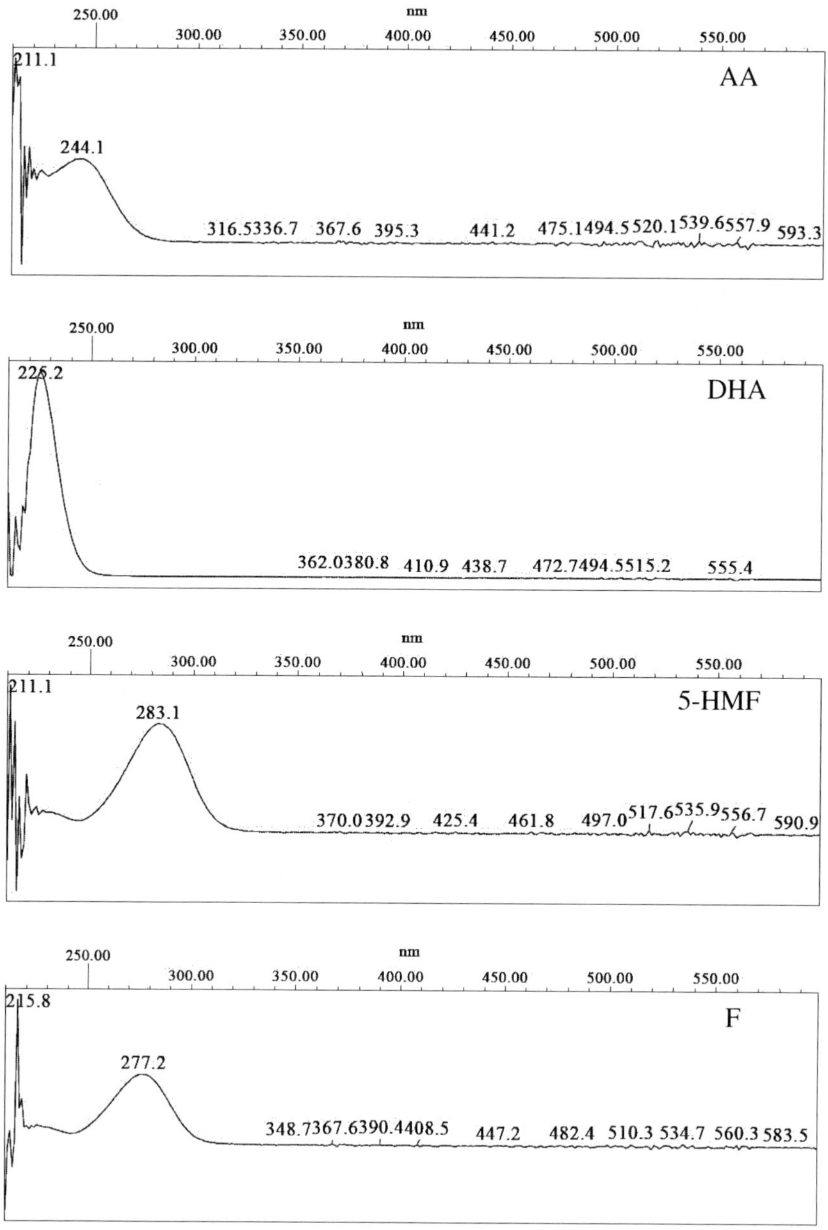

Figure 4. The spectra obtained from four standard compounds (ascorbic acid, AA; dehydroascorbic acid, DHA; 5-hydroxymethylfurfural, 5-HMF; furfural, F) separated by HPLC and detected with a photodiode array detector.

Method validation: The quantitative features of the proposed method were studied under the optimum conditions. The calibration curves were found to be linear in the range of 0.1-100 μg/mL for ascorbic acid, 5-HMF and furfural and in the range of 0.1-10 μg/mL for DHA. Excellent linearity was obtained between peak area (y) and the corresponding concentration (x) on the standard curve of ascorbic acid, DHA, 5-HMF and furfural. The multiple determination coefficients obtained were higher than 0.9990. The results are shown in Table 3.

Table 3. Linearity and sensitivity for ascorbic acid, DHA, 5-HMF and furfural

Compound	Linear range (μg/mL)	Regression equation	r^2	LOD (μg/mL)	LOQ (μg/mL)
Ascorbic acid	0.1-100	y = 87430 x + 37996	0.9998	0.020	0.070
DHA	0.1-10	y = 669537 x + 70256	0.9992	0.050	0.200
5-HMF	0.1-100	y = 209752 x + 117254	0.9991	0.006	0.020
Furfural	0.1-100	y = 200127 x + 122868	0.9992	0.010	0.040

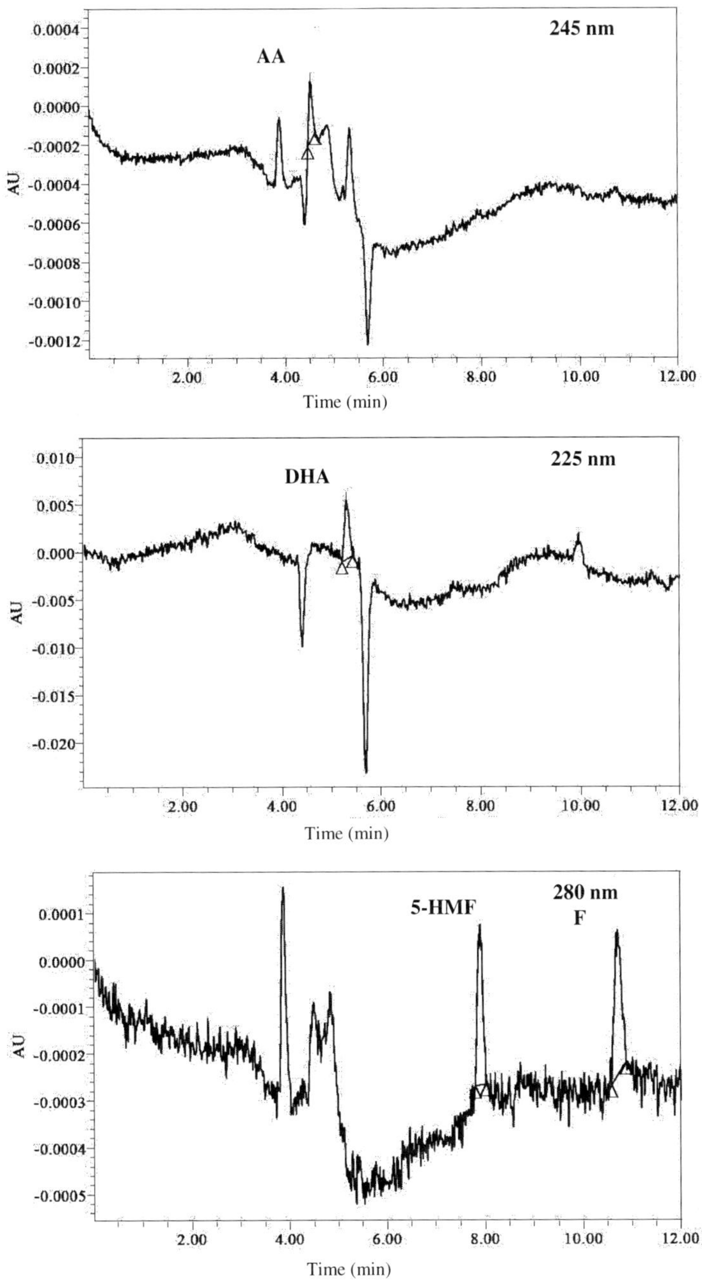

Figure 5. HPLC chromatograms of ascorbic acid (AA), DHA, 5-HMF and furfural (F) at the LOD.

Both LOD and LOQ under the chromatographic conditions were determined by injecting a series of each of the standard solution until the S/N for each compound was 3 for LOD and 10 for LOQ. The LOD was found to be 0.006 µg/mL and 0.05 µg/mL for 5-HMF and DHA, respectively, while the LOQ was 0.02 µg/mL and 0.20 µg/mL for 5-HMF and DHA, respectively. Both LOD and LOQ values of the studied compounds are also summarized in Table 3. Figure 5 and 6 show the typical HPLC chromatograms of these compounds at their LOD and LOQ, respectively.

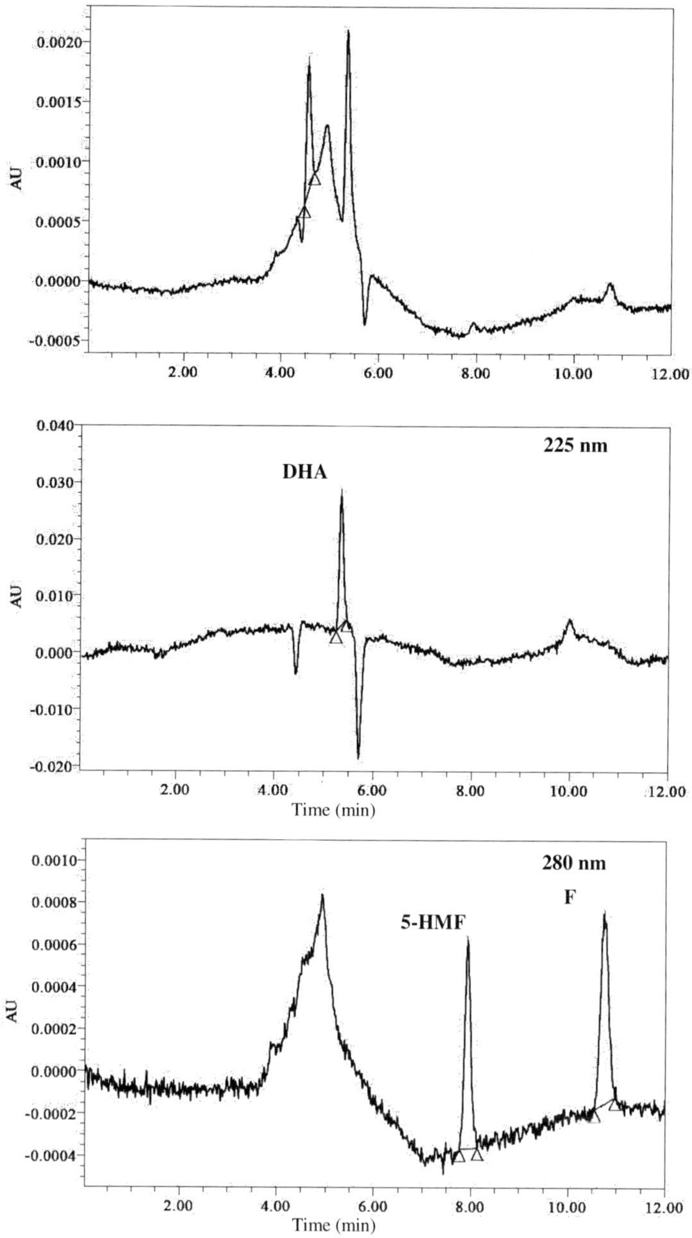

Figure 6. HPLC chromatograms of ascorbic acid (AA), DHA, 5-HMF and furfural (F) at the LOQ.

Table 4. Precision data for retention times and peak areas of these compounds studied

Compound	Precision (%RSD)[a]			
	Intra-day (n = 10)		Inter-day (n = 3×5)	
	t_R	Peak area	t_R	Peak area
Ascorbic acid	0.17	2.87	0.44	4.17
DHA	0.48	2.87	0.69	2.85
5-HMF	0.13	3.12	0.35	0.98
Furfural	0.14	2.76	0.34	1.99

[a] RSD (%) = (SD/mean)×100.

The precision of the proposed method was evaluated by analyzing the standard solutions of ascorbic acid, DHA, 5-HMF and furfural at concentration of 5.0 μg/mL and expressed as RSD of repetitive injections for both peak area and retention time.

Intra-day precision was deduced from ten times per day (n = 10). The RSD of the retention time was lower than 0.48%, ranging from 0.13-0.48% and the RSD of peak area was lower than 3.12%, ranging from 2.76-3.12%. Inter-day precision was calculated from five times a day for three consecutive days (n = 3×5). The RSD of the retention time was lower than 0.69%, ranging from 0.34-0.69% and the RSD of peak area was lower than 4.17%, ranging from 0.98-4.17%. Table 4 summarizes the RSD of both retention time and peak area.

Table 5. Recovery of ascorbic acid (AA), DHA, 5-HMF and furfural (F) from chilli samples (n = 3)

Sample	Compound	Spiked amount (mg/100 g)	Found (mg/100 g)	Recovery (%)	RSD[a] (%)
RP1	AA	5.0	4.94	98.83	0.22
		25.0	24.35	97.38	0.34
	DHA	2.5	2.15	85.95	3.65
		25.0	21.91	87.63	1.07
	5-HMF	7.5	6.23	83.12	3.85
		25.0	20.26	81.02	1.06
	F	7.5	6.85	91.28	3.86
		25.0	23.09	92.35	0.53
RP2	AA	5.0	4.62	92.34	1.45
		25.0	22.55	90.21	3.47
	DHA	2.5	2.26	90.44	1.75
		25.0	23.09	92.36	0.35
	5-HMF	7.5	6.65	88.67	1.75
		25.0	20.88	83.53	1.60
	F	7.5	7.05	93.96	1.76
		25.0	22.76	91.05	2.06
RP3	AA	5.0	4.38	87.64	2.27
		25.0	22.66	90.63	1.14
	DHA	2.5	2.03	81.11	3.14
		25.0	20.89	83.57	0.02
	5-HMF	7.5	6.45	86.01	3.13
		25.0	22.52	90.09	0.07
	F	7.5	6.32	84.33	3.31

Table 5. (Continued)

Sample	Compound	Spiked amount (mg/100 g)	Found (mg/100 g)	Recovery (%)	RSD[a] (%)
		25.0	21.07	84.28	2.45
RP4	AA	5.0	4.35	86.92	3.54
		25.0	22.57	90.28	0.03
	DHA	2.5	2.37	94.79	0.42
		25.0	23.58	94.31	0.24
	5-HMF	7.5	6.44	85.93	0.08
		25.0	23.33	93.31	0.30
	F	7.5	6.89	91.89	0.08
		25.0	23.78	95.14	0.15
RP5	AA	5.0	4.84	96.80	0.81
		25.0	24.04	96.18	0.34
	DHA	2.5	2.26	90.41	0.28
		25.0	22.15	88.61	0.40
	5-HMF	7.5	6.66	88.84	0.09
		25.0	22.01	88.02	0.41
	F	7.5	7.60	101.37	0.18
		25.0	25.61	102.43	0.01
RF5	AA	5.0	4.27	85.48	0.92
		25.0	21.21	84.83	1.53
	DHA	2.5	2.22	88.95	2.05
		25.0	21.79	87.14	1.98
	5-HMF	7.5	6.78	90.41	2.36
		25.0	21.89	87.55	0.35
	F	7.5	7.48	99.73	2.15
		25.0	24.37	97.49	1.17
OP1	AA	5.0	4.37	87.47	2.92
		25.0	23.15	92.59	1.39
	DHA	2.5	2.41	96.21	0.25
		25.0	24.26	97.03	0.08
	5-HMF	7.5	7.10	94.63	0.06
		25.0	23.91	95.64	0.08
	F	7.5	7.63	101.68	0.01
		25.0	24.85	99.42	0.01
GP1	AA	5.0	4.05	81.00	4.11
		25.0	20.92	83.69	3.47
	DHA	2.5	2.07	82.86	1.08
		25.0	20.25	81.01	1.46
	5-HMF	7.5	7.49	99.84	3.26
		25.0	24.58	98.32	0.26
	F	7.5	6.82	90.91	1.32
		25.0	22.82	91.28	1.28
GP2	AA	5.0	4.29	85.84	0.77
		25.0	21.08	84.31	1.06
	DHA	2.5	2.07	82.75	0.68
		25.0	20.89	83.57	2.41
	5-HMF	7.5	6.32	84.33	0.69
		25.0	21.35	85.39	0.29
	F	7.5	6.08	81.12	0.96
		25.0	20.87	83.46	1.84

Sample	Compound	Spiked amount (mg/100 g)	Found (mg/100 g)	Recovery (%)	RSD[a] (%)
GP3	AA	5.0	4.20	84.00	1.98
		25.0	21.91	87.62	0.99
	DHA	2.5	2.08	83.03	4.50
		25.0	21.30	85.18	0.46
	5-HMF	7.5	6.45	85.99	1.64
		25.0	21.06	84.23	1.24
	F	7.5	6.20	82.60	1.78
		25.0	21.30	85.21	0.72
GP4	AA	5.0	4.61	92.14	4.89
		25.0	23.68	94.71	4.48
	DHA	2.5	2.43	97.27	2.10
		25.0	24.40	97.60	0.20
	5-HMF	7.5	7.18	95.68	0.02
		25.0	23.58	94.31	0.23
	F	7.5	7.78	103.72	0.08
		25.0	26.27	105.09	4.85
GP5	AA	5.0	4.45	89.00	0.73
		25.0	22.09	88.37	0.52
	DHA	2.5	2.38	95.21	1.14
		25.0	23.45	93.81	0.98
	5-HMF	7.5	6.77	90.24	0.09
		25.0	22.04	88.14	0.22
	F	7.5	7.80	103.98	0.16
		25.0	25.92	103.68	0.48
GP6	AA	5.0	4.41	88.22	1.51
		25.0	22.64	90.57	1.75
	DHA	2.5	2.33	93.28	1.97
		25.0	22.87	91.49	0.02
	5-HMF	7.5	6.70	89.36	0.06
		25.0	23.23	92.91	0.21
	F	7.5	7.50	100.04	0.23
		25.0	24.90	99.06	0.04
GP7	AA	5.0	4.29	85.77	1.99
		25.0	21.44	85.75	0.06
	DHA	2.5	2.38	95.16	0.60
		25.0	23.85	95.41	0.12
	5-HMF	7.5	6.81	90.84	0.28
		25.0	22.95	91.79	0.84
	F	7.5	7.61	101.43	0.04
		25.0	25.84	103.37	0.52
GF7	AA	5.0	4.01	80.11	0.87
		25.0	20.87	83.49	0.06
	DHA	2.5	2.31	92.23	0.62
		25.0	23.85	95.41	0.23
	5-HMF	7.5	6.58	87.70	0.29
		25.0	22.71	90.85	0.85
	F	7.5	7.58	101.10	0.04
		25.0	25.60	102.39	0.53
GF8	AA	5.0	4.25	85.01	1.22
		25.0	21.29	85.16	0.02
	DHA	2.5	2.17	86.94	0.08

Table 5. (Continued)

Sample	Compound	Spiked amount (mg/100 g)	Found (mg/100 g)	Recovery (%)	RSD[a] (%)
		25.0	21.29	85.15	0.31
	5-HMF	7.5	6.51	86.79	0.05
		25.0	21.76	87.04	1.26
	F	7.5	7.82	104.28	0.57
		25.0	26.23	104.91	0.19
FC	AA	5.0	4.87	97.47	1.81
		25.0	24.21	96.83	1.20
	DHA	2.5	2.14	85.51	0.20
		25.0	21.70	86.80	0.64
	5-HMF	7.5	6.57	87.63	0.11
		25.0	21.99	87.96	0.24
	F	7.5	7.82	104.28	0.57
		25.0	26.23	104.91	0.19
VC	AA	5.0	4.82	96.34	1.83
		25.0	23.64	94.57	1.23
	DHA	2.5	2.06	82.59	0.21
		25.0	20.97	83.88	0.66
	5-HMF	7.5	6.34	84.50	0.11
		25.0	21.76	87.03	0.25
	F	7.5	7.33	97.73	0.60
		25.0	23.77	95.08	0.21

[a] RSD (%) = (SD/mean)×100.

Recovery test was used to evaluate the accuracy of the proposed method and determined by adding the mixed standard solution with different concentrations to the known amounts of sample. Then the resultant samples were extracted and analyzed by the proposed method. The percentage recoveries were calculated according to the following equation: [(detected amount – original amount)/spiked amount] × 100. As shown in Table 5, the developed analytical method was reproducible with good recovery in the range of 80.11-105.09% (RSD <4.89%).

2.2. Sample Preparation

Mixture of 3% (w/v) MPA with 8% (v/v) HOAc was prepared by dissolving 3.0 g of MPA in 100 mL of de-ionized water containing 8.0 mL of HOAc. The solution was prepared everyday and kept in the dark and cold (4°C). A 20% (w/v) TCA was prepared by dissolving 20.0 g of TCA in 100 mL of de-ionized water. A 3% (w/v) MPA was prepared by dissolving 3.0 g of MPA in 100 mL of de-ionized water. A 5% (v/v) HOAc was prepared by dilution 5.0 mL of HOAc to 100 mL with de-ionized water. A 1% (v/v) HCl was prepared by dilution 1.0 mL of concentrated HCl to 100 mL with de-ionized water. Carrez I solution was prepared by dissolving 15.0 g of potassium hexacyanoferrate (II) trihydrate in 100 mL of de-ionized water [53]. Carrez II solution was prepared by dissolving 30.0 g of zinc acetate in 100 mL of de-ionized water [53].

The extraction was performed under the optimized conditions. Optimal extraction conditions were achieved by sequentially varying the experiment parameters, one at the time, while all the other parameters remained fixed. The following extraction conditions were altered: types of solvents (mixture of 3% (w/v) MPA with 8% (v/v) HOAc, 20% (w/v) TCA, 3% (w/v) MPA, 5% (v/v) HOAc, 1% (v/v) HCl and Carrez (I and II) in 0.2 M HOAc); amount of solvent (1, 2, 3, 4 and 5% (v/v) HOAc); extraction methods (vortex mixing, shaking, standing and magnetic stirring extractions); extraction times (0.5, 1, 3 and 5 min); and centrifugation times (10, 15, 20 and 30 min), in order to determine the conditions that would allow for maximum extraction efficiency (high recovery). Following the extraction, fresh hot peppers were accurately weighed, after being washed with de-ionized water, rinsed free of seeds and cut into small pieces. Sample (1 g) was weighed into a test tube. Then 5 mL of solvent was added. The mixture was extracted and the sample extract was clarified by centrifuging at 5000 rpm and then filtered through a 0.45 μm filter membrane prior to analysis. The efficiency of recovery was performed by addition of a known amount of the mixed standards into the sample before the extraction.

In order to optimize the extraction conditions for achievement of quantitative extraction, various parameters involved the extraction procedure such as type of solvent (mixture of 3% (w/v) MPA with 8% (v/v) HOAc, 20% (w/v) TCA, 3% (w/v) MPA, 5% (v/v) HOAc, 1% (v/v) HCl and Carrez (I and II) in 0.2 M HOAc), amount of solvent (1, 2, 3, 4 and 5% (v/v) HOAc), extraction method (vortex mixing, shaking, standing and magnetic stirring extractions) and extraction time (0.5, 1, 3 and 5 min), and centrifugation time (10, 15, 20 and 30 min) were optimized.

Type of solvent: In the first part of this study, the effect of different solvents on the extraction efficiency was evaluated while keeping other parameters constant. The results for the former being expressed as percentage of recovery as shown in Figure 7. The recoveries of ascorbic acid, DHA, 5-HMF and furfural in the sample extracted by 5% (v/v) HOAc were very high among them compared to those of using the other solvents. Moreover, it was found that the recovery of ascorbic acid was the highest because of co-elution with solvent peak when using 20% (v/v) TCA as the extraction solvent. Therefore, 5% (v/v) HOAc was a choice as the most suitable solvent for extraction.

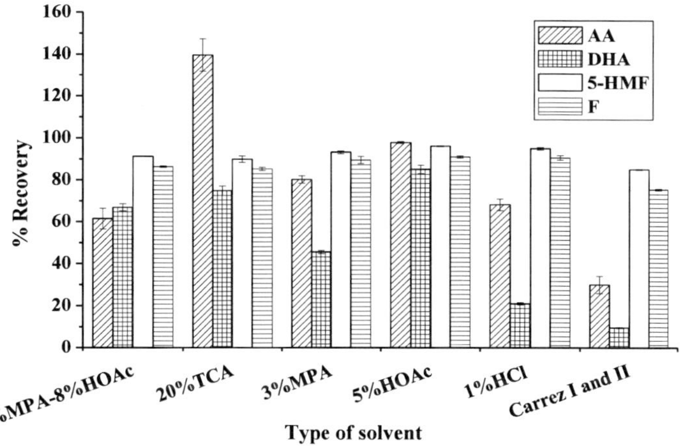

Figure 7. Effect of solvents on the percentage of recovery of ascorbic acid (AA), DHA, 5-HMF and furfural (F) in chilli samples.

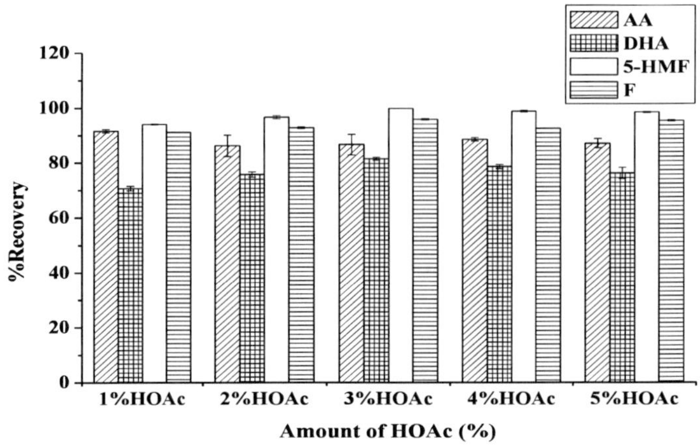

Figure 8. Effect of concentration of HOAc on the percentage of recovery of ascorbic acid (AA), DHA, 5-HMF and furfural (F) in chilli samples.

Amount of solvent: After choosing the optimal extraction solvent, which was HOAc, the effect of amount of HOAc (1, 2, 3, 4 and 5%) on the extraction of ascorbic acid, DHA, 5-HMF and furfural from hot pepper sample was then investigated. As shown in Figure 8, an increasing in the amount of HOAc from 1 to 3% moderately increased the recoveries of DHA, 5-HMF and furfural in the extracts but when increasing from 3 to 5% the recoveries of these compounds were reduced. While the amount of HOAc changed, the recovery of ascorbic acid did not change significantly. Therefore, the suitable amount of solvent was found to be 3% (v/v) acetic acid.

Extraction method: To determine the effect of extraction method on the recovery of ascorbic acid, DHA, 5-HMF and furfural, hot pepper sample was extracted with various simple methods (vortex mixing, shaking, standing, and magnetic stirring extractions). The results are summarized in Figure 9. From results, it is suggested that a vortex mixing gave rather high recovery better than the other extraction ones. Thus, the vortex mixing method was used throughout the experiments.

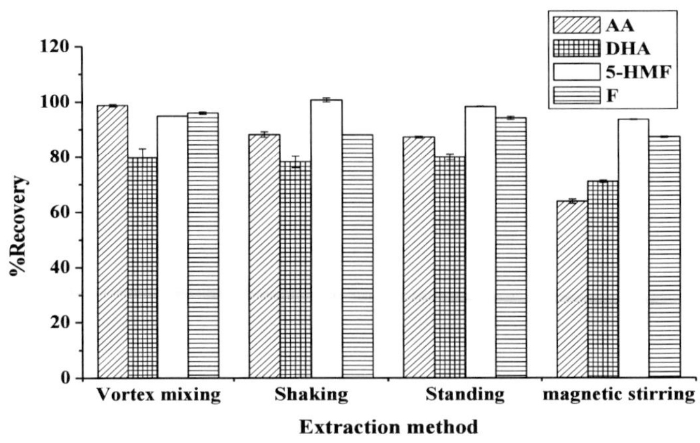

Figure 9. Effect of extraction methods on the percentage of recovery of ascorbic acid (AA), DHA, 5-HMF and furfural (F) in chilli samples.

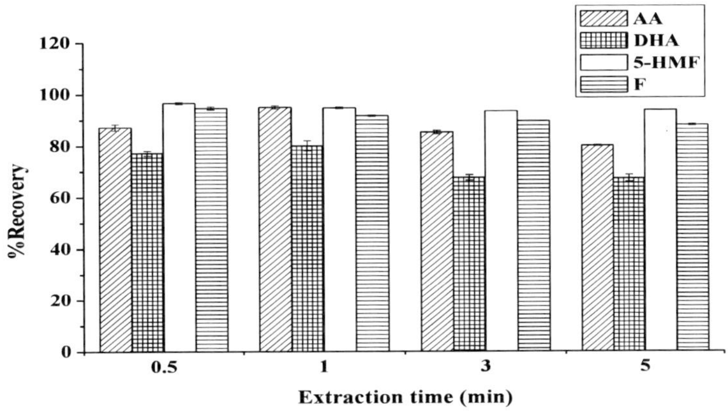

Figure 10. Effect of extraction times on the percentage of recovery of ascorbic acid (AA), DHA, 5-HMF and furfural (F) in chilli samples.

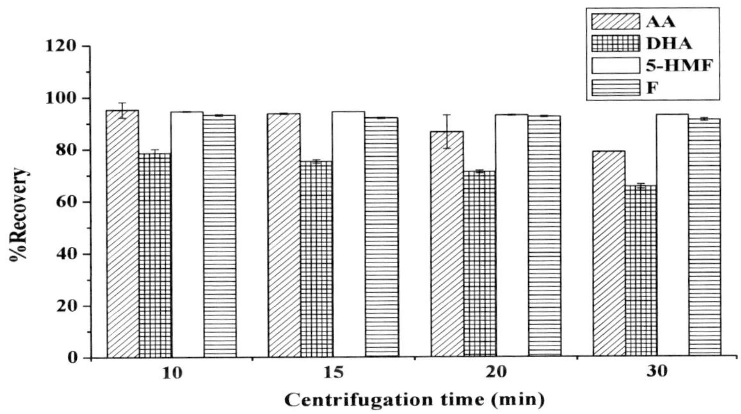

Figure 11. Effect of centrifugation times on the percentage of recovery of ascorbic acid (AA), DHA, 5-HMF and furfural (F) in chilli samples.

Extraction time: In order to investigate extraction time, samples were extracted with 5 mL 3% (v/v) acetic acid for 0.5, 1, 3 and 5 min. As shown Figure 10, the results showed that all analytes were almost completely extracted within 1 min. It has been the presence of light, time and temperature affects the ascorbic acid and DHA retention in the extracts [20]. Thus, the optimum time for extraction of ascorbic acid, DHA, 5-HMF and furfural in hot pepper sample was 1 min.

Table 6. Optimum conditions for the extraction of ascorbic acid, DHA, 5-HMF and furfural in chilli samples

Parameter	Optimum condition
Extraction solvent	HOAc
Amount of solvent	5% (v/v)
Extraction method	Vortex mixing
Extraction time	1 min
Centrifugation time	10 min

Centrifugation time: For centrifugation time, the extract was clarified by centrifuging at 5000 rpm for 10, 15, 20 and 30 min. The results are summarized in Figure 11. Different recoveries were also obtained depending on the duration of the sample centrifugation. The recoveries of ascorbic acid, DHA, 5-HMF and furfural decreased with higher centrifugation time. The recoveries of all compounds were the highest at 10 min ranged from 78.57 to 95.21%, therefore, the optimum centrifugation time was 10 min. The optimum conditions for the extraction of ascorbic acid, DHA, 5-HMF and furfural in hot pepper sample are summarized in Table 6.

2.3. Analysis of Ascorbic Acid, DHA, 5-HMF and Furfural in Real Samples

The proposed method was applied for the determination of ascorbic acid, DHA, 5-HMF and furfural in hot pepper, wine and fruit juice samples. Figure 12 and 13 show chromatograms obtained from the hot pepper and fruit juice samples, respectively. The target compounds in the samples were identified by comparing their retention times and absorption spectra with their authentic compounds.

The developed HPLC method was applied to analyze the four analytes in sixteen samples. In all hot pepper samples, ascorbic acid was found with high contents up to 99.58 mg/100 g. DHA was found in some samples in the range of 0.16-2.75 mg/100 g. 5-HMF was not detectable in any hot pepper samples, and furfural was found only in GP5 sample (0.06 mg/100 g). These results are summarized in Table 7. The contents of ascorbic acid, DHA, 5-HMF and furfural found in both wine and fruit juice samples are shown in Table 8. The contents of ascorbic acid, DHA, 5-HMF and furfural were found to be 0.47-46.47, 0.91-7.93, 0.20-12.41 and 0.01-0.70 mg/100 mL, respectively. Ascorbic acid was almost found in the sample RW1, WW1, FJ2, FJ3 and FJ4 whereas DHA was also found in all samples, except for RW1, WW1 and FJ4. 5-HMF was also found in all samples, whereas furfural was found only in the sample FJ2.

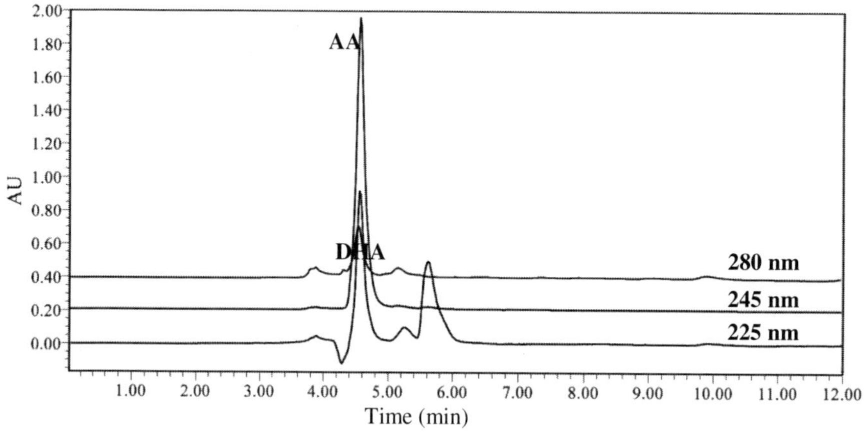

Figure 12. The superimposed typical HPLC chromatograms of the chilli extract at each wavelength (ascorbic acid, AA and dehydroascorbic acid, DHA).

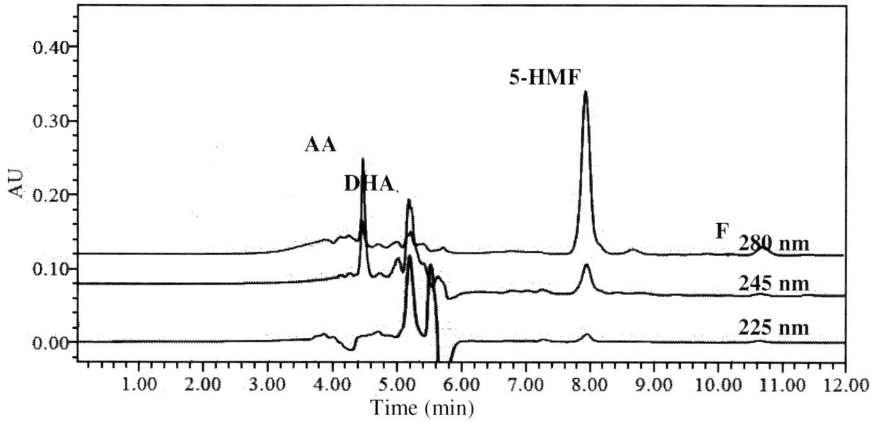

Figure 13. The superimposed typical HPLC chromatograms of fruit juice sample at each wavelength (ascorbic acid, AA; dehydroascorbic acid, DHA; 5-hydroxymethylfurfural, 5-HMF and furfural, F).

Table 7. The contents of ascorbic acid (AA), DHA, 5-HMF and furfural (F) in hot chilli samples

Sample	Content[a] (mg/100 g)			
	AA	DHA	5-HMF	F
RP1	66.08±3.47	N.D.[b]	N.D.	N.D.
RP2	86.72±1.27	N.D.	N.D.	N.D.
RP3	23.46±0.53	N.D.	N.D.	N.D.
RP4	64.17±2.41	0.21±0.00	N.D.	N.D.
RP5	99.58±4.29	0.61±0.00	N.D.	N.D.
RF5	73.85±3.44	0.48±0.05	N.D.	N.D.
OP1	26.17±2.35	0.16±0.02	N.D.	N.D.
GP1	49.48±0.53	N.D.	N.D.	N.D.
GP2	24.09±1.06	N.D.	N.D.	N.D.
GP3	12.63±0.54	N.D.	N.D.	N.D.
GP4	22.56±0.02	0.21±0.01	N.D.	N.D.
GP5	15.57±1.84	0.23±0.01	N.D.	0.06±0.00
GP6	6.45±0.29	0.45±0.02	N.D.	N.D.
GP7	20.96±1.26	0.25±0.01	N.D.	N.D.
GF7	13.47±1.10	0.13±0.05	N.D.	N.D.
GF8	1.31±0.01	0.74±0.05	N.D.	N.D.
FC	N.D.	2.75±0.07	N.D.	N.D.
VC	N.D.	N.D.	N.D.	N.D.

[a] Content = mean ± SD (n=3).
[b] N.D.: not detected.

Table 8. The contents of ascorbic acid (AA), DHA, 5-HMF and furfural (F) in wine and fruit juice samples

Sample	Content[a] (mg/100 mL)			
	AA	DHA	5-HMF	F
RW1	0.89±0.02	N.D.[b]	4.27±0.01	0.02±0.00
RW2	N.D.	0.91±0.07	0.48±0.00	0.01±0.00
RW3	N.D.	1.70±0.00	0.20±0.01	0.13±0.01
WW1	0.38±0.02	N.D.	12.41±0.07	0.12±0.00
WW2	N.D.	3.45±0.07	2.45±0.00	0.13±0.00
WW3	N.D.	1.01±0.07	1.28±0.00	0.01±0.00
FJ1	N.D.	3.52±0.01	0.28±0.00	0.02±0.00
FJ2	7.45±0.02	7.93±0.19	1.70±0.00	N.D.
FJ3	46.47±1.14	3.62±0.13	1.57±0.01	0.10±0.00
FJ4	0.47±0.02	N.D.	4.17±0.00	0.70±0.00

[a] Content = mean ± SD (*n*=3).
[b] N.D.: not detected.

2.4. Determination of Total Ascorbic Acid

Standard solution of DHA was prepared as described in 3.3.1.1(2). 0.1 M phosphate buffer (pH 7.0) was prepared by dissolving 1.3609 g of potassium dihydrogen phosphate in 100 mL deionized water and adjusted to pH 7.0 with 0.1 M sodium hydroxide. DTT solution (1000 μg/mL) was prepared by dissolving 0.0100 g of DTT in 10 mL deionized water and transferred to vial and kept in cool at 4°C until use.

The reduction of DHA to ascorbic acid was performed in phosphate buffer pH 7.0 at ambient temperature. Phosphate buffer (0.1 M) was mixed with DTT solution at variable final concentrations (50, 70, 100 and 120 μg/mL) in an Appendrop vial (1.0 mL). The reaction was started by adding 100 μL of DHA (100 μg/mL) into vial. The mixture was kept in the dark for 1-30 min and injected into HPLC. 800 μL of phosphate buffer (0.1 M) pH 7.0 was mixed with 100 μL of DTT solution (1000 μg/mL) in an Appendrop vial (1.0 mL). Then 100 μL of the sample extract was added into the mixture and kept in the dark for 5 min prior to analysis.

Optimization reduction of dehydroascorbic acid: The reduction of DHA to ascorbic acid was catalyzed by DTT, but the amount of DTT required to complete the reaction depends on how much DHA is present in the sample [34]. The DTT concentration was optimized by the reaction of a 10 μg/mL DHA solution with the solution at various concentrations (50, 70, 100 and 120 μg/mL) of the DTT in the dark at ambient temperature. The complete conversion was achieved within 15, 10, 3 and 2 min for 50, 70, 100 and 120 μg/mL DTT, respectively (Figure 14). Because it is important to reduce the consumption of expensive chemicals for analysis, a concentration of 100 μg/mL DTT (which gave a reaction time sufficiently short) was selected as optimal conditions. Figure 15 shows that the reduction procedure is able to make the complete conversion of DHA.

Analysis of total ascorbic acid in real samples: Total ascorbic acid of hot chilli, wine and fruit juice samples was determined by the proposed method. Figure 16 and 17 show HPLC chromatograms obtained from the samples used.

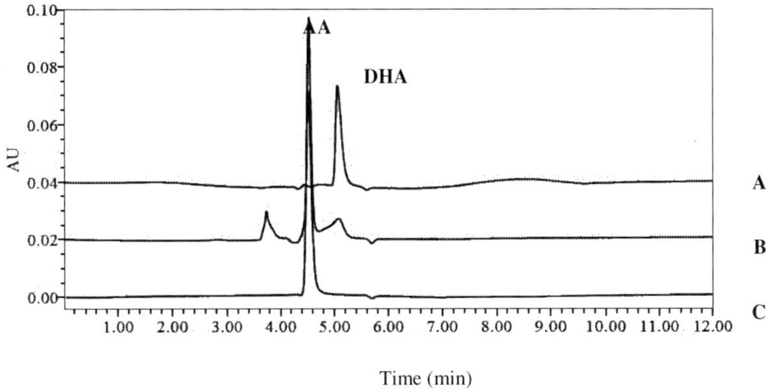

Figure 14. The superimposed typical HPLC chromatograms of standard solutions (ascorbic acid, AA and dehydroascorbic acid, DHA) at 245 nm. 10 μg/mL DHA (A), after the reduction treatment of 10 μg/mL DHA with DTT (B) and 10 μg/mL ascorbic acid (C).

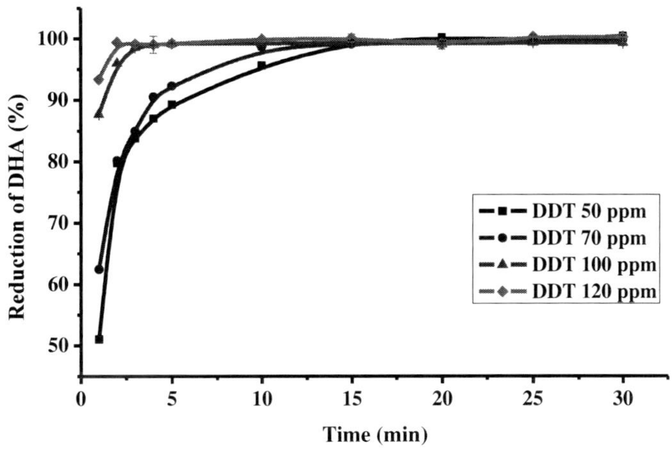

Figure 15. Conversion (%) of 10 μg/mL DHA in the presence of ascorbic acid (10 μg/mL) at different DTT concentrations.

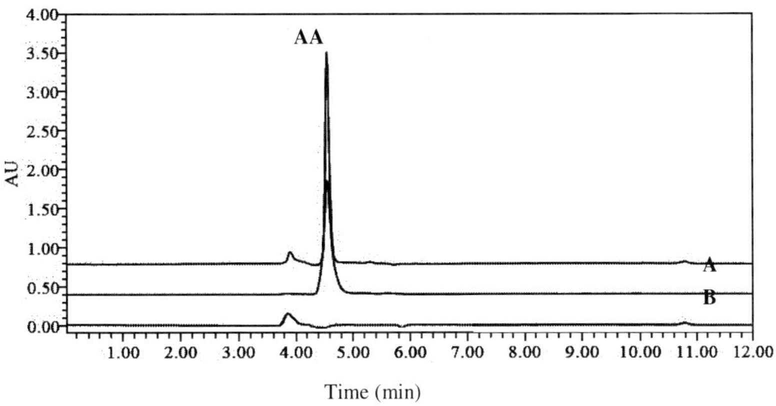

Figure 16. The superimposed typical HPLC chromatograms of the chilli extract after the reduction with DTT (A), chilli extract (B) and control (DTT 100 μg/mL in 0.1 M phosphate buffer pH 7.0) (C) at 245 nm (ascorbic acid, AA).

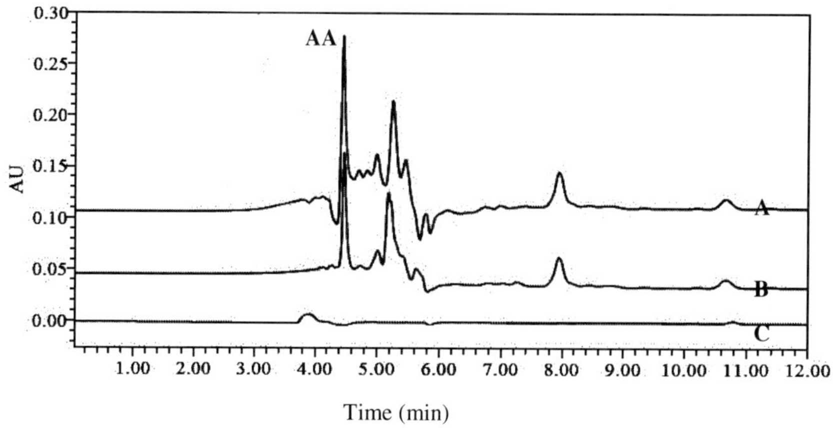

Figure 17. The superimposed typical HPLC chromatograms of fruit juice sample after the reduction with DTT (A), fruit juice sample (B) and control (DTT 100 μg/mL in 0.1 M phosphate buffer pH 7.0) (C) at 245 nm (ascorbic acid, AA).

Table 9. The contents of total ascorbic acid in hot chilli samples

Sample	Total ascorbic acid[a] (mg/100 g)
RP4	64.43±0.45
RP5	100.49±0.03
OP1	26.37±2.43
GP4	22.79±0.38
GP5	15.81±0.12
GP6	6.93±0.00
GP7	21.24±0.07
GF8	2.59±0.52
FC	2.87±0.04
VC	N.D.[b]

[a] Content = mean ± SD (*n*=3).
[b] N.D.: not detected.

Table 10. The contents of total ascorbic acid in wine and fruit juice samples

Sample	Total ascorbic acid[a] (mg/100 mL)
RW1	0.89±0.01
RW2	0.91±0.01
RW3	1.90±0.13
WW1	0.39±0.01
WW2	3.71±0.15
WW3	1.00±0.03
FJ1	3.49±0.03
FJ2	15.66±0.24
FJ3	50.35±0.63
FJ4	0.48±0.03

[a] Content = mean ± SD (*n*=3).

All samples of hot pepper, wine and fruit juice were determined under the optimal conditions. The results are shown in Table 9 and 10. The contents of total ascorbic acid in hot pepper samples were 2.59-100.49 mg/100 g, while the contents of total ascorbic acid in wine and fruit juice samples were 0.48-50.35 mg/100 mL.

The results obtained from direct measurement of DHA by the proposed method and determination of DHA as the difference between the total ascorbic acid after DHA reduction with DTT and ascorbic acid content of the original sample (indirect method) are presented in Table 11 and Table 12. Figure 18 and 19 show the contents of DHA in samples from direct and indirect methods. Student's t-test was used to compare those obtained from both direct and indirect methods. The Student's *t*-test showed that no statistically significant difference at 95% confidence level was found between the two methods.

Table 11. Comparison of direct and indirect methods for the determination of DHA in hot chilli extracts

Sample	Content of DHA[a] (mg/100 g)		*t*-test
	Direct method	Indirect method	
RP4	0.21±0.00	0.26±0.06	1.50<2.78 no significant
RP5	0.61±0.00	0.92±0.23	2.38<2.78 no significant
OP1	0.16±0.02	0.20±0.08	0.85<2.78 no significant
GP4	0.21±0.01	0.23±0.06	0.63<2.78 no significant
GP5	0.23±0.01	0.24±0.06	0.41<2.78 no significant
GP6	0.45±0.02	0.48±0.29	0.19<2.78 no significant
GP7	0.25±0.01	0.28±0.04	1.37<2.78 no significant
GF8	0.74±0.05	1.28±0.53	1.77<2.78 no significant
FC	2.75±0.07	2.87±0.04	2.57<2.78 no significant
VC	N.D.[b]	N.D.	-

[a] Content = mean ± SD (*n*=3). [b] N.D.: not detected. - No data.

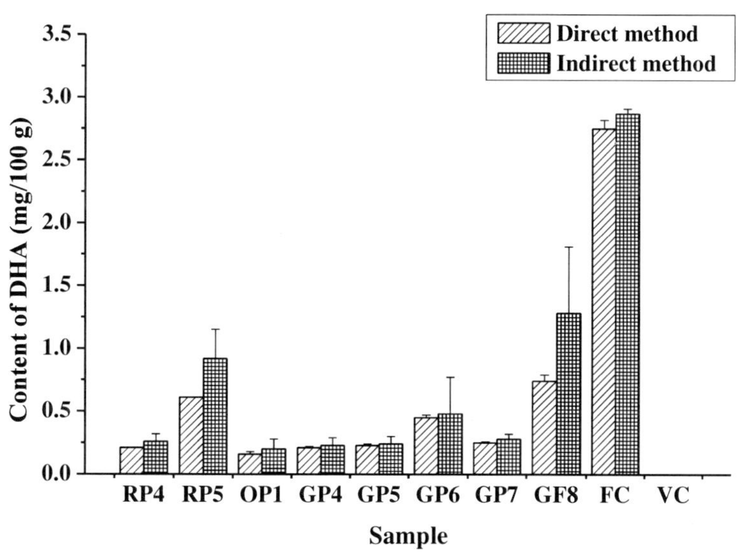

Figure 18. The contents of DHA in hot chilli samples from direct and indirect methods.

Table 12. Comparison of direct and indirect methods for determination of DHA in wine and fruit juice samples

Sample	Content of DHA[a] (mg/100 mL)		*t*-test
	Direct method	Indirect method	
RW1	N.D.[b]	N.D.	-
RW2	0.91±0.07	0.91±0.01	0.02<2.78 no significant
RW3	1.70±0.00	1.90±0.13	2.71<2.78 no significant
WW1	N.D.	N.D.	-
WW2	3.45±0.07	3.84±0.03	2.67<2.78 no significant
WW3	1.01±0.07	1.00±0.03	0.26<2.78 no significant
FJ1	3.52±0.01	3.44±0.05	1.33<2.78 no significant
FJ2	7.93±0.19	8.21±0.22	1.67<2.78 no significant
FJ3	3.62±0.13	3.88±1.78	0.25<2.78 no significant
FJ4	N.D.	N.D.	-

[a] Content = mean ± SD (n=3). [b] N.D.: not detected. - No data.

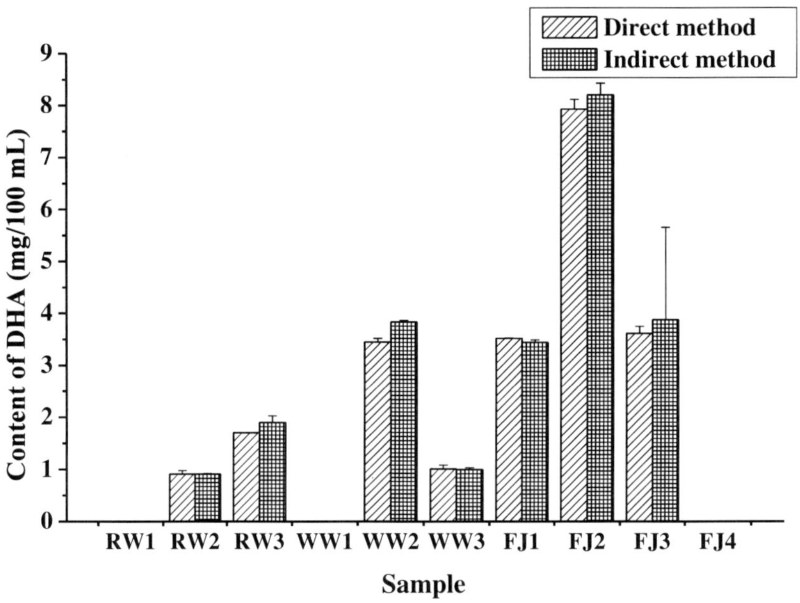

Figure 19. The contents of DHA in wine and fruit juice samples from direct and indirect methods.

2.5. Measurement of Browning Index in Chilli Peppers

The measurement of browning index was carried out according to the method described by Göğüş and Eren [57]. Before measurement of browning index in hot pepper, the sample was washed with deionized water, rinsed free of seeds and cut into small pieces. Sample (1 g) was added with de-ionized water (10 mL) and stirred continuously with a magnetic stirrer for 15 min in order to extract the color. Then it was filtered with Whatman No.42 filter paper. The absorbance of the filtrate was read at 420 nm. The value obtained was considered as the non-enzymatic browning index.

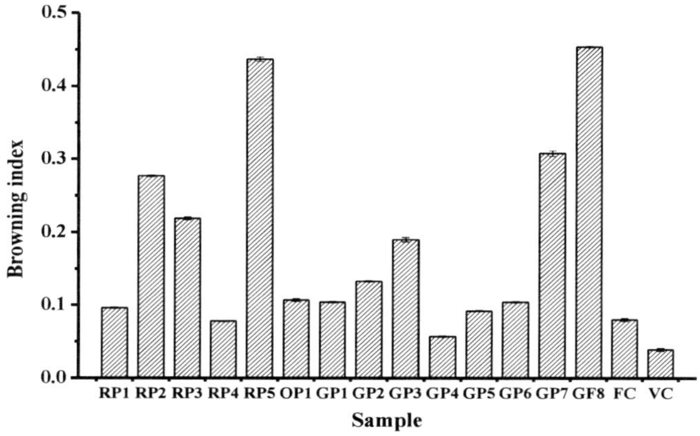

Figure 20. Browning index (Abs 420 nm) of hot chilli samples.

The browning index is one of the parameters that indicate as the non-enzymatic browning of the hot pepper samples. The browning index of the extracts varied from 0.039 to 0.454. Values obtained are shown in Figure 20.

Degradation of ascorbic acid and formation of browning: Four samples of hot pepper (RP4, RP5, GP5 and GP7) were subjected to determine the ascorbic acid content and the browning index for 7 days. In during refrigerated storage, gradual degradation of the ascorbic acid was observed in hot pepper samples. Figure 21 shows the changes of ascorbic acid contents in hot pepper samples. Initially, the ascorbic acid contents of RP4, RP5, GP5 and GP7 samples were 74.83±0.45, 109.71±1.04, 42.55±0.28 and 40.39±0.51 mg/100 g, respectively. The ascorbic acid contents in RP4 sample were rapidly dropped about 40% of the initial value during the first 2 days and continuously decreased from 2 to 7 day storage. The ascorbic acid contents in RP5 sample were rapidly decreased during the first 3 days but constantly maintained from 3 to 5 day storage and continuously decreased up to 7 day storage. The ascorbic acid contents of GP5 and GP7 samples were continuously decreased during the 7 days, especially the ascorbic acid contents of GP5 sample was more rapidly destructed than those of GP7 sample. Finally, the ascorbic acid contents of RP4, RP5, GP5 and GP7 samples were found to be 16.28±0.23, 63.19±0.69, 8.26±0.31 and 8.40±0.29 mg/100 g, respectively. Nonenzymatic browning index is determined with an absorbance measurement at 420 nm. The browning index was the characteristic change in hot chilli color during refrigerated storage. Figure 22 shows the changes of browning index in hot pepper samples. Initially, the browning index of RP4, RP5, GP5 and GP7 samples were 0.027, 0.036, 0.029 and 0.041, respectively. The browning index of RP4 sample was continuously increased during the first 5 days and rapidly increased up to 7 day storage. The browning index of RP5 sample was rapidly increased during the first 2 days but constantly maintained from 2 to 4 day storage and continuously increased up to 7 day storage. The browning indexes of GP5 and GP7 samples were continuously increased during the 7 days, especially the browning index of GP5 sample was more rapidly increased than those of GP7 sample. Finally, browning indexes of RP4, RP5, GP5 and GP7 samples were found to be 0.065, 0.083, 0.085 and 0.094, respectively.

However, as the ascorbic acid has degraded, it results in the production of reactive carbonyl compounds which act as intermediates in non-enzymatic browning effect in the samples [5].

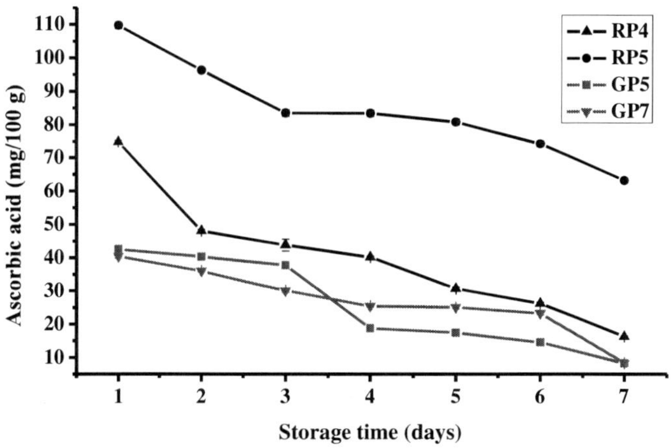

Figure 21. Changes in ascorbic acid contents of hot chilli samples.

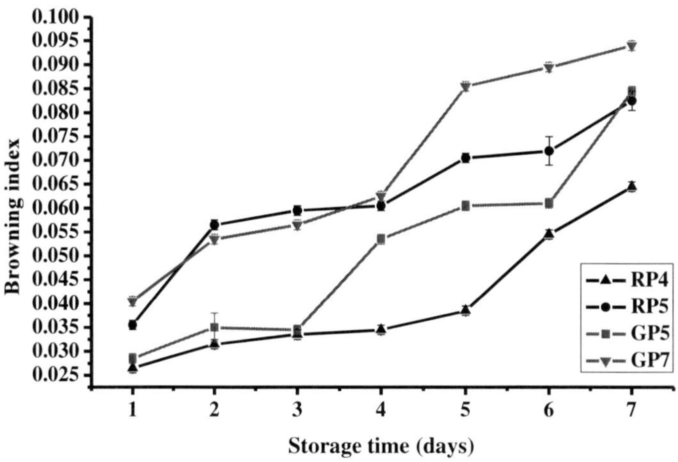

Figure 22. Browning index of hot chilli samples.

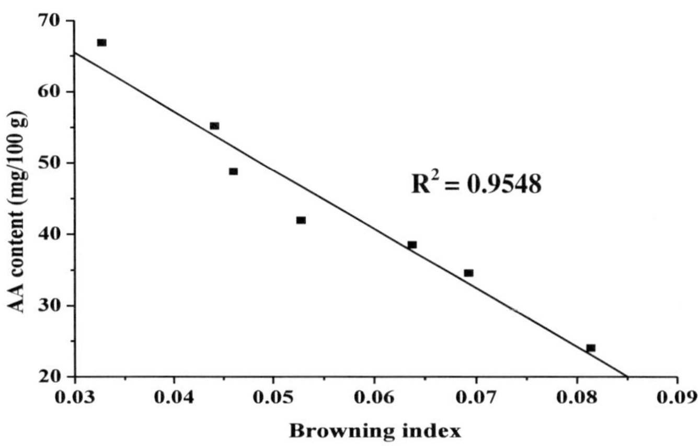

Figure 23. Correlation between the ascorbic acid content and the browning index of hot chilli pericarp samples ($n = 4$) for 7 days.

Although these reactive intermediates have not been identified in this study, they have been observed to undergo further reactions leading to the formation of brown pigments. Figure 23 show the correlation between the ascorbic acid content and the browning index of hot pepper pericarp samples (n = 4) for 7 days. It was observed that the ascorbic acid degradation was negative correlated to the browning formation. The ascorbic acid content in hot pepper samples decreased with increasing storage time. On the other hand, browning index of the samples also increased depending on the storage time.

CONCLUSION

This study reported the analytical methods aiming to use HPLC for the determination of ascorbic acid, dehydroascorbic acid (DHA), 5-hydroxymethylfurfural (5-HMF) and furfural in chilli fruits including wine and fruit juice samples. The developed method was applied for the determination of ascorbic acid, DHA, 5-HMF and furfural in sixteen samples of hot chilli and ten samples of wine and fruit juice.

The contents of ascorbic acid and DHA in hot chilli sample were ranged from 1.31-99.58 and 0.13-2.75 mg/100 g, respectively. 5-HMF was not detectable in any studied samples and F was found only in one sample (0.06 mg/100 g). The contents of ascorbic acid, DHA, 5-HMF and furfural in wine and fruit juice samples were ranged from 0.38-46.47, 0.91-7.39, 0.20-12.41 and 0.01-0.70 mg/100 mL, respectively. However, the contents of DHA in these samples were determined using the proposed method in comparison with the determination of DHA as the differences in the total ascorbic acid after DHA reduction with dithiothreitol and the ascorbic acid content of the original sample.

The contents of DHA using the proposed method were close to those of using the reference method. Thus, the proposed method was suitable for analysis of ascorbic acid, DHA, 5-HMF and furfural in the samples. This method could also be used to evaluate their quality of chilli fruits, wines and fruit juices during processing and storage.

Moreover, the relationship between the degradation of ascorbic acid and the formation of browning was studied. The browning index was determined with an absorbance measured at 420 nm. Four samples of hot chilli were determined for both ascorbic acid content and browning index during 7 days. From the results, the ascorbic acid content in the chilli samples decreased while the browning index of the samples increased with an increasing storage time.

ACKNOWLEDGMENTS

This research work was supported financially by the Centre for Innovation in Chemistry (PERCH-CIC), Commission on Higher Education, Ministry of Education, Thailand and the Hitachi Scholarship Foundation, Tokyo, Japan.

REFERENCES

[1] Eskin, N.A.M. (1990). *Biochemistry of Foods*. New York: Academic Press.

[2] Martinez, M.V. and Whitaker, J.R. (1995). The biochemistry and control of enzymatic browning. *Trends in Food Science and Technology*, 6, 195-200.

[3] Yuan, J.P. and Chen, F. (1999). Simultaneous separation and determination of sugars, ascorbic acid and furanic compounds by HPLC-dual detection. *Food Chemistry*, 64, 423-427.

[4] Simsek, A., Poyrazoglu, E.S., Karacan, S. and Velioglu, S. (2007). Response surface methodological study on HMF and fluorescent accumulation in red and white grape juices and concentrates. *Food Chemistry*, 101, 987-994.

[5] Roig, M.G., Bello, J.F., Rivera, Z.S. and Kennedy, J.F. (1999). Studies on the occurrence of non-enzymatic browning during storage of citrus juice. *Food Research International*, 32, 609-619.

[6] Burdurlu, H.S., Koca, N. and Karadeniz, F. (2006). Degradation of vitamin C in citrus juice concentrates during storage. *Journal of Food Engineering*, 74, 211-216.

[7] Shinoda, Y., Murata, M., Homma, S. and Komura, H. (2004). Browning and decomposed products of model orange juice. *Bioscience Biotechnology and Biochemistry*, 68, 529-536.

[8] Shinoda, Y., Komura, H., Homma, S. and Murata, M. (2005). Browning of model orange juice solution: factors affecting the formation of decomposition products. *Bioscience Biotechnology and Biochemistry*, 69, 2129-2137.

[9] Yuan, J.P. and Chen, F. (1998). Degradation of ascorbic acid in aqueous solution. *Journal of Agricultural and Food Chemistry*, 46, 5078-5082.

[10] Rojas, A.M. and Gerschenson, L.N. (1997). Ascorbic acid destruction in sweet aqueous model systems. *LWT- Food Science and Technology*, 30, 567-572.

[11] Śanchez-Mata, M.C., Cámara-Hurtado, M., Díez-Marqués, C. and Torija-Isasa, M.E. (2000). Comparison of high-performance liquid chromatography and spectrofluorimetry for vitamin C analysis of green beans (*Phaseolus vulgaris* L.). *European Food Research and Technology,* 210, 220-225.

[12] Solomon, O., Svanberg, U. and Sahlström, A. (1995). Effect of oxygen and fluorescent light on the quality of orange juice during storage at 8°C. *Food Chemistry*, 53, 363-368.

[13] Hui, Y.H., Nip, W.K., Nollet, L.M.L., Paliyath, G. and Simpson, B.K. (Eds.) (2006). *Food Biochemistry & Food Processing*. Iowa: Blackwell.

[14] Pino, J., González, M., Ceballos, L., Yah, A.C., Aguirre, J.T., Moreno, L.L. and Duch, E.S. (2007). Characterization of total capsaicinoids, colour and volatile compounds of Habanero chilli pepper (*Capsicum chinese* Jack.) cultivars grown in Yucatan. *Food Chemistry,* 104, 1682-1686.

[15] Oboh, G., Puntel, R.L. and Rocha, J.B.T. (2007). Hot pepper (*Capsicum annuum*, Tepin and *Capsicum chinese*, Habanero) prevents Fe^{2+}-induced lipid peroxidation in brain – *in vitro*. *Food Chemistry*, 102, 178-185.

[16] Zerdin, K., Rooney, M.L. and Vermuë, J. (2003). The vitamin C content of orange juice packed in an oxygen scavenger material. *Food Chemistry*, 82, 387-395.

[17] Kim, S., Lee, K.W., Park, J. Lee, H.J. and Hwang, I.K. (2006). Effect of drying in antioxidant activity and changes of ascorbic acid and colour by different drying and

storage in Korean red pepper (*Capsicum annuum*, L.). *International Journal of Food Science and Technology,* 41, 90-95.

[18] Cortés, C., Esteve, M.J. and Frígola, A. (2008). Color of orange juice treated by High Intensity Pulsed Electric Fields during refrigerated storage and comparison with pasteurized juice. *Food Control*, 19, 151-158.

[19] Li, H., Guo, A. amd Wang, H. (2008). Mechanisms of oxidative browning of wine. *Food Chemistry,* 108, 1-13.

[20] Hernández, Y., Lobo, M.G. and González, M. (2006) Determination of vitamin C in tropical fruits: A comparative evaluation of methods. *Food Chemistry*, 96, 654-664.

[21] Deepa, N., Kaur, C., Singh, B. and Kapoor, H.C. (2006). Antioxidant activity in some red sweet pepper cultivars. *Journal of Food Composition and Analysis*, 19, 572-578.

[22] Deepa, N., Kaur, C., George, B., Singh, B. and Kapoor, H.C. (2007). Antioxidant constituents in some sweet pepper (*Capsicum annuum* L.) genotypes during maturity. *LWT- Food Science and Technology*, 40, 121-129.

[23] Scala, K.D. and Crapiste, G. (2008). Drying kinetics and quality changes during drying of red pepper. *LWT- Food Science and Technology*, 41, 789-795.

[24] Ayranci, E and Tunc, S. (2004). The effect of edible coatings on water and vitamin C loss of apricots (*Armeniaca vulgaris* Lam.) and green peppers (*Capsicum annuum* L.). *Food Chemistry*, 87, 339-342.

[25] Noroozifar, M. and Motlagh, M.K. (2003). Application of potassium chromate-diphenylcarbazide in the quantitative determination of ascorbic acid by spectrophotometry. *Turkish Journal of Chemistry*, 27, 717-722.

[26] Arya, S.P., Mahajan, M. and Jain, P. (2001). Spectrophotometric determination of vitamin C with iron(II)-4-(2-pyridylazo)resorcinol complex. *Analytica Chimica Acta*, 427, 245-251.

[27] Pfendt, L.B., Vukašinović, V.L., Blagojević, N.Z. and Radojevìć, M.P. (2003). Second order derivative spectrophotometric method for determination of vitamin C content in fruits, vegetables and fruit juices. *European Food Research and Technology*, 217, 269-272.

[28] Aydoğmuş, Z. and Cetin, S.M. (2002). Determination of ascorbic acid in vegetables by derivative spectrophotometry. *Turkish Journal of Chemistry*, 26, 697-704.

[29] Grudpan, K., Kamfoo, K. and Jakmunee, J. (1999). Flow injection spectrophotometric or conductometric determination of ascorbic acid in a vitamin C tablet using permanganate or ammonia. *Talanta*, 49, 1023-1026.

[30] Simonne, A.H., Simonne, E.H., Eitenmiller, R.R., Mills, H.A. and Green, N.R. (1997). Ascorbic acid and provitamin A contents in unusually colored bell peppers (*Capsicum annuum* L.). *Journal of Food Composition and Analysis*, 10, 299-311.

[31] Wang, R., Liu, Z., Cai, R. and Li, X. (2002). A new spectrofluorometric method for the determination of ascorbic acid based on its activating effect on a hemoglobin-catalyzed reaction. *Analytical Sciences*, 18, 977-980.

[32] Dilgin, Y. and Nişli, G. (2005). Fluorimetric determination of ascorbic acid in vitamin C tablets using methylene blue. *Chemical and Pharmaceutical Bulletin*, 53, 1251-1254.

[33] Carvajal, M., Martinez, M.R., Sánchez, F.M. and Alcaraz, C.F. (1997). Effect of ascorbic acid addition to peppers on paprika quality. *Journal of the Science of Food and Agriculture*, 75, 442-446.

[34] Gökmen, V., Kahraman, N., Demir, N. and Acar, J. (2000). Enzymatically validated liquid chromatographic method for the determination of ascorbic and dehydroascorbic acids in fruit and vegetables. *Journal of Chromatography A*, 881, 309–316.
[35] Yahia, E.M., Padilla, M.C. and Aguilar, G.G. (2001). Ascorbic acid content in relation to ascorbic acid oxidase activity and polyamine content in tomato and bell pepper fruits during development, maturation and senescence. *LWT- Food Science and Technology*, 34, 452-457.
[36] Aguilar, G.A.G., Zavala, J.F.A., Cruz, S.R., Félix, E.A. and Cinco, M.E.D. (2004). Effect of temperature and modified atmosphere packaging on overall quality of fresh-cut bell peppers. *LWT- Food Science and Technology*, 37, 817-826.
[37] Navarro, J.M., Flores, P., Garrido, C. and Martinez, V. (2006). Changes in the contents of antioxidant compounds in pepper fruits at different ripening stages, as affected by salinity. *Food Chemistry*, 96, 66-73.
[38] Topuz, A. and Ozdemir, F. (2007). Assessment of carotenoids, capsaicinoids and ascorbic acid composition of some selected pepper cultivars (*Capsicum annuum* L.) grown in Turkey. *Journal of Food Composition and Analysis*, 20, 596-602.
[39] Raghu, V., Platel, K. and Srinivasan, K. (2007). Comparison of ascorbic acid content of *Emblica officinalis* fruits determined by different analytical methods. *Journal of Food Composition and Analysis*, 20, 529-533.
[40] Serrano, I.O., Jover, T.H. and Belloso, O.M. (2007). Comparative evaluation of UV-HPLC methods and reducing agents to determine vitamin C in fruits. *Food Chemistry*, 105, 1151-1158.
[41] Furusawa, N. (2001). Rapid high-performance liquid chromatographic identification/quantification of total vitamin C in fruit drinks. *Food Chemistry*, 12, 27-29.
[42] Kall, M.A. and Andersen, C. (1999). Improved method for simultaneous determination of ascorbic acid and dehydroascorbic acid, isoascorbic acid and dehydroisoascorbic acid in food and biological samples. *Journal of Chromatography B*, 730, 101-111.
[43] Cohen, E., Birk, Y., Mannheim, C.H. and Saguy, I.S. (1998). A rapid method to monitor quality of apple juice during thermal processing. *LWT- Food Science and Technology*, 31, 612-616.
[44] Gomis, D.B., Alvarez, M.D.G., Naredo, L.S. and Alonso, J.J.M. (1991). High-performance liquid chromatographic determination of furfural and hydroxymethylfurfural in apple juices and concentrates. *Chromatographia*, 32, 45-48.
[45] Gökmen, V. and Acar, J. (1999). Simultaneous determination of 5-hydroxymethylfurfural and patulin in apple juice by reversed-phase liquid chromatography. *Journal of Chromatography* A, 847, 69-74.
[46] Teixidó, E., Santos, F.J., Puignou, L. and Galceran, M.T. (2006). Analysis of 5-hydroxymethylfurfural in foods by gas chromatography-mass spectrometry. *Journal of Chromatography A,* 1135, 85-90.
[47] Ferrer, E., Alegría, A., Courtois, G. and Farré, R. (2000). High-performance liquid chromatographic determination of Maillard compounds in store-brand and name-brand ultra-high-temperature-treated cows' milk. *Journal of Chromatography A*, 881, 599-606.

[48] Mendoza, M.R., Olano, A. and Villamiel, M. (2002). Determination of hydroxymethylfurfural in commercial jams and in fruit-based infant foods. *Food Chemistry,* 79, 513-516.
[49] Gökmen, V. and Şenyuva, H.Z. (2006). Improved method for the determination of hydroxymethylfurfural in baby foods using liquid chromatography-mass spectrometry. *Journal of Agricultural and Food Chemistry*, 54, 2845-2849.
[50] Alcázar, A., Jurado, J.M., Pablos, F., González, A.G. and Martín, M.J. (2006). HPLC determination of 2-furaldehyde and 5-hydroxymethyl-2-furaldehyde in alcoholic beverages. *Microchemical Journal*, 82, 22-28.
[51] Rocha, S.M., Coimbra, M.A. and Delgadillo, I. (2004). Occurrence of furfuraldehydes during the processing of *Quercus suber* L. cork. Simultaneous determination of furfural, 5-hydroxymethylfurfural and 5-methylfurfural and their relation with cork polysaccharides. *Carbohydrate Polymers*, 56, 287-293.
[52] Merás, I.D., Mansilla, A.E. and López, F.S. (1995). Simultaneous kinetic spectrophotometric determination of 2-furfuraldehyde and 5-hydroxymethyl-2-furfuraldehyde by application of a modified Winkler's method and partial least squares calibration. *The Analyst*, 120, 2567-2571.
[53] Ameur, L.A., Trystram, G. and Aragon, I.B. (2006). Accumulation of 5-hydroxymethyl-2-furfural in cookies during the backing process: Validation of an extraction method. *Food Chemistry*, 98, 790-796.
[54] Doxastakis, G., Papageorgiou, M., Mandalou, D., Irakli, M., Papalamprou, E., D'Agostina, A., Resta, D., Boschin, G. and Arnoldi, A. (2007). Technological properties and non-enzymatic browning of white lupin protein enriched spaghetti. *Food Chemistry,* 101, 57-64.
[55] Yuan, J.P. and Chen, F. (1998). Separation and identification of furanic compounds in fruit juices and drinks by high-performance liquid chromatography photodiode array detection. *Journal of Agricultural and Food Chemistry*, 46, 1286-1291.
[56] Meydav, S., saguy, I. and Kopelman, I.J. (1977). Browning determination in citrus products. *Journal of Agricultural and Food Chemistry*, 25, 602-604.
[57] Göğüş, F. and Eren, S. (1998). Effect of temperature and pH on nonenzymic browning in minced dried pepper during storage. *Turkish Journal of Engineering and Environmental Sciences,* 22, 33-38.
[58] Burdurlu, H.S. and Karadeniz, F. (2003). Effect of storage on nonenzymatic browning of apple juice concentrates. *Food Chemistry*, 80, 91-97.
[59] Vercet, A. (2003). Browning of white chocolate during storage. *Food Chemistry*, 81, 371-377.
[60] Iwe, M.O., Zuilichem, D.J., Stolp, W. and Ngoddy, P.O. (2004). Effect of extrusion cooking of soy–sweet potato mixtures on available lysine content and browning index of extrudates. *Journal of Food Engineering*, 62, 143-150.
[61] Bozkurt, H. and Erkmen, O. (2005). Effects of salt, starter culture and production techniques on the quality of hot pepper paste. *Journal of Food Engineering*, 69, 473-479.
[62] Zhang, Y.L and Zhang, R.P. (2008). Study on the mechanism of browning of pomegranate *(Punica granafum* L. *cv.* Ganesh) peel in different storage conditions. *Agricultural Sciences in China*, 7, 65-73.

[63] Lindsay, S. (1992). *High Performance Liquid Chromatography*. Sussex: John Wiley & Sons.

[64] Miller, J. C. and Miller, J. N. (1993). *Statistics for Analytical Chemistry*. 3nd. New York: Ellis Horwood. S.E.

In: Natural Heterocycles
Editors: K. L. Ameta and Saksit Chanthai

ISBN: 978-1-63463-424-3

Chapter 7

NATURAL COUMARIN MOTIFS: ANTICANCER AGENTS

Sunil U. Tekale[1], Sushama S. Kauthale[1], Rajita D. Ingle[1], Sanjay B. Ubale[1], Satish U. Deshmukh[1], K. L. Ameta[2] and Rajendra P. Pawar[1*]

[1]Department of Chemistry, Deogiri College, Aurangabad (MS) India
[2]Department of Chemistry, FASC, Mody Institute of Technology and Science, Rajasthan, India

ABSTRACT

Coumarins constitute one of the most important classes of naturally occurring compounds in the realm of natural products and synthetic organic chemistry. Coumarin and its derivatives are well recognized for many biological activities such as anticoagulant, anticancer, antioxidant, antimicrobial, enzyme inhibitors etc. The present chapter highlights biological potential of naturally occurring coumarins; mainly anticancer activity.

1. INTRODUCTION

Coumarin (2H-chromen-2-one or benzopyran-2-one) (1) belongs to benzopyrone class of organic compounds. Chemically coumarins are a group of naturally occurring lactones.

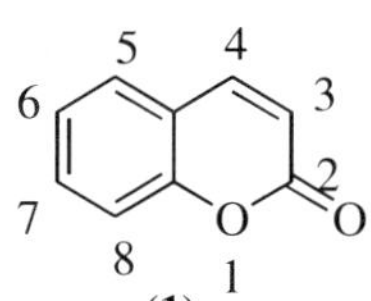

(1)

* Email: rppawar@yahoo.com Telephone No. +91-(0240) 2334577, Fax No. +91 (240) 2334430.

These are the molecules of plant kingdom; *rutaceae* and *umbelliferae* being the rich sources and also detected in microorganisms and animal sources. The name coumarin comes from the word *coumarou* (French: tonka bean); one of the sources from which it was first isolated as a natural product in 1820.

They are present in high concentrations in essential oils mainly cinnamon bark oil, lavender oil, cassia leaf oil *etc.* [1]. Coumarin also occurs in fruits such as bilberry, cloudberry, green tea *etc.* [2]. It is presumed to be produced by plants as a defense chemical to discourage predation. Its derivatives have been used in perfumes since 1882. It is used as a starting material in pharmaceutical industry for the synthesis of different synthetic anticoagulants. Coumarins act as vitamin K antagonists [3]. Coumarin and its derivatives are known to stimulate macrophages to degrade extracellular albumen thus allowing faster resorption of edematous fluids.

The coumarin ring system present in natural products displays interesting pharmacological properties. These are the plant secondary metabolites whose biological activity varies with the substituents. They are widely used as antibacterials [4], anti-inflammatory agents [5], anti-coagulants [6], pesticides [7] *etc.* Thus simple coumarins and their analogues are of great interest as therapeutic agents in medicinal and pharmaceutical chemistry.

The wide spectrum applications of coumarin based heterocyclic compounds have intrigued organic and medicinal chemists to explore the natural coumarins or synthetic analogs for their applicability as drugs. The bacteriostatic [8] and anti-tumour activity [9] of these scaffolds makes them attractive for further derivatisation and evaluation as novel therapeutic agents.

2. Biological Significance of Coumarins

Coumarin is the core skeleton of many naturally occurring as well as artificially synthesized bioactive molecules [10]. The diverse bioactivity of coumarins is utilized in many drugs which are marketed as anticoagulants, anti-inflammatory and anticoagulants. Acenocoumarol (2) is a marketed coumarin based drug commercially sold under the brand name- sintrom or sinthrome that functions as an anticoagulant and vitamin K antagonist [11]. Coumatetralyl (3) is an anticoagulant and vitamin K antagonist type. From the chronic animal studies; it has carcinogenic or teratogenic effects [12]. Dicoumarol (4) is a naturally occurring anticoagulant and vitamin K deplete of plant and fungal origin. It is also used as an inhibitor of reductases [13].

Phenprocoumon (5); commercially sold under the trade names like marcoumar, marcumar or falithrom is a long term oral anticoagulant drug [14]. It inhibits coagulation by blocking the synthesis of coagulation factors and is used for the prophylaxis and treatment of thromboembolic disorders. Tioclomarol (6) is an anticoagulant and vitamin K antagonist. It is used as a rodenticide which is useful for the control of rodents [15]. Warfarin (7); commercially sold under the brand names like coumadin, marevan, uniwarfin, warf or jantoven is an anticoagulant. It is used for the prevention of thrombosis and thromboembolism [16]. Initially in 1948 it was regarded as a pesticide against rats and mice.

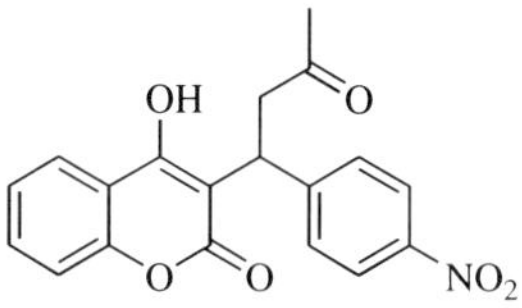

(2)

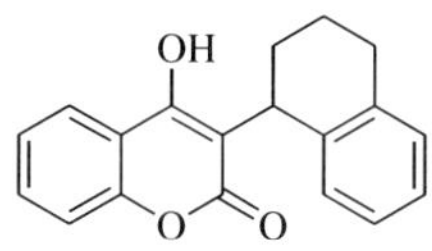

(3)

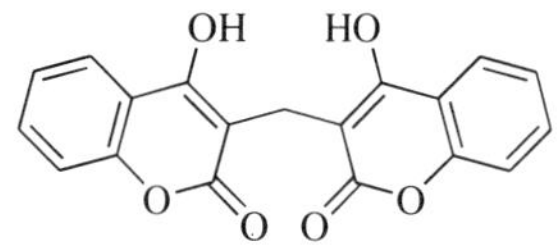

(4)

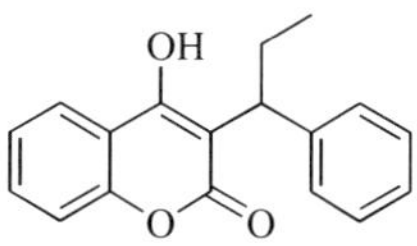

(5)

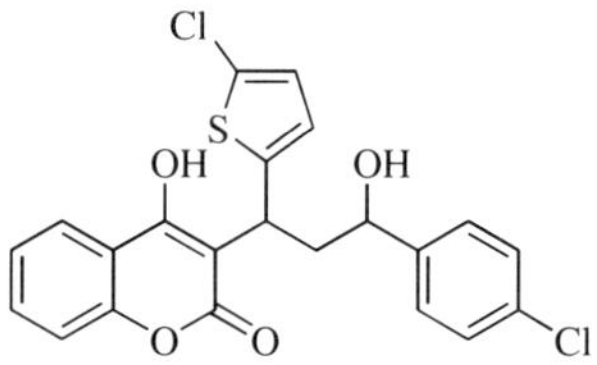

(6)

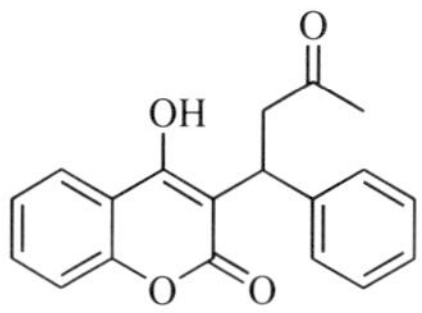

(7)

Anticoagulant coumarins

The marine alkaloids ningalin B (8) and lamellarin D (9) possessing good HIV-1 inhibition activity [17], and cytotoxicity [18] also contain coumarin cores in their structures. Umbelliprenin (10) has promising cancer chemo preventive activity [19]. Auraptene (11) is a natural bioactive monoterpene coumarin firstly isolated from the genus *Citrus*. It is helpful in the prevention of many degenerative diseases such as cancers of skin, liver etc. [20].

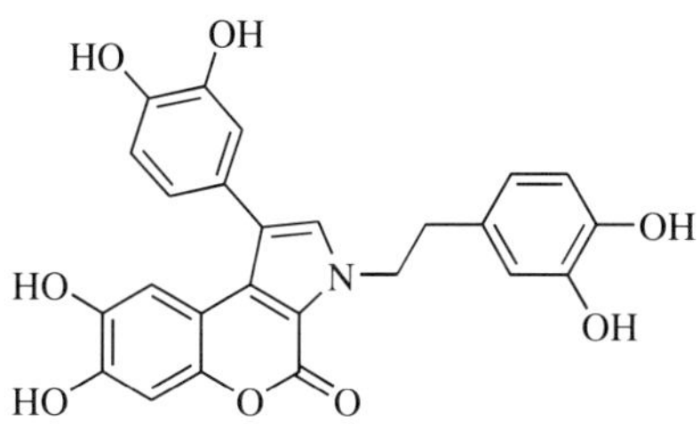

(8)

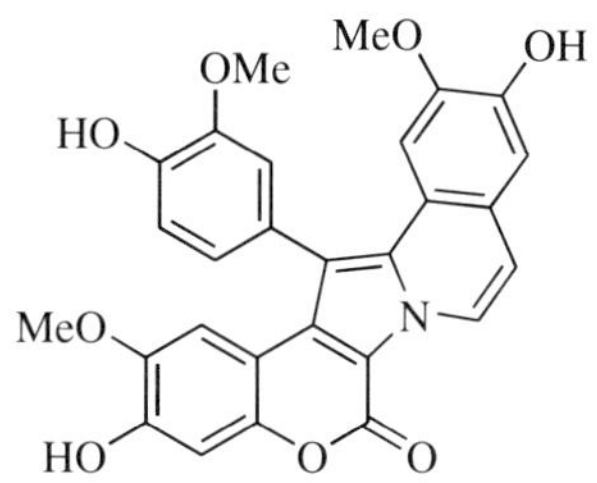

(9)

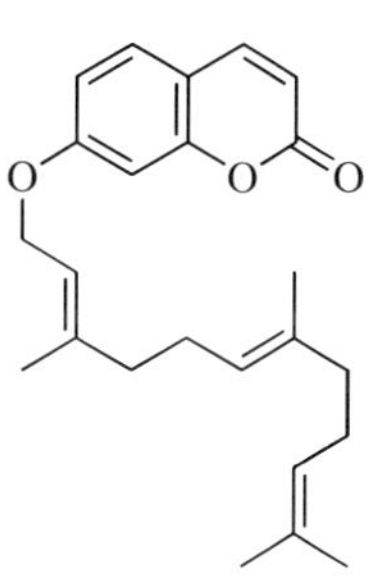

(10)

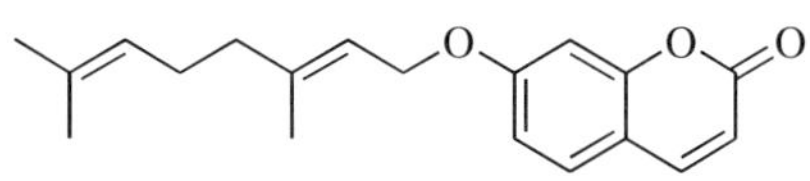

(11)

Anticancer coumarins

Flocoumafen (12) is an anticoagulant and vitamin K antagonist. It is commercially employed as a rodenticide [21]. Brodifacoum (13) is a vitamin K antagonist, anticoagulant poison and one of the commonly used pesticides [22]. Coumaphos (14) has neurological problems in bees and hence has ectoparasiticidal activity which is useful for farm and domestic animals to control mites, fleas and flies.

Ensaculin (15) is a coumarin based drug useful in the treatment of dementia by working on a number of receptor systems [23]. In human being; it causes side effect of orthostatic hypotension (low blood pressure).

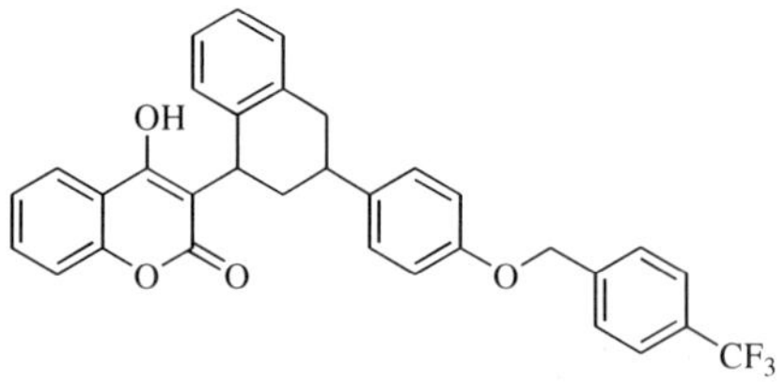

(12)

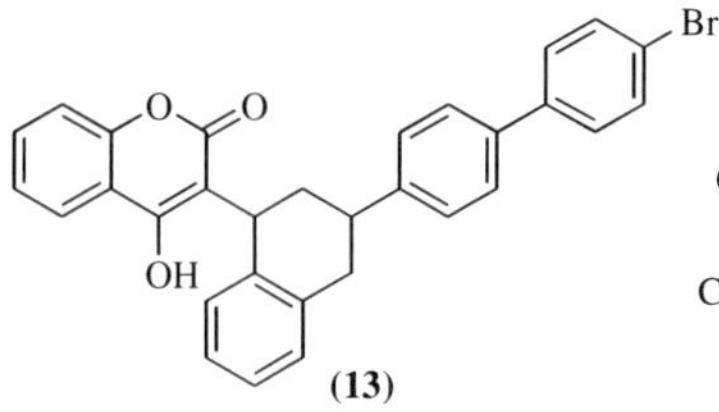

(13)

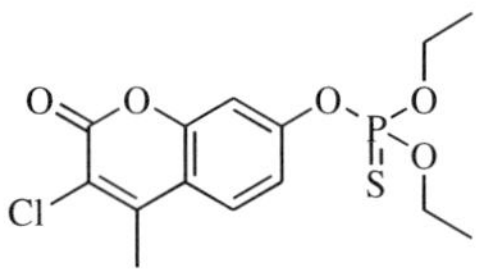

(14)

Pesticidal coumarins

Aminocoumarin (16) is an antibiotic that inhibits the DNA gyrase enzyme which is involved in the cell division in many bacteria [24]. Aminocoumarin can be isolated from the *Streptomyces* species. Hymecromone (17) drug is used in bile therapy. It is used as choleretic and antispasmodic drugs [25]. Herniarin (18) which can be isolated from *Herniariaglabra*, *Ayapanatriplinervis etc* is well known for antigenotoxic and antioxidant activity [26]. Batoprazine (19) is a phenylpiperazine drug acting as serenic or antiaggressive agent and 5-HT_{1B} receptor agonist [27]. Carbocromen (20) is a vasodilator [28].

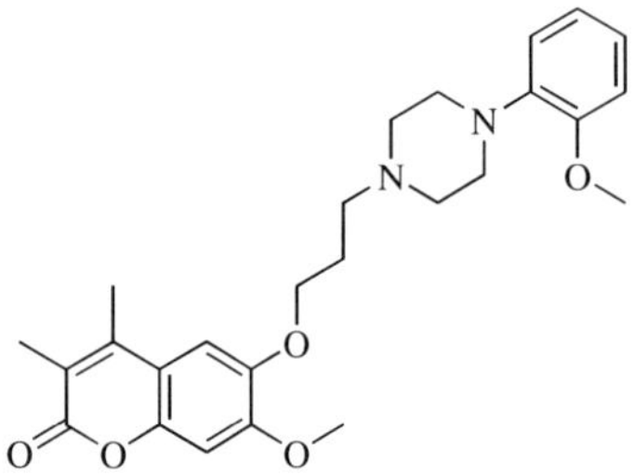

(15)

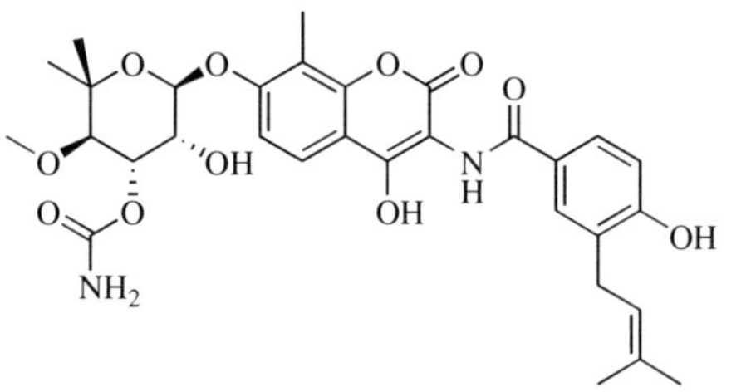

(16)

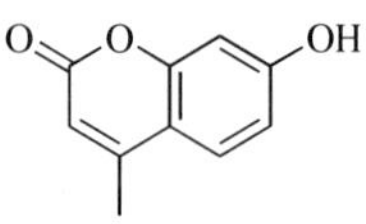

(17)

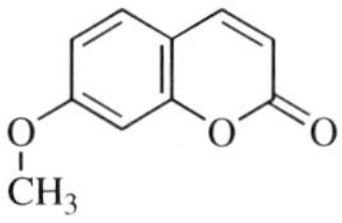

(18)

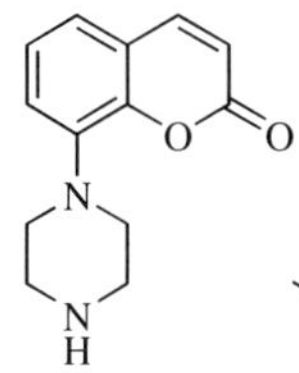

(19)

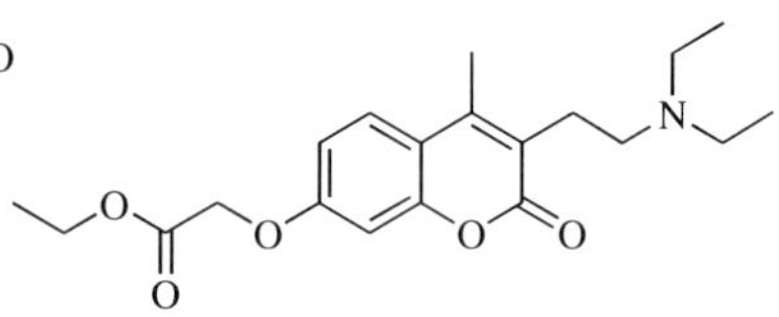

(20)

3. Extraction, Isolation and Anticancer Evaluation of Some Natural Coumarins

Cancer is a broad group of diseases involving unregulated cell growth. In the current era, cancer has become the most threatening disease to mankind all over the world. In fact; it has turned to be the world's biggest killer. Anti-cancer drugs help to prevent cancer by interrupting the cell division of the uncontrollably growing cells.

However many of the anticancer agents although effective have several side effects on normal proliferating tissues. Very commonly radiotherapy is also combined with chemotherapy. Studies have shown that coumarins not only treat cancer but also treat the side effects of radiotherapy [29-30].This part of the present chapter deals with the isolation and application of certain naturally occurring coumarins as anticancer agents.

The *ferula* species is one of the most popular species for the treatment of various disorders due to presence of certain coumarins in them [31]. Valiahdi et al. reported the isolatin and investigation of some coumarins - umbelliprenin (21), herniarin (22), 7-isopentenyloxycoumarin (23), galbanic acid (24), diversin (25), farnesiferol A (26) from the *ferula* species which were studied for their cytotoxicity on ovarian carcinoma (CH1), lung cancer (A549) and melanoma (SK-MEL-28) cell lines calorimetrically using MTT assay. Umbelliprenin, 7-isopentenyloxycoumarin and herniarin were synthesized artificially whereas galbanic acid, farnesiferol A, diversin were isolated from the roots of *F. szowitsiana* [32], *F. persica* [33], *F. diversivittata* [34] respectively.

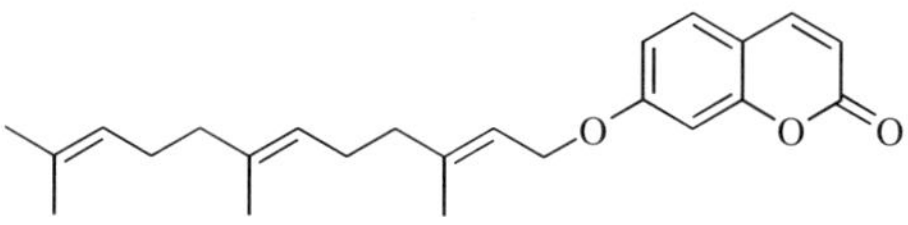

Umbelliprenin

(21)

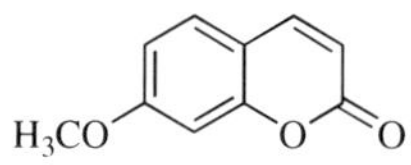

Hemiarin

(22)

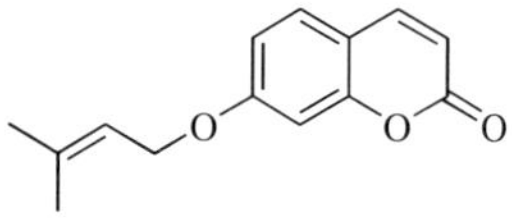

7-Isopentenlyoxycoumarin

(23)

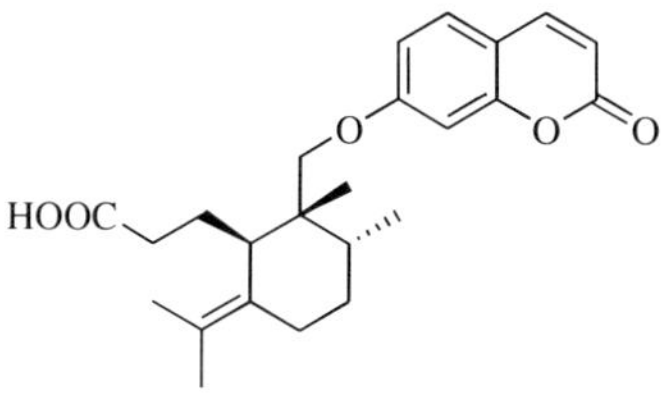

Galbanic acid

(24)

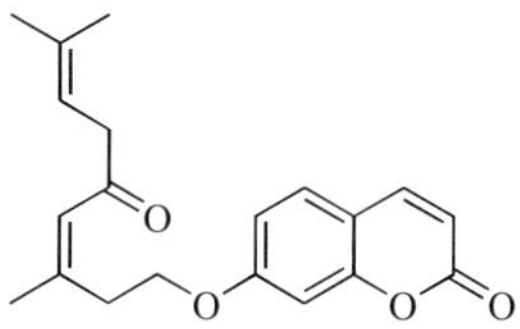

Diversin

(25)

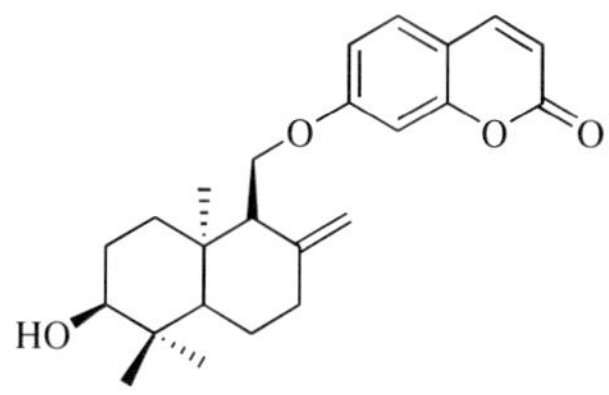

Farnesiferol A

(26)

Good to moderate cytotoxic activity results were obtained with IC_{50}values in the micromolar range (Table 1).

Saeed et al. documented the isolation of potentially six coumarins - 5,6,7-trimethoxycoumarin (27), 6-methoxy-7,8-methylenedioxycoumarin (28), bergapten (29), decursinolangelate (30), decursin (31) and nodakenetin (32) isolated from the roots of *Angelica glauca* collected from the hilly areas of Swart (Northern hilly state of Pakistan) [35].

Table 1. Cytotoxicity of coumarins isolated from *Ferula* species on three human cancer cells

Coumarin	IC_{50} [μM]		
	A549	SK-MEL-28	CH1
Umbelliprenin	133±6	58.1 ± 15.2	37.2± 2.7
Herniarin	> 250	> 250	> 62,5
7-isopentenyloxycoumarin	> 250	> 250	> 62.5
Galbanic acid	89.1±4.4	76.1 ± 5.4	46.3± 2.8
Diversin	> 250	> 250	> 62,5
Farnesiferol A	53.7±5.1	213 ± 17	25.7± 1.2

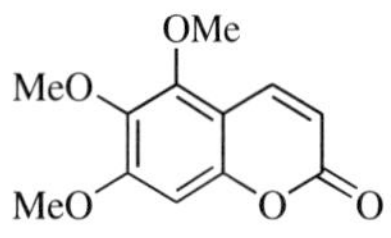

(27)

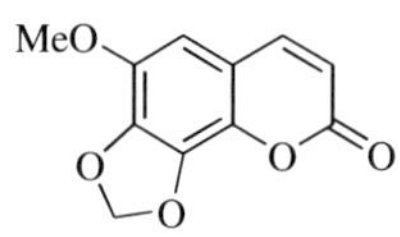

(28)

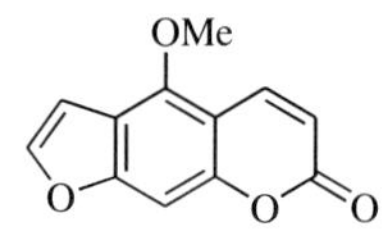

(29)

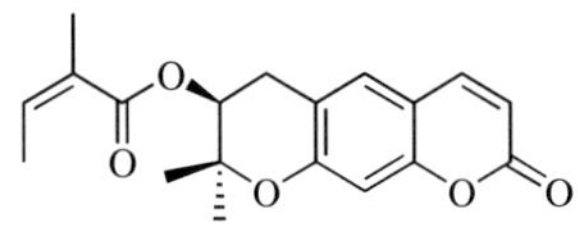

(30)

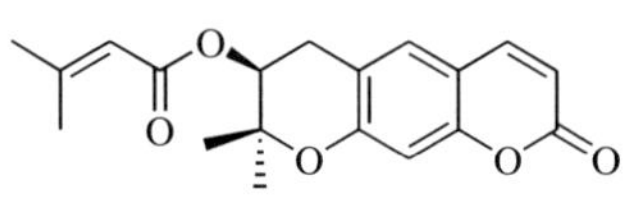

(31)

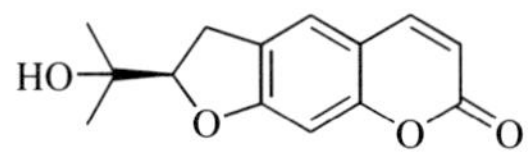

(32)

Extraction and isolation of coumarins (27)-(32): Air dried root powder of *A. glauca* was extracted thrice with methanol in a soxhlet and the combined extracts were concentrated under reduced pressure to obtain a residue. The residue was suspended in water and then fractionated successively with equal volumes of CH_2Cl_2/n-BuOH and the fractions were separately evaporated in vacuum. The DCM soluble fraction was then purified by column chromatography with gradient elution using cyclohexane/ethyl acetate to get the compounds (27), (28) and (29). The n-BuOH fraction after chromatography on silica gel with a gradient elution of $CHCl_3$/MeOH afforded compounds (30), (31) and (32).

Cytotoxicity: Decursinolangelate (30) was observed to be stronger cytotoxic agent than the remaining coumarins. 5, 6, 7-trimethoxycoumarin (27) showed intermediate cytotoxicity whereas 6-methoxy-7, 8-methylenedioxycoumarin (28), bergapten (29) and nodakenetin (32), exhibited less against brine shrimp larvae (Table 2).

Table 2. Cytotoxicity of coumarins (27) - (32) by brine shrimp assay

Dose (mg/mL)	27	28	29	30	31	32	Colchicine
400	-	18[#]/30*	15/30	-	-	-	26/30
200	16/30	12/30	14/30	-	-	16/30	24/30
100	14/30	10/30	11/30	11/30	-	14/30	23/30
80	13/30	9/30	10/30	10/30	22/30	12/30	20/30
60	12/30	8/30	9/30	9/30	21/30	10/30	19/30
40	10/30	5/30	8/30	8/30	20/30	9/30	18/30
20	8/30	2/30	5/30	6/30	19/30	8/30	15/30
10	6/30	1/30	4/30	4/30	15/30	7/30	10/30
5	2/30	0/30	0/30	3/30	13/30	4/30	6/30
2.5	-	-	-	2/30	10/30	0/30	2/30
1.25	-	-	-	1/30	2/30	-	0/30
LC_{50} mg/mL	128.291	243.038	268.866	12.084	9.456	249.289	28.297
S.D.	0.172	0.094	0.127	0.096	0.046	0.127	0.054
χ^2	1.040	1.221	2.283	4.829	4.422	3.333	2365
U.C.L.	328.408	446.847	696.483	18.846	12.404	784.321	39.225
L.C.L.	77.537	164.764	157.194	7.679	7.087	129.387	19.913

LC_{50}: lethal conc. causing death of 50% brine shrimps; SD: standard deviation, U.C.L.: upper confident limit, L.C.L.: lower confident limit; [#]number of dead brine shrimps; * total number of shrimps used.

Two coumarins - dentatin (33) and 5-hydroxydentatin (34) were isolated along with some cabazoles from the roots of *Clausena Harmandiana* collected from Phuwiang District, Khon Kaen Province, Thailand by T. Thongthoom. The compounds after isolation and characterization by spectral techniques were evaluated for several biological activities such as anti-TB, antimalarial, antifungal and cytotoxic activities [36].

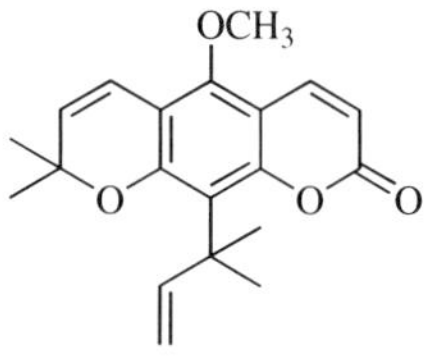

(33)

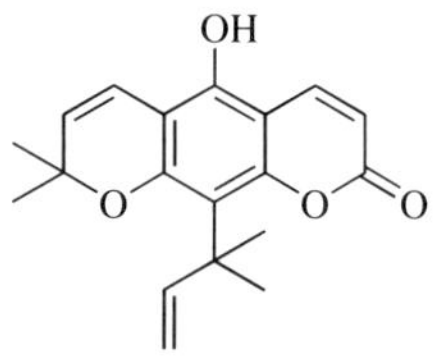

(34)

Extraction and isolation of coumarins (33) and (34): A dry powder obtained from the roots of *C. Harmandiana* was initially stirred at room temperature for one day and the hexane extract was collected by removal of the solvent in vacuo which was subsequently purified by column chromatography on silica gel by gradient elutions with several solvents such as hexane, EtOAc and MeOH with increasing polarity to afford 16 different fractions. Among the collected fractions during column chromatography; the filtrate from the crystallization of 8th fraction after crystallization afforded the coumarin dentatin (33). Recrystallization of fraction F11.7 by EtOAc/hexane afforded 5-hydroxydentatin (34).

Cytotoxicity Assay: Cytotoxicity of the isolated coumarins as performed by colorimetric method on vero cell line using ellipticine as the standard indicated that the compound (34) exhibited cytotoxicity against vero cells having IC_{50} values of 17.60 μg/mL (Table 3).

Two glycosides containing coumarins namely 6-methoxy-5, 7-dihydroxy-3, 4-dihydrocoumarin-8-*C*-glucopyranoside (35) and 5-vinyl-6, 7-dimethoxy-3, 4-dihydrocoumarin-8-*C*-glucopyranoside (36) were isolated from the aerial parts of *Diceratella elliptica* from the Jonsell growing in Egypt along with four different flavonoids. These isolated coumarins were found to be potent against three human carcinoma cell lines; liver (HEPG2), cervix (HELA) and colon (HCT116) [37].

Table 3. Cytotoxicity of coumarins (33) and (34)

Compound	Anti-TB (MIC, µg/mL)	Antimalerial (IC_{50}, µg/mL)	Antifungal (IC_{50}, µg/mL)	Cytotoxicity against vero cells (IC_{50}, µg/mL)
(33)	25.0	Inactive	Inactive	Inactive
(34)	25.0	1.32	Inactive	17.60

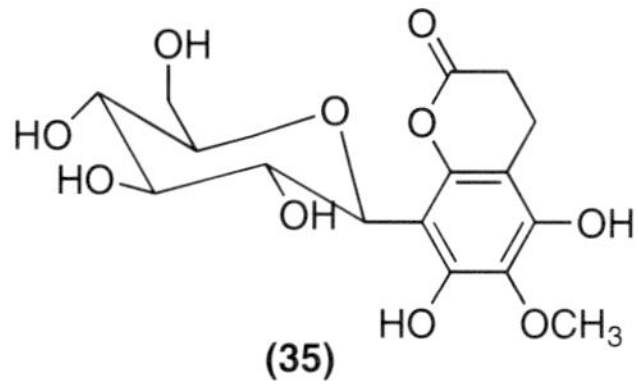

(35)

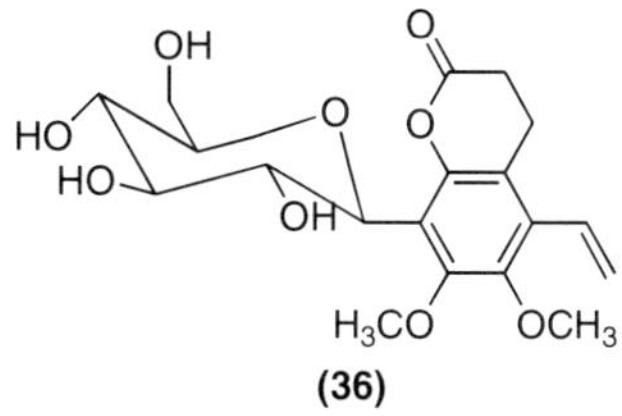

(36)

Extraction and isolation of coumarins (35) and (36): Arel parts of the *Diceratella elliptica* species were dried in air and defatted with pet ether. These were then extracted thrice in 70% ethanol/water under reflux condition.

Table 4. *In vitro* cytotoxicity of coumarins (35) and (36)

Compound	Carcinoma cell lines (HCT116) (IC_{50}, μg/mL)	Liver colon (HEPG2), (IC_{50}, μg/mL)	Cervix (HELA) (IC_{50}, μg/mL)
(35)	13.5	21.1	15.6
(36)	17.9	20.7	14.7

The combined extracts were evaporated under vacuum and the residue thus obtained was purified over a polyamide column starting with water then decreasing the polarity by increasing the concentration of methanol. Among the 28 fractions collected in this manner; the 1st fraction (20% MeOH:H_2O) was chromatographed on silica gel column using CH_2Cl_2: MeOH (2:3) as the eluent giving two major fractions FI-1 and FI-2. The FI-1 fractions and FI-2 after chromatography on PC using BAW twice followed by purification on sephadex LH-20 column using methanol afforded coumarins (35) and (36).

Cytotoxicity study: In vitro cytotoxic study of these compounds compared with the standard doxorubcin revealed that the coumarin (35) shows good activity against liver carcinoma cell line whereas the compound (36) exhibited good activity against colon and cervix cell lines (Table 4).

Mammea E/BB cyclo D coumarin (37) was isolated by Laphookhieo et al. along with five known compounds from the seeds of *Mammea siamensis* [38].

Extraction and isolation of coumarin (37): Powdered seeds of *M. siamensis* were extracted with dichloromethane successively for five days. The combined filtrates on evaporation under reduced pressure resulted in an extract which was chromatographed by QCC and eluted with hexane-EtOAc mixtures to give seven different fractions. The second fraction after purification by RP-18 CC with acetone: H_2O (3:1) and then RP-18 preparative TLC with acetone: H_2O (3:1) afforded the coumarin (**37**). Unfortunately the coumarin was found to be inactive against the different cancer cell lines, MCF-7 (breastadeno carcinoma), HeLa (human cervical cancer), HT-29 (colon cancer) and KB (human oral cancer).

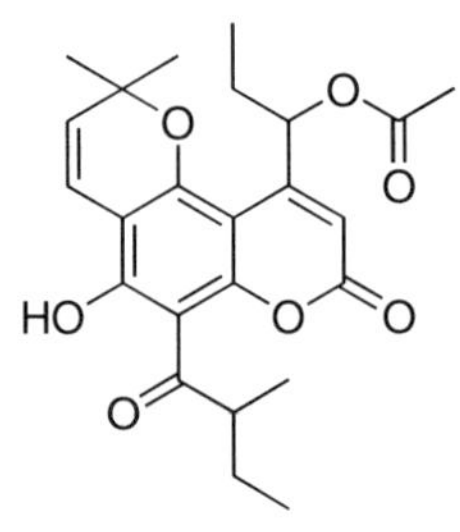

(**37**)

4. Recent Synthetic Approaches for Coumarin Derivatives

As a result of broad spectrum activities of coumarins and their limited natural resources; development of coumarins, their derivatives for chemotherapy becomes essential. Although many plants are rich in coumarins; environmental conditions influence their occurrence in diverse parts of the plant. Hence the laboratory synthesis of these compounds becomes essential. This portion of the present chapter is related with the recent advances in the synthesis of coumarin based heterocyclic compounds:

Y. Yamamoto et al. documented the synthesis of coumarins (40) by the copper acetate catalyzed coupling of arylpropionic acid methyl esters having MOM protected OH group (38) at ortho position with boronic acids (39) in methanol as the solvent [39].

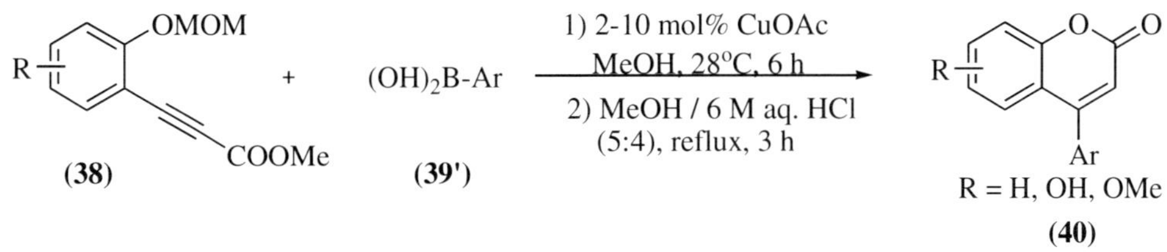

Coumarins (42) can be synthesized by the $Hf(OTf)_4$ catalyzed intramolecular Friedel-Crafts alkenylations of aromatic compounds with alkyl- and aryl-substituted alkynes (41) in ionic liquids [40].

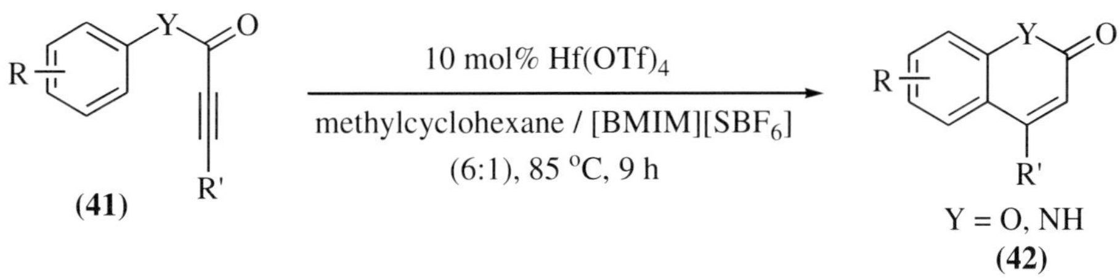

Palladium acetate catalyzed cyclocarbonylation of 2-vinylphenols (43) was reported by J. Ferguson for the synthesis of coumarins (44) [41].

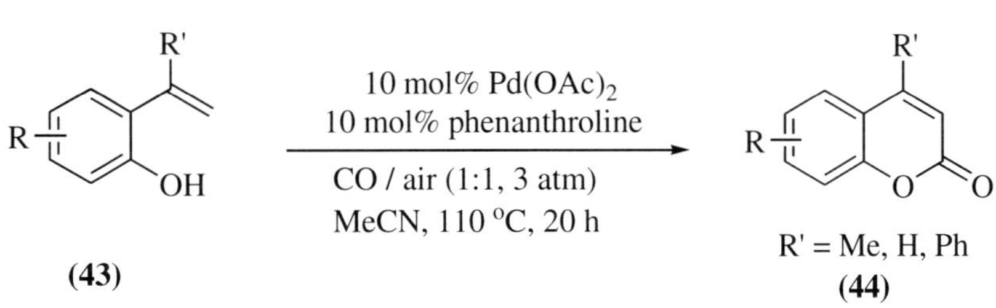

The ionic liquid 1-butyl-3-methylimidazonium tetrafluoroborate $[BMIM][BF_4]$ was employed for the reaction between salicylaldehyde (45) and active methylene compounds for the synthesis of coumarins (46) [42].

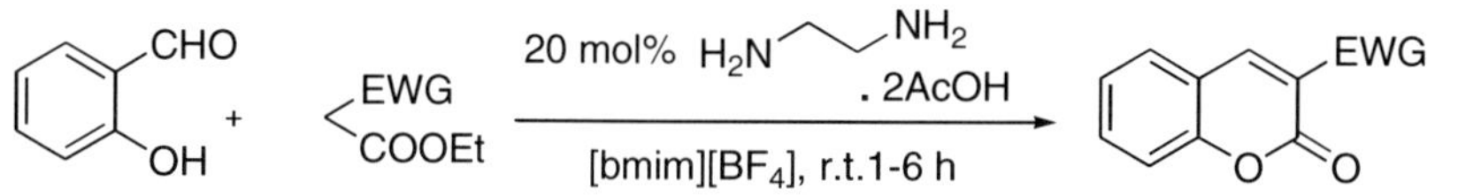

(45) (EWG = COMe, COOEt) (46)

Coumarins (49) can be obtained by triphenyl phosphine catalyzed reaction between 2-hydroxybenzaldehydes (47) and dialkylacetylenedicarboxylate (48) [43].

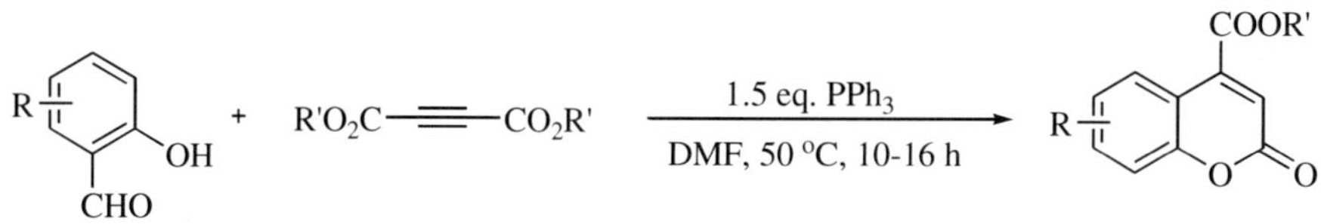

(47) (48) (49)

The coumarins (51) were obtained from the intramolecular cyclization of ortho hydroxyl ketones (50) in the presence of trifluoroacetic acid [44].

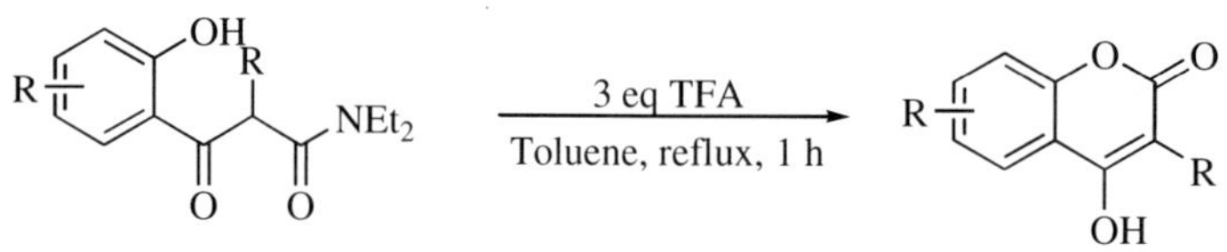

(50) (51)

Sashidhara et al. synthesized coumarins (54) by the coupling of salicylaldehydes (52) and substituted phenyl acetic acids (53) in the presence of TCT and NMM in refluxing DMF [45].

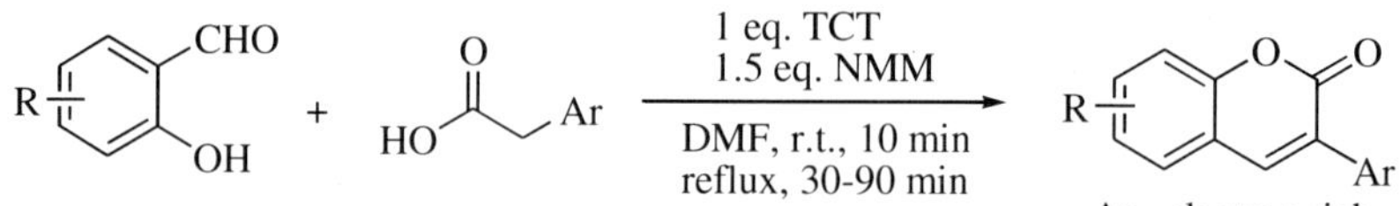

(52) (53) Ar= electron rich (54)

$BiCl_3$ was used as a catalyst for the synthesis of coumarins (57) by the Pechmann condensation reaction of phenols with β-keto esters (56) under solvent-free conditions [46]. A similar synthesis was documented by Potdar et al. in Lewis acidic chloroaluminate ionic liquid [47].

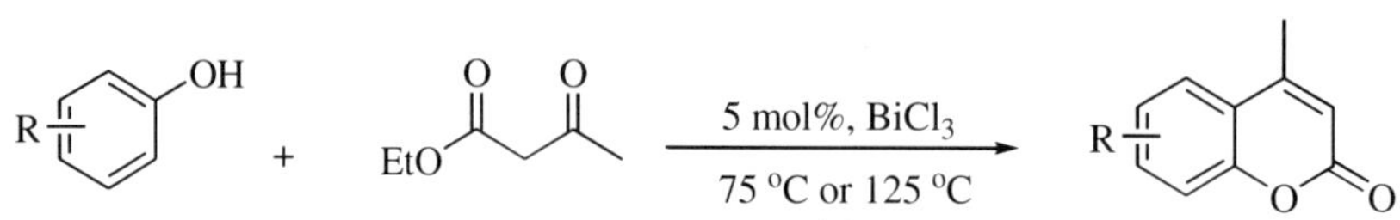

(R = OH, Me, $COCH_3$ etc)

(55) (56) (57)

A green organocatalyst containing proline and hydroxyproline material was recently reported by Sangsuwan et al. [48] for the synthesis of coumarins (62) and chromenes (63) under solvent-free conditions.

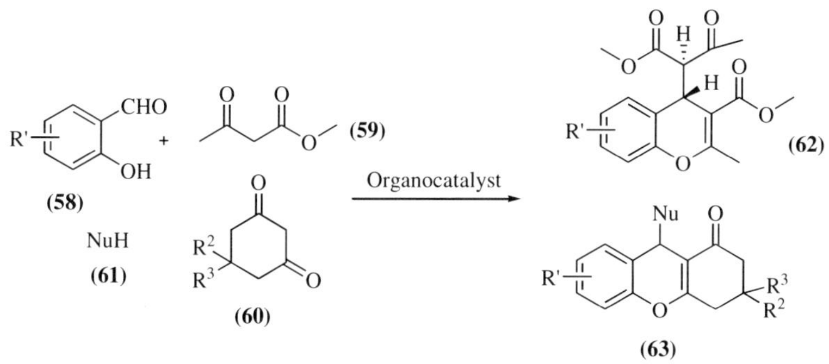

Conclusion

In summary; coumarins have attracted significant attention as naturally occurring cytotoxic agents of great therapeutic value. Besides their anticancer activity, anti-inflammatory and anticoagulant activities are equally important. Biological properties of coumarins make them suitable for further backbone derivatisation and screening as novel therapeutic agents in future. These facts highlight the future of these compounds for chronic inflammation diseases such as asthma and rheumatoid arthritis and cancer. Developing new analytical methods for natural and non-natural coumarin compounds is important in the screening of natural products with biological activity as well as in the studies of drug metabolism.

References

[1] Chouchi, D. and Barth, D. (1994) Rapid identification of some coumarin derivatives in deterpenated citrus peel oil by gas chromatography. *J. of Chromatography A,* 672(1-2):177-183.

[2] Olmedo, D., Rodriguez, N., Vasquez, Y., Solis, P.N., Lopez-Perez, J.L., Feliciano A.S. and Gupta M.P. (2007) *Natural Product Research*, 21(7), 625-631.

[3] Helgeland, L. (1980) A new coumarin from the fruits of Coutareahexandra. The action of vitamin K and coumarin anticoagulants. *Biochemical Education,* 8(3), 66-69.

[4] De Souza, S.M., Delle, M.F. and Smania, A. (2005). Antibacterial activity of coumarins. *Jr. Z Naturforsch C,* 60(9-10):693-700.

[5] Christos, A.K., Dimitra, and Hadjipavlou-Litina, J. (2005) Synthesis and anti-inflammatory activity of coumarin derivatives. Journal of Medicinal Chemistry, 48 (20), 6400-6408.

[6] Ya, L. Garazd, E.M., Kornienko, L.N., Maloshtan, M.M., Garazd and Khilya, V.P. (2005) Synthesis and anticoagulant activity of 3,4-cycloannelated coumarin D-glycopyranosides. *Chemistry of Natural Compounds,* 41(5), 508-512.

[7] Batra, T., Singh, R.V., Sangwan, N.K., Malik, M.S. and Malik, O.P. (1989) Studies on pesticides based on coumarin IV: Synthesis and antifungal activity of twelve 4-(substituted phenoxy)methyl-6-methyl coumarins. *Pesticide Science,* 25(1), 53-58.

[8] Fischer, F.C., Doorne, H.V., Lim, M.I., Svendsen, A.B. (1996) Bacteriostatic activity of some coumarin derivatives. *Phytochemistry*, 15(6):1078-1079.

[9] Chen, Z., Bi, J. and Su, W. (2013) Synthesis and antitumor activity of novel coumarin derivatives via a three-component reaction in water. *Chinese Journal of Chemistry,* 31(4) 507-514.

[10] Slapsyte, G., Dedonyte, V., Lazutka, J.R., Mierauskiene, J., Morkunas, V., Kazernaviciute, R., Pukalskas, A. and Venskutonis P.R. (2013) Evaluation of the biological activity of naturally occurring 5, 8-dihydroxycoumarin. *Molecules,* 18 (4):4419-4436.

[11] Arboix, M., Frati, M.E. and Laporte, J-R. (1984) The potentiation of acenocoumarol anticoagulant effect by amiodarone. *British Journal of Clinical Pharmacology*, 18, 355-360.

[12] Morgan, A.M. (2006) Teratogenic effect of the coumarinic anticoagulant rodenticide, racuminin white rats. *J. Egypt. Soc. Toxicol.* 34, 5-14.

[13] Cullen, J.J., Hinkhouse, M.M., Grady, M. et al. (2003) Dicumarol inhibition of NADPH: Quinone oxidoreductase induces growth inhibition of pancreatic cancer via a superoxide-mediated mechanism. *Cancer Research,* 63, 5513-5520.

[14] Haustein, K.O. and Hemost, S.T. (1999) Pharmacokinetic and pharmacodynamic properties of oral anticoagulants, especially phenprocoumon. 25(1):5-11.

[15] Totfi, H., Dreyfuss, M.F., Marquet, P., Debord, J., Merle, L. and Lachtre, G.A. (1996) Screening procedure for the determination of 13 oral anticoagulants and rodenticides. *Journal of Analytical Toxicology*, 20, 93-100.

[16] Leclerc, J.R., Geerts, W.H., Desjardins, L., Laflamme, G.H., Lesperance, B., Demers, C., Kassis, J., Cruickshank, M., Whitman, L. and Delorme F. (1996) Prevention of venous thromboembolism after knee arthroplasty. A randomized, double-blind trial comparing enoxaparin with warfarin. *Ann. Intern. Med.,* 124(7):619-26.

[17] Fan,G., Li, Z., Shen, S., Zeng, Y., Yang, Y., Xu, M., Bruhn, T., Bruhn, H., Morschhauser, J., Bringmann, G. and Lin, W. (2010) Baculiferins A-O, O-sulfated pyrrole alkaloids with anti-HIV-1 activity, from the chinese marine sponge *iotrochotabaculifera. Bioorganic and Medicinal Chemistry*, 18(15):5466-5474.

[18] Boger, D.L., Soenen , D.R., Boyce , C.W., Hedrick , M.P. and Jin, Q. (2000) Total Synthesis of Ningalin B utilizing a heterocyclic azadiene Diels-Alder reaction and discovery of a new class of potent multidrug resistant (MDR) reversal agents. Journal of the Organic Chemistry, 65 (8), 2479-2483.

[19] Iranshahi, M., Sahebkar, A., Takasaki, M., Konoshima, T. and Tokuda, H. (2009) Cancer chemopreventive activity of the prenylated coumarin, umbelliprenin, in vivo. *European Journal of Cancer Prevention*, 18(5):412-415.

[20] Genovese, S., Epifano and Auraptene, F. (2011) A natural biologically active compound with multiple targets. *Current Drug Targets,* 12(3), 381-386.

[21] Newton, I., Wyllie, I., Gray, A. and Charles, Eadsforth, V. (1994) The toxicity of the rodenticide flocoumafen to barn owls and its elimination via pellets. *Pesticide Science*,

[22] Patocka, J., Petroianu, G., Kuca, K. (2013) Toxic potential of superwarfarin: Brodifacoum. *Militray Medical Science Letters*, 82(1), 32-38.

[23] Hoerr, R., Noeldner, M. (2002) Ensaculin (KA-672 HCl): A multitransmitter approach to dementia treatment. *CNS Drug Rev.,* 8(2):143-58.

[24] Falconer, S.B., Czarny, T.L. and Brown, E.D. (2011) Antibiotics as probes of biological complexity. *Nature Chemical Biology,* 7, 415- 423.

[25] Garrett, E.R., Venitz, J., Eberst, K., Cerda, J.J. (1993) Pharmacokinetics and bioavailabilities of hymecromone in human volunteers. *Biopharm Drug Dispos,* 14(1):13-39.

[26] Rezaee, R., Behravan, Behravan, E.J., Soltani, F., Naderi, Y., Emami, B., Iranshahi, M. (2013) Antigenotoxic activities of the natural dietary coumarin sumbelliferone, herniarin and 7-isopentenyloxy coumarin on human. *Drug and Chemical Toxicology -* 37(2):144-148.

[27] Olivier, B., Oorschot, R.V. (2005) A 5-HT1B receptors and aggression: A review. *European Journal of Pharmacology,* 526, 207-217.

[28] Kjekshus, J.K., Simonsen, S., Bohmer, T. (1978) Effect of carbocromen induced coronary vasodilatation on myocardial metabolism in coronary artery disease. *Clinical Cardiology*, 1 (2):74-79.

[29] Lacy, A. and O'Kennedy, R. (2004) Studies on coumarins and coumarin-related compounds to determine their therapeutic role in the treatment of cancer. *Current Pharmaceutical Design*, 10, 3797-3811.

[30] Jain, P.K. and Joshi, H. (2012) Coumarin: Chemical and pharmacological profile. *Journal of Applied Pharmaceutical Science,* 2 (6): 236-240.

[31] Valiahdi, S.M., Iranshahi, M. and Sahebkar, A. (2013) Cytotoxic activities of phytochemicals from *Ferula* species. *DARU Journal of Pharmaceutical Sciences,* 21:39.

[32] Iranshahi, M., Arfa P., Ramezani M., Jaafari M.R., Sadeghian H., Bassarello C., Piacente S., Pizza C. (2007) Sesquiterpene coumarins from Ferula Szowitsiana and in vitro antileishmanial activity of 7-prenyloxycoumarins against promastigotes. *Phytochemistry,* 68:554-561.

[33] Iranshahi, M., Amin, G. and Shafiee, A. (2004) A new coumarin from *Ferula persica. Pharm. Biol.,* 42:440-442.

[34] Iranshahi, M., Hosseini, S.T., Shahverdi, A.R., Molazade, K., Khan, S.S., Ahmad, V.U. (2008) Diversolides A-G, guaianolides from the roots of *Ferula diversivittata. Phytochemistry*, 69, 2753-2757.

[35] Saeed, M.A. and Sabir, A.W. (2008) Irritant and cytotoxic coumarins from *Angelica glauca* Edgew roots. *Journal of Asian Natural Products Research*, 10(1), 49-58.

[36] Thongthoom, T., Yenjai, C. and Phaosiri, C. (2009) Biological activities of carbazoles and coumarins from the roots of *clausena harmandiana. Pure and Applied Chemistry International Conference,* 401-403.

[37] Marzouk, M.M., Elkhateeb, A., Ibrahim, L.F., Hussein, S.R. and Kawashty, S.A. (2012) Two cytotoxic coumarin glycosides from the aerial parts of *Diceratella Elliptica* (DC.) Jonsell growing in *Egypt Rec. Nat. Prod.,* 6:3, 237-241.

[38] Laphookhieo, S., Promnart, P., Syers, J.K., Kanjana-Opas, A., Ponglimanontc, C. and Karalai, C. (2007) Coumarins and xanthones from the seeds of *mammea siamensis. Journal of the Brazilean Chemical Society,* 18(5), 1077-1080.

[39] Yamamoto, Y. and Kirai, N. (2008) Synthesis of 4-arylcoumarins via Cu-catalyzed hydroarylation with arylboronic acids. *Organic Letters,* 10, 5513-5516.

[40] Song, C.E., Jung, D.-U., Choung, S.Y., Roh, E.J. and Lee, S.G. (2004) Dramatic enhancement of catalytic activity in an ionic liquid: Highly practical Friedel-Crafts alkenylation of arenes with alkynes catalyzed by metal. triflates. *Angew. Chem. Int. Ed.*, 43, 6183-6185.

[41] Ferguson, J., Zeng, F. and Alper, H. (2012) Synthesis of coumarins via Pd-catalyzed oxidative cyclocarbonylation of 2-vinylphenols. *Organic Letters*, 14, 5602-5605.

[42] Su, C., Chen, Z.C., Zhen and Q.G. (2003) Organic reactions in ionic liquids: Knoevenagel condensation catalyzed by ethylenediammoniumdiacetate. *Synthesis*, 555-559.

[43] K.C. Majumdar, I. Ansary, S. Samanta and B. Roy (2011) Aromatic electrophilic substitution vs. intramolecular Wittig reaction: Vinyltriphenylphosphonium salt mediated synthesis of 4-carboxyalkyl-8-formyl coumarins. *Synlett*, 694-698.

[44] Kalinin, A.V., Da Silva, A.J.M., Lopes, C.C., Lopes, R.S.C. and Snieckus, V. (1998) Directed ortho metalation-cross coupling links. Carbamoyl rendition of the Baker-Venkataraman rearrangement. Regiospecific route to substituted 4-hydroxycoumarins. *Tetrahedron Lett.*, 39, 4995-4998.

[45] Sashidhara, K.V., Palnati, G.R. and Avula, S.R.A. Kumar (2012) Efficient and general synthesis of 3-aryl coumarins using cyanuric chloride. *Synlett,* 23, 611-621.

[46] De, S.K. and Gibbs, R.A. (2005) An efficient and practical procedure for the synthesis of 4-substituted coumarins. *Synthesis*, 1231-1233.

[47] Potdar, M.K., Mohile, S.S., and Salunkhe, M.M. (2001) Coumarin syntheses via Pechmann condensation in Lewis acidic chloroaluminate ionic liquid. *Tetrahedron Lett.,* 42, 9285-9287.

[48] Sangsuwan, R., Sangher, S., Aree, T., Mahidol, C., Ruchirawat, S. and Kittakoop, P. (2014) An organocatalyst from renewable materials for the synthesis of coumarins and chromenes: three-component reaction and multigram scale synthesis. *RSC Advances,* 4, 13708-13718.

In: Natural Heterocycles
Editors: K. L. Ameta and Saksit Chanthai
ISBN: 978-1-63463-424-3

Chapter 8

NATURAL FLAVONOIDS: ASSETS FOR LIFE

***Shipra Bhardwaj*[1] *and K. L. Ameta*[2,*]**
[1]Government P.G. College, Kota, Rajasthan, India
[2]Department of Chemistry, FASC, Mody Institute of Technology & Science,
Lakshmangarh, Rajasthan

ABSTRACT

Flavonoids are essential constituents of plants and are meant to cure a number of problems in human beings as well as plants themselves. Flavonoids offer anti-inflammatory, anti-cancerous, anti-fungal, anti-bacterial, ion balancing and many more properties. Nearly 8000 flavonoids have been extracted by various conventional as well as new techniques and have been characterized by various methods like chromatographic analysis, spectrometric methods etc. In addition, flavonoids have been characterized for their uses in different fields like medical, health and other. Their uses have been tested on plants, animals and human beings in vivo as well as in vitro. The present chapter gives a broad summary of few such flavonoids with their respective classification, tests, extraction methods, characterization procedure, properties and uses.

INTRODUCTION

Flavonoids are also known as bioflavonoids and indigenously occurred in fruits (apple, blue berry, strawberry, raspberry, and pears), vegetables (cabbage, parsley, tomato, onion, and black beans), nuts, seeds, stems, flowers, tea, wine, honey etc. The word flavonoid has been derived from Latin word "Flavus" that means yellow. Generally flavonoids possess yellow, blue, and/or green colours. Due to their attractive colours they act as visual signals for pollinating insects and can represent a defence system against herbivores harmful to the plants [1].

Flavonoids also act as catalysts in the photosynthesis and as regulators of ion channels in phosphorylation [2]. Flavonoids are said to be the most common pigments found in plants and are placed in the order just next to chlorophyll and carotenoids. They also function as stress

* Email: klameta77@hotmail.com

protectants in plant cells by scavenging reactive oxygen species produced by the photosynthetic electron transport system [3]. Plants are also being protected from ultra violet radiations and from UV generated reactive oxygen species [4]. Recently, nearly 8000 natural flavonoid have been characterized. As far as chemical structure is concerned, flavonoids are poly phenolic compounds which posses' at least two phenolic subunits. These are planar molecules found in almost all plants, formed from the aromatic amino acids like phenylalanine, tyrosine and malonates [5]. The basic flavonoid structure is the flavan nucleus constructed from 15 carbon atoms arranged in three rings (C6–C3–C6), which are labelled A, B, and C. Their structures vary due to hydroxylation, methoxylation, prenylation, or glycosylation of different rings.

1.1. Classification of Flavonoids

Flavonoids have been classified broadly in three major classes according to IUPAC nomenclature [6, 7].

1. Flavonoids or bioflavonoids with basic back bone of 2-phenyl-1,4-benzopyrone (Figure 1).

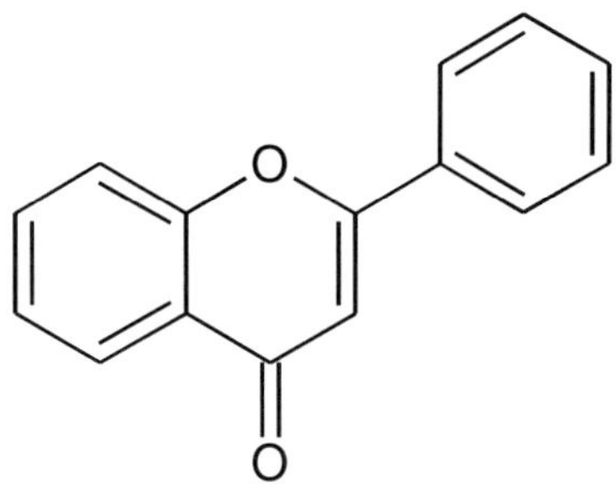

Figure 1. Flavonoid.

2. Isoflavonoids having basic back bone of 3-phenyl-1,4-benzpyrone (Figure-2).

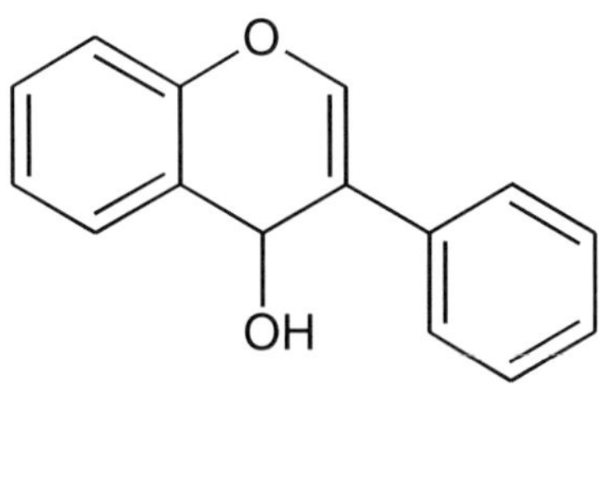

Figure 2. Isoflavonoid.

3. Neoflavonoids having basic back bone of 4-phenyl-1,4-pyrone which is also known as 4-phenyl couarine (Figure-3).

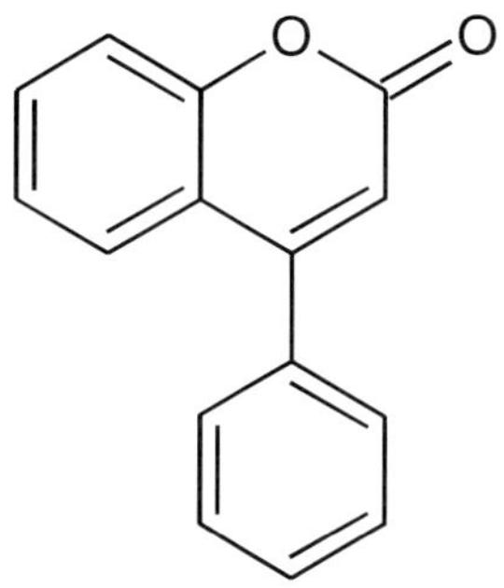

Figure 3. Neoflavonoid.

Flavonoids have also been classified depending upon functional groups attached as-

1. Anthoxanthins which are further classified into two sub groups-

a) Flavones with ketonic groups and has basic skeleton as given in Figure-4. Examples of such flavonoids are Luteolin, Apigenin, Tangeritin etc.

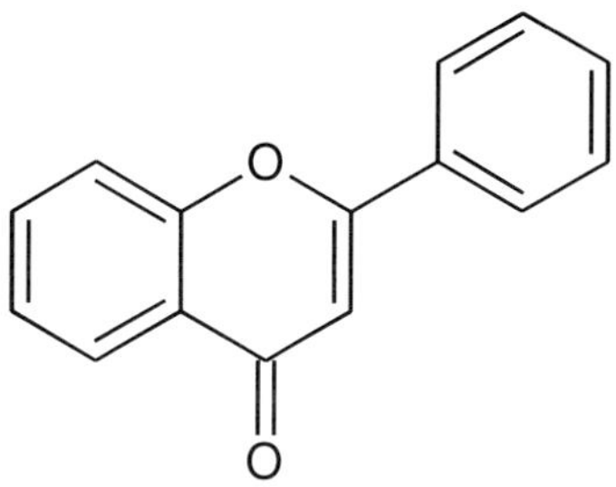

Figure 4. Flavone.

b) Flavonols are 3- hydroxy phenyl chromen-4-one with the basic skeleton given below in Figure-5. Few examples of this class are Quercetin, Kaempferol, Myricetin, Fisetin, Galangin etc.

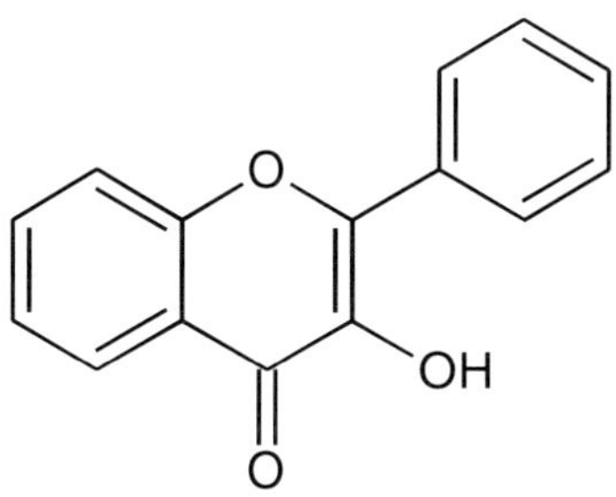

Figure 5. Flavonol.

2. Flavonones have 2,3- dihydro ketone as backbone. The skeleton is shown in Figure-6. Some examples of this class are Hesperetin, Homoeriodictyol etc.

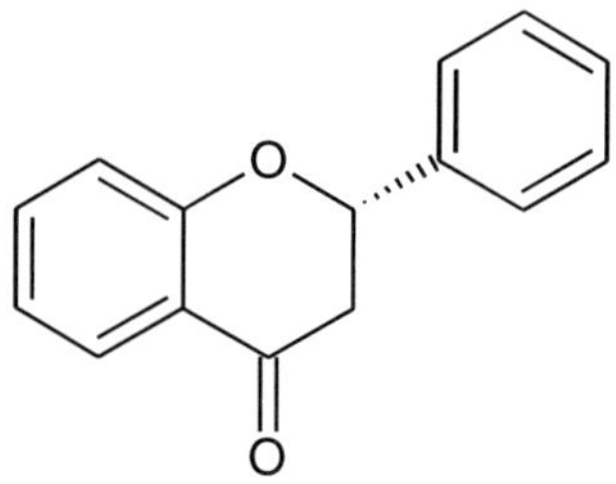

Figure 6. Flavonone.

3. Flavanonols are compounds having -OH groups with 2,3-dihydro ketones. The basic structure is given in Figure-7. Few examples are Taxifolin, Dihydrokaempferol etc.

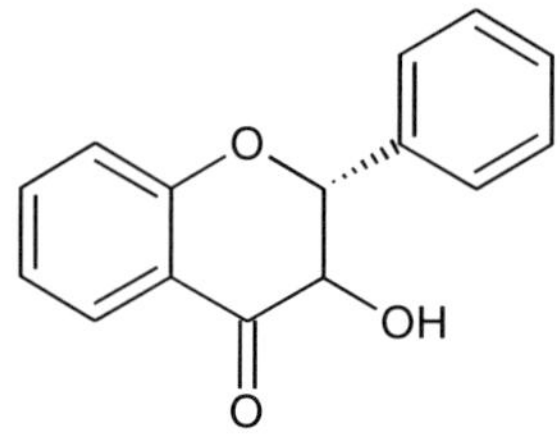

Figure 7. Flavanonol.

4. Flavans have -OH group instead of ketone as functional group. The backbone structure is given in Figure-8. The -OH group may be placed at 2- or 3- or both 2- and 3- positions. Thus it is a 2,3-dihydro structure. Catechin, Gallocatechin, Theaflavin etc. are some examples of this class.

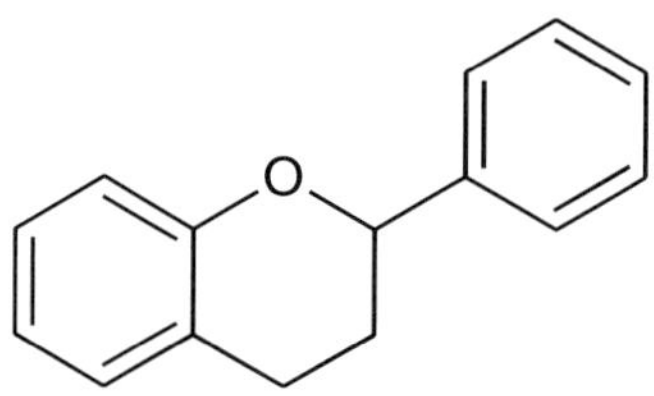

Figure 8. Flavan.

5. Anthocyanidins have flavylium as basic skeleton given in Figure-9. Few examples of this class are Cyanidin, Delphinidin etc.

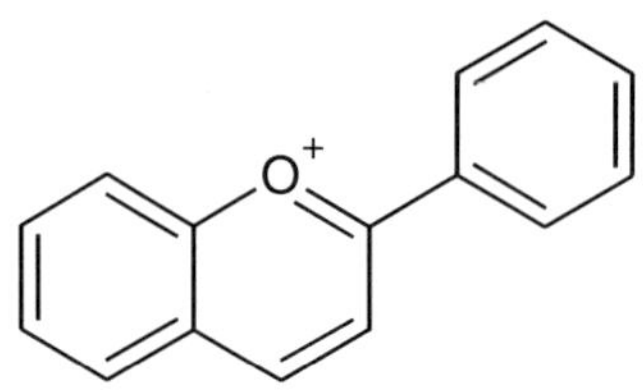

Figure 9. Flavylium ion.

1.2. Classification of Isoflavonoids

Isoflavonoids have been classified in following classes-

1. Isoflavones are the class having ketone as functional group and -OH groups are not present in the basic skeleton which is given in Figure-10. Some examples of this class are Genistein, Daidzein etc.

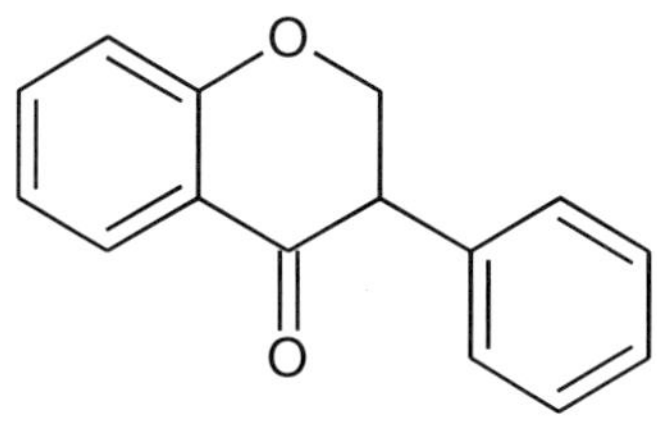

Figure 10. Isoflavone.

2. Isoflavanes are a type of polyphenolic compounds having backbone structure of isoflavan and is given in Figure-11. Some of its examples are Equol, Ionchocarpane, Laxiflorane etc.

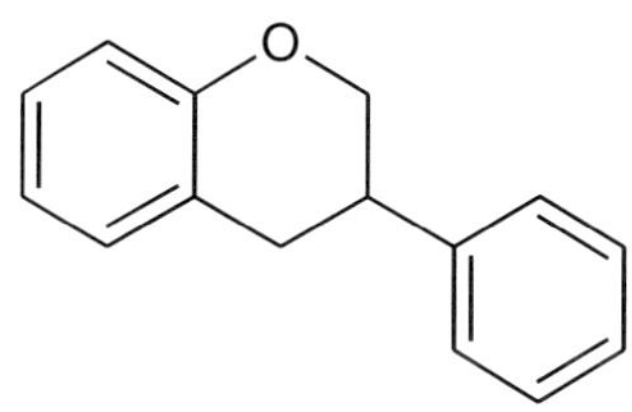

Figure 11. Isoflavane.

3. Isoflavandiols [8] are compounds containing two -OH groups may be on same ring or on different rings. Examples of this class are Sequol, Equol etc.
4. Isoflavenes are a type of isoflavonoid which have a double bond present in etherial ring. The backbone contains structure given in Figure-12. Some examples are Glabrene, Haginin D etc.

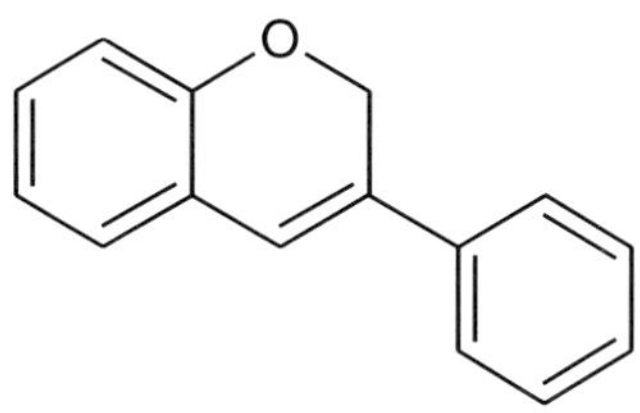

Figure 12. Isoflavene.

5. Coumestans are types of isoflavonoids having fused rings and are oxidation products of Pterocarpan [9]. The backbone of these compounds is given in Figure-13. Examples of this class are Coumestrol, Wedelolactone, Pilcadin etc.

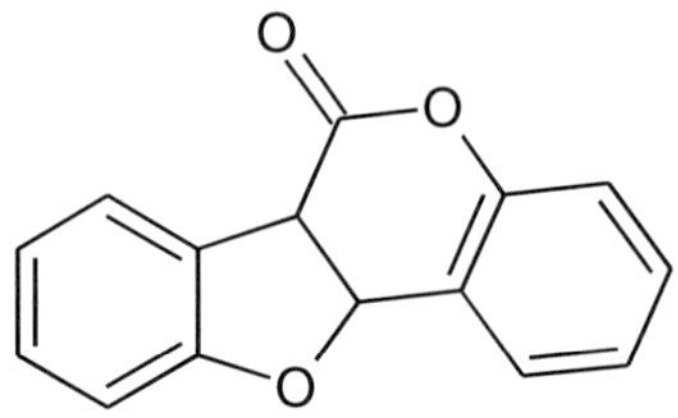

Figure 13. Coumestan.

6. Pterocarpans in the class of isoflavonoids have general backbone structure given in Figure-14. Some examples of this class are Bitucarpin A and B, Erybraedin A and B, Trifolirhizen etc. This class is synthesized by coupling of B ring at 4-one position of isoflavones.

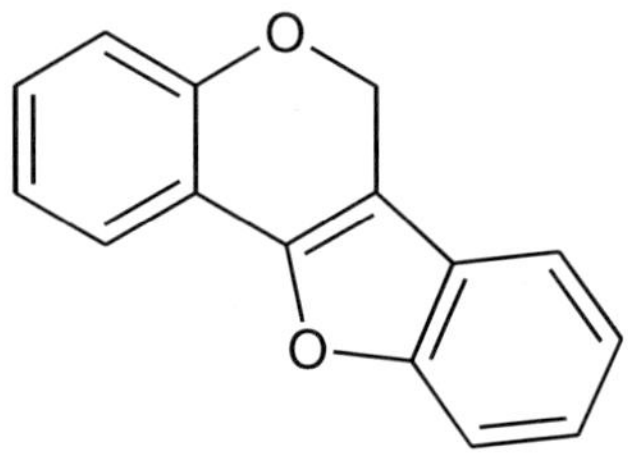

Figure 14. Petrocarpans.

1.3. Classification of Neoflavonoids

Neoflavonoids are derivatives of polyphenolic compounds [10]. They have a backbone of neoflavan as given in Figure-15.

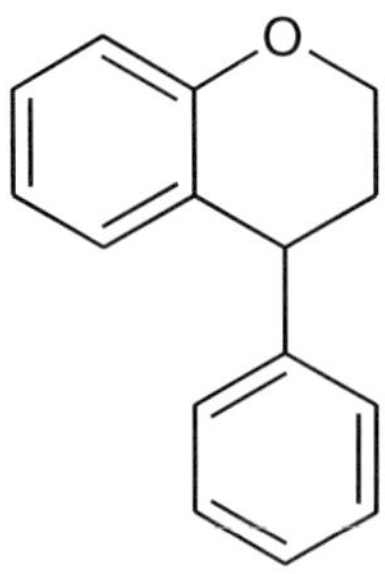

Figure 15. Neoflavonoid.

These neoflavonoids have been classified into two classes:

1. Neoflavones [11] are derivatives of 4-phenyl couarin. These have backbone as given in Figure-16. The first isolated neoflavone was Calcophylloide in 1951 [12].
2. Neoflavenes contain 4- phenyl chromen backbone as given in Figure-16. An example of this class is Dalbergichromene [13]. Some other examples are Coutareagenin, Dalbergin, Nivetin etc.

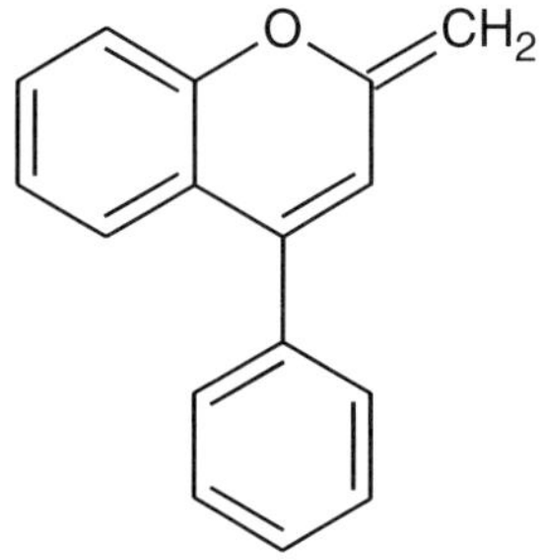

Figure 16. Skeleton of Neoflavone and Neoflavene.

1.4. Properties of Flavonoids

Numerous flavonoids have been found to act as anti diarrheal agents [14]. Coca beans containing polyphenols inhibit intestinal Cl^-secretion thus reducing diarrhoea. The structural related flavonoids like quercetin and luteolin were found blocking completely the Cl^- currents. The process was found to follow michealis menten kinetics. Flavonoid also act as anti-cancerous agents [15, 16]. Flavonoids have been used as antiviral and as antifungal agents [17, 18]. Antibacterial effect of flavonoids was reported by Cushnie et al., [19] and Manner et al. [20]. Bacterial biofilm in the work was defined as a grown community surrounded by a self-produced thick layer of extracellular polymeric substances known as matrix or slime thus keeping the bacteria protected from out skirt disturbances like harsh environmental conditions or chemicals. An extensive study was made on 500 natural and synthetic flavonoids that can counter act on S. Aureus biofilms (Staphylococcus). A detailed mechanism was proposed for treatment of this bacterium before and after formation of biofilm [21]. Plant extract rich in flavonoids have been recovered from species of Hypericum [22], Capsella and Chrolaena [23] and reported to possess antibacterial activity. Tsuchiya et al. [24] in their studies indicated that 2'4' or 2'6'-dihydroxylation of B ring and 5,7-dihydroxylation of A ring in flavanone structure was important for anti-methicillin resistant staphylococcus aureus activity.

A number of flavonoids act as antioxidants [16] for human body. Anti-allergic [25] and anti-inflammatory activities [16, 21] were proposed by Yamamoto et al. Green tea contains three flavonoids which are Catechins, catechin gallates and proanthocyanidins. Catechin is found up to 30% weigh of dry leaves. It was observed that antioxidative and free radical scavenging properties of polyphenolic components [26, 27] of tea can help in cure of cardiovascular disease [28], strokes [29], obesity [30] and cancer [31]. Inhibition of topoisomerase enzymes by flavonoids was reported in year 2009 [32] and 2008 [33]. Flavonoids are also found active to induce DNA mutations in the mixed linkage leukemia gene [34].

Flavonoids found to have direct antibacterial activities or synergistic activities with antibiotics or suppressing effect on bacteria. Green tea Catechins have been reported to offer antimicrobial activities [35]. A number of in vivo studies (in test tubes or peter dish outside the body of living organism) has been carried out [36, 37, 38]. It has been observed that due to acute and chronic stress reactive oxygen species were developed which may destroy the mucosa of intestine. Flavonoid rich grape seed extract has been found to act *in vivo* as an antioxidant in rats protecting their gastrointestinal mucosa [39]. Flavonoids due to their tendency to inhibit reactive oxygen and nitrogen compounds, can affect the anti-inflammatory mechanism. Also they have been found to inhibit the pro-inflammatory activity of enzymes which produce free radicals that are responsible for oxidation [40, 41]. Inflammation is caused by a number of diseases.

Dietary flavonoids have been found to be effective as preventers of cancer [42], celiac disease [43], diabetes [44] and cardiovascular problems [45] wherein procynidins have been found to act on as anti-inflammatory in one or other way. They exhibited a role in secretion of anti-inflammatory mediators, activation of inflammatory enzymes together with inhibition of gene transcription [46].

Flavonoids have been found to act as anti-cancerous agents when taken in diet. Cancer prevention, a review was proposed by Romagnolo et al., [47]. A number of studies suggest the role of flavonoids in cardiovascular disease. Non oxidant role of dietary flavonoids was suggested by Higdon et al., [48].

Action of flavonoids have been reported and extensively studied in type 2 diabetes cardiovascular disease [49], inflammation [50] and atherosclerosis [51]. Flavonoids have also been found to act in inhibiting coagulation, thrombus formation or platelet aggregation, reducing oxidative stress, arterial blood pressure and hyper tension, modifying vascular inflammatory mechanism, blood lipid levels and improving endothelial and capillary functions.

In addition, flavonoids have also been studied as one of the major constituent in plants to produce colours like yellow, red, and/or blue. Besides all these properties, flavonoids act as chemical messengers, cell cycle inhibitors, physiological inhibitors, in ultra violet filtration, floral pigmentation, and nitrogen fixation in plants. Flavonoids also show inhibitory action on diseases causing organisms and may also played a role in treatment of allergy, asthma, cataract, gout, atopic dermatisis, varicose veins, stomach ulcer, migraine, haemorrhoids, and periodontal disease [52].

1.5. Tests for Detection of Flavonoids

1.5.1. Reaction of ring A of flavonoids carried out with p-dimethyl amino cinnamaldehyde and assessed colorimetrically. This was then compared with vanillin procedure [53].

1.5.2. 5mg of compound was dissolved in water, warmed and filtered. To this 10% of aqueous solution of NaOH was added and yellow color was obtained which was found to disappear on addition of dilute HCl [54]. This confirms the presence of flavonoids.

1.5.3. Flavonoids are extracted in ethanol. To this the addition of few fillings of magnesium was followed by addition of concentrated HCl. Appearance of pink or red colour confirms the presence of flavonoids [55].

1.6. Extraction of Some Flavonoids

When the evidences came in of utilities of flavonoids, then began the research in field of extraction, biological applications and effect of these flavonoids. Till now a number of research papers have indicated the extraction methods of flavonoids and their presence in a huge variety in nature. Epimedium, also known as barrenwort or bishops hat, is a flowering plant of berberidaceae family and was said to contain flavonoids [56]. Isolation and extraction of flavonoids from this plant by super critical fluid extraction method (SFE) was carried out by Liu et al., [57]. It was reported earlier that beside other constituents, flavonoids are the major constituents of Epimedium [58, 59]. Leaves of 15 Epimidium species were used as medicines in Chinese tradition. The SFE method of extraction is found superior over other techniques because this method uses super critical carbon di oxide which has low critical point (a point of phase diagram where liquid and gas phase of a substance have same density and so are not distinguished), non-toxicity, low vapour loss, easily removable, inert nature etc. properties. Due to these properties less amount of supportive solvent was used and so concentration was not required. Optimum conditions for extraction of flavonoids like temperature, pressure, extraction time, amount of ethanol as modifier etc. were established [56]. The extracted material was estimated quantitatively for flavonoids by UV-visible spectrophotometer at 270nm. The total extraction was optimized at conditions 70mL of 70% ethanol as modifier, 30MPa pressure, 60°C temperature and 30 minutes of static extraction.

Brevicornin a new flavonol was extracted from plant Epimedium Brevicornum [60]. Its structure was determined as 5,7-trihydroxy-4'-methoxy-8-(3"methoxyisopentyl)flavones. Epimedin B, Epimedin C, Beta-anhydroicaritin and Tricin were also extracted from same plant [61].Extraction and isolation of flavonoids present in the methanolic extract of leaves of Acanthospermium Hispidium DC was carried out by Temidaya [62]. It was observed that leaves of this plant contain flavonoids like flavanones, isoflavones, flavones and flavonols. Thus it was suggested that it can be used for treatment of tuberculosis and also can act as anti-poisoning agent. Flavonoids were extracted by n-hexane, chloroform, ethyl acetate and n-butanol. Test with NaOH suggested the presence of flvanones in chloroform, presence of isoflavones in ethyl acetate, presence of flavones in n-butanol and presence of flavonols and flavones in water. UV-visible and IR investigations with chromatographic analysis of obtained products were carried out.

Flavones from Marchantia Convolute were extracted by Xiao et al., [63]. Flavonoids in plants were characterised by their 1H and 13C NMR studies [64]. Rhizomes of Cyclosorusacuminatus were found to contain acetylated flavanone glycosides [65]. A common method used for estimation as well as extraction is chromatography and specially thin layer chromatography [66]. Separation of two flavonoids and more was also carried out by column chromatography and their structures were confirmed by UV, IR, Mass and NMR spectroscopic analysis [67]. The fruit of Lagenaria Siceraria (bottle guard) was used for this purpose for the first time. It was found with medicinal uses and worked as cardio protective, cardio tonic and aphrodisiac agent. It was also found useful in allergies, inflammatory disorders, bronchial asthma, rhinitis, bronchitis etc.

A number of methods including conventional and modern, were used to extract flavonoids from Spearmint leaves (Mentha Spicata L.) [68]. Catechin, Epicatechin, Rutin, Myricetin, Luteolin, Apigenin and Naringenin flavonoids were extracted by conventional soxhlet extraction (CSE) method and super critical carbon di oxide (SC-CO_2) extraction method. Many factors like time, temperature, pressure etc. were varied to get optimum conditions for maximum extraction through both processes. It was observed that SC-CO_2 extraction gave yield of 60.57 mg/g at 60°C, 200 bar and in duration of 60 minutes where as 70% ethanol soxhlet extraction gave a superior yield of 257.67mg/g. Further the extract was characterized by HPLC for quantification of major flavonoid compound.

The leaf extract from any plant is found useful as stabilizer for edible oils as well as posses' antioxidant properties. The super critical fluid extraction with CO_2 is a method where gases above their critical point extract selectively soluble components from raw material [69]. Few draw backs of the super critical fluid extraction method by CO_2 limits its uses. One of them is its non-polar nature which does not allow the extraction of polar component at certain temperature and pressure. Rutin, Quercitrin, Hesperidin, Myricetin, Morin, Luteolin, Quercetin, Apigenin, Kaempferol, Baicalein, Caffeic acid and Catechol were characterize by HPLC after hydrolysis in Japanese seaweeds [70]. Super critical fluid extraction of flavonoids from Hops (Humulus Lupulus L.) was investigated [71]. The maximum extraction conditions were 50°C, 25MPa, solvent material ratio 1:1, ethanol 80% and extracted amount was 7.8mg/g. Obtained flavonoid Xanthohumol was characterized by HPLC-MS analysis.

Analysis of flavonoids in Ixeris Laevigata Var. Oldhami by HPLC was carried out [72]. Lin et al., [73] carried out super critical fluid extraction of flavonoids from Scutellariae Radix. The extraction was compared with parallel conventional extraction methods. Analysis of products was carried out by TLC and GC-MS methods. Amount of Baicalin was 137.6mg/g, Baicalein was 8.6mg/g and of Wogonin was 2.2mg/g by SFE method. Optimum conditions for maximum extraction were super critical CO_2-MeOH-H_2O (20:2, 1:0.9), 50°C and 200 bar.

Super critical CO_2 extraction of flavonoids from Citrus Grandis was carried out [74]. It is also known as Pomelo Peel. The highest extraction conditions were temperature 80°C, pressure 39MPa, and time 49 minutes in presence of 85% ethanol as modifier. The yield was 2.37%. Naringin flavonoid was extracted from peel of Citrus Paradisi by super critical fluid extraction method [75]. The highest yield was 14.4g/Kg. The yield of super critical extraction method was found higher than the extraction carried out by reflux and soxhlet methods. Extraction of flavonoids from Opuntia Milpa alta was carried out by Cai et al. [76]. It was observed that maximum yield of flavonoids (5.55mg/g) was obtained at 90°C with 80% ethanol(v/v). The main components were characterized by Liquid chromatography-mass spectrometry(LC-MS) and Ultra violet visible spectrometry and were found as Isorhamnetin 3-O-(2,6-dirhamnosyl)glucoside and Isorhamnetin 3-O-D-rutinoside. The extraction was carried out by soxhlet method. The flavonoids were separated on to a column of AB-8 resin by reverse phase chromatography. Effect of various factors like volume fraction of ethanol, temperature, extraction time, effect of solvents etc. were studied on the yield of flavonoids.

Arcoleo et al., [77] first reported in1961, the presence of Isorhamnetin flavonoid in flowers of O. Ficus Indica. Gas chromatography-mass spectrometric determination of new flavonoid triglycosides from leaves of Cerbera Manghas L. was carried out in 1980's [78]. Two new flavonol triglycosides named Manghaslin and Clitorin were isolated from the leaves. The structure of Manghaslin was elucidated as Quercetin-3-O-L-rhamnosyl-(1-2)-O-[L-rhamnosyl-(1-6)]D-glucoside and of Clitorin was elucidated as Kaempferol-3-O-L-rhamnosyl-(1-2)-O-[L-rhamnosyl-(1-6)]D-glucoside. The elucidation was carried out by GC-MS studies.

Isolation of Nicotiflorin and Rutin from leaves of Cerbera Manghas was also reported [79]. Presence of flavonol skeleton was confirmed by UV spectrum and presence of –OH group (3350cm^{-1}), carbonyl group (1660cm^{-1}) and aromatic ring(1600cm^{-1}) were confirmed by IR spectrum.

Flavonol and dihydroflavonol glycosides of Echinocereus triglochidiatus Var. Gurneyi was discussed by Miller et al. [80] whereas Markham explained the techniques of flavonoid identification [81]. Flavonoides from methanol extract of Paepalanthus Chiquilensis Herzog were characterised by HPLC-ESI-IT-MS method [82]. The method employed was high performance liquid chromatography coupled with electrospray negative ionisation multistage ion trap mass spectrometry. It was used to identify 24 flavonoids. Total flavonoids in Bulgarina fruits and vegetables was studied by Marenova et al., [83]. Flavonoids in the fruits of Opuntia Ficus-Indica Var. Saboten were investigated by Jeong et al., [84]. Flavonol glycosides were analysed by LC-UV and thermospray LC-MS method in thirteen species of Epilobium belonging to Onagraceae family [85].

1.7. Properties of Some Flavonoids

Phenolic compounds like flavonoids are found to possess anti-oxidant activities due to their redox properties. Generally these phenolic compounds are expected to neutralize lipid free radicals and prevent decomposition of hydroperoxide into free radicals [86, 87]. A number of flavonoids were extracted naturally or were being synthesized because of their tremendous medicinal uses. Properties and medical uses of flavonolignans were studied in 1996 [88]. Gangwal et al., [89] investigated antioxidant free radical scavenging potential of Citruullus Colocynthis (L.) Schrad. Chemistry, metabolism and structure activity relationships of flavonoid antioxidants in methanolic fruit extract was studied by Heim et al. [90]. Measurement of oxidative process was carried out in vitro and in vivo both conditions. It was suggested that multiple hydroxyl groups confer upon the molecule substantial antioxidant, chelating and pro oxidant activity. Methoxy group due to steric effect increases lipophilicity where as a double bond and a ketonic group generates a more stable flavonoid radical through conjugation and electron delocalization which increases activity of flavonoids.

Antioxidant properties of flavonoids extracted from Pomelo peel were investigated [74]. It was observed during investigations that flavonoids extracted from Pomelo peel were found good scavengers of stable DPPH radicals (hydroxyl,1,1-diphenyl-2-picrylhydrazyl) and ABTS(2,2'-azino-bis(3-ethylbenzthiazoline-6-sulphonic acid). The flavonoids extracted by SC-CO_2(super critical carbon dioxide) method were found more active as compared to conventional solvent extracted (CSE) flavonoids.

Antioxidant and vasorelaxant activities of flavonoids from Amygdalus Lycioides Var. Lorrida was studied by Babaei et al., [91]. The aerial parts of the roots of plant were used for treatment of diabetes. Six flavonoids Quercetin 3-O-Rhamnoside, Luteolin 7-O-Rhamnoside, Isorhamnetin 3-O-Rutinoside, Kaempferol 3-O-Rhamnoside, Apigenin and Naringenin were extracted and characterized by UV, MS and NMR spectroscopy.

Relation between structure of 42 flavonoids and their antioxidant and antiradical activity was studied [92]. It was found that flavonols with free –OH group at C-3 position of flavonoid skeleton showed high inhibitory activity to beta-carotene oxidation. Also the activity of C-4 –OH group was modified by presence of C-3 free –OH groups.

Antioxidant and pro oxidant behaviour of flavonoids was studied by Cao et al., [93]. It was observed that flavonoids including flavones, isoflavones and flavonones act as antioxidants against peroxyl and hydroxyl radicals and act as pro oxidants in presence of Cu^{2+} ions. It was also observed that both pro oxidant and antioxidant properties depend upon hydroxyl substituent on the flavonoid structure.

A comparative study of chemosensetizing effects of flavonoids was investigated by Jeong et al., [94]. It was observed that flavonoids having 3,5,7,3',4' –OH groups were found to show higher antioxidant properties as compared to flavonoids having 5,7,3',4' –OH groups. Flavonoids with $-OCH_3$ groups showed reduced antioxidant properties. Inhibitory activities of flavonoids from Prunus Davidiana was carried out [95]. It was observed that methanol extraction of the bark of plant exerts inhibitory activities on 1,1-diphenyl-2-picryl hydrazyl radicals(DPPH), total reactive oxygen species and peroxynitrites (ONOO-). The antioxidant properties of seven flavonoids from this plant were investigated. A review was also written on the antioxidant properties of plant flavonoids [96]. Scavenging of reactive oxygen species by a novel glucurinated flavonoid was studied by Bergman et al., [97]. It was observed that the natural antioxidants extracted from spinach may affect the scavenging of reactive oxygen species. This new glucurinated flavonoid was obtained from a well-known water soluble antioxidant NAO.

A review summarized the bioactivity of flavonoids [98]. The effect of flavonoids on eicosanoid and nitric oxide generating enzymes and the effect on expression of proinflammatory gene was discussed. A review on flavonoids as anti-inflammatory agents was discussed by Garcia-Lafuente et al. [99]. The anti-inflammatory plant flavonoids and cellular action mechanism was studied by Kim et al., [100] while Tunon et al., [101] studied potential of flavonoids as anti-inflammatory agents. A review on mechanism of action of flavonoids as anti-inflammatory [102] showed a number of in vivo and in vitro activities like antimicrobial, antiviral, antiulcerogenic, cytotoxic, anti-neoplastic, mutagenic, antioxidant, anti-hepatotoxic etc. Flavonoids are found to inhibit a number of enzymes like aldose reductase, Ca(+2)-ATPase etc. and also have regulatory role on different hormones like androgens etc. The present review focussed on mechanism of action of potential anti-inflammatory flavonoids.

Biological properties of flavonoids pertaining to inflammation was investigated by Manthey [103]. In the review a class of flavonoids were suggested to be effective in treatment of chronic venous insufficiency (CVI). Micro circulation of Daflon 500mg in chronic venous insufficiency was examined by Struckmann et al., [104]. Deflon 500mg contains 450mg Diosmin and 50mg flavonoid. The therapeutic efficacy was discussed in the work. Chemistry and biological activity of flavonoids was discussed by Kumar et al., [105]. Flavonoids in food and their health benifits was discussed by Yao et al. [106] while health benifits and molecular mechanism of flavonoids was studied byXiao et al., [107]. The antitumor activities of flavonoids was studied [108] and a variety of effects like cancer cell growth inhibition and kinase activity inhibition, apoptosis induction etc. were found in the plant flavonoids.

Flavonoids as promising anticancer agents was given by Ren et al., [109] while Flavonoid Luteolin and its properties were discussed by Lopez-Lazaro [110]. Dietary sources of Luteolin were carrots, peppers, celery, olive oil, peppermint, thyme etc. Luteolin was found to inhibit angiogenesis, apoptosis, prevents carcinogenesis, reduces tumor growth in vivo etc. and the mechanisms involved in the biological activities of Luteolin were STAT3, IGF1R, HER2 etc.

Luteolin as a cancer protecting flavonoid was also projected by Lin et al., [111]. Whereas properties of citrus flavonoids were discussed by Benavente-Garcia et al., [112]. Hyocholesterolemic effects of a flavonoid rich extract of Hypericum Perforatum L. in rats was studied bt Zou et al., [113]. The study was carried out in vivo and in vitro and it was observed that oral administration of extract (75mg/Kg of body weight and 150mg/Kg of body weight) significantly lowered the levels of cholesterol, triglyceride and lipoprotein cholesterol level in serum. Also it increased the activity of super oxide dismutase in serum and liver. Thus it proved the antioxidant enzyme activity. Antioxidant activities of flavonoids extracted from red fruit (Pandanus Conoideus Lam) was studied by Rohman et al., [114]. The work was related to antioxidant activities using in vitro method of 2,2-diphenyl-1-picrylhydrazyl (DPPH) radical scavenging, reducing power and metal chelating activities and to determine the phenolic and flavonoid contents of ethyl acetate extract which was found useful in reducing Fe^{3+} to Fe^{2+} ions. Extraction of flavonoids was carried out in methanol, chloroform and ethyl acetate.

Extraction from roots and aerial parts of Polycarpaea Corymbosa Lam. was found to show anti-oxidantproperties [115]. Antioxidant activity, phenol and flavonoid contents of Acacia Sinuata (Lourr) Merr. was carried out by Anand et al., [116]. Extraction was carried out in methanol and water and methanol extract was found to show significant activity than aqueous extract. Flavonoid contents were 143.19-423.12mg/100g of crude extract. Phytochemical analysis and evaluation of antioxidant property of leaf extract of Ficus was investigated [117]. Free radical scavenging and total antioxidant capacity of Ethulia Conyzoides was studied by Aliyu et al., [118]. Antioxidant property was investigated over 1,1-diphenyl-2-picrylhydrazyl radical(DPPH) in methanol extract. The phytochemical screening revealed the presence of tannins and flavonoids whose synergistic effect was found responsible for the high antioxidant activity. It was stated that methanol extract of E. Conyzoids possess interesting antioxidant activity which may provide protection against free radical induced damage to biomolecules. Improved antioxidant activity of Apigenin was studied by Liu et al., [119]. For this 5-prenyloxy apigenin was synthesized, characterized by HPLC-UV-MS/ES and was found to show higher antioxidant properties than Apigenin. Flavonoids Quercetin, Myrcetin, Kaemferol and Catechin as antioxidants were discussed by Pekkarinen et al., [120].

Although lots of researches regarding extraction and properties of flavonoids is being carried out, there is yet scope in the field as nature changes continuously and generates new aspects for research. Till now lot many researches are going around opening new avenues for scientists.

REFERENCES

[1] Mazza, G.; Miniati, E. Anthocyanins in Fruits, Vegetables and Grains, *CRC Press, Boca Raton, FL,* 1993.

[2] Pietta, P. G.; Simonetti, P.; in: Packer, L.; Hiramatsu, M.; Yoshikawa, T. (Eds), Antioxidant Food Supplements in Human Health, *Academic Press, San Diego,* 1999, 283 – 308.

[3] Harborne, J. B.; Flavonoids: Advances in Research since 1986, *Chapman and Hall, London*, 1994, 589 – 618.
[4] Shirley, B. W., *Trends Plant. Sci.,* 1996, 31, 377 – 382.
[5] Harborne, J.B.; In: Cody, V.; Middleton, V. Plants flavonoids in biology and medicine, *Alan R. Liss., New York,* 1986, 15-24.
[6] Mc Naught, A.D.; Wilkinson, A. Flavonoids (isoflavonoids and neoflavonoids), *IUPAC Compendium of Chemical Terminology (2 ed.), Oxford: Blackwell Scientific*, 1997.
[7] Online corrected by Nic, M.; Jirat, J., Kosata, B., Flavonoids (isoflavonoids and neoflavonoids), *IUPAC Compendium of Chemical Terminology (2 ed.), The Gold Book*, 2012.
[8] Rufer, C.E.; Glatt, H.; Kulling, S.E. Structural elucidation of hydroxylated metabolities of the isoflavan equol by GC/MS and HPLC/MS, *Drug Metabolism and Disposition*, 2005.
[9] Tuskaev, V.A. Synthesis and biological activity of Coumestan derivatives (review), *Pharmaceutical Chemistry Journal*, 2013, 47(1), 1-11.
[10] Ollis, W.D. The neoflavonoids, a new class of natural products, *Cellular and Molecular Life Sciences,* 1966, *22*(12), 777-783.
[11] Linuma, M.; Tanaka, T.; Hamada, K.; Mizuno, M.; Asai, F.; Reher, G.; Kraus, L. Revised structure of neoflavone in Coutarea hexandra, *Phytochemistry,* 1987,*26(11),* 3096-3097.
[12] Neoflavones-Natural distribution and spectral and biological properties, *Chemistry of Natural Products,* 2003, 39(1), 54-121.
[13] Mukhrjee, S.K.; Saroia, T.; Seshadri, T.R. Dalbergichromene-A new neoflavonoid from stem bark and heartwood of Dalbergia Sissoo, *Tetrahedron,* 1971,27(4), 799-803.
[14] Schuier, M.; Sies, H.; Illek, B.; Fisher, H.; Coca-related flavonoids inhibit CFTR-mediated chloride transport across T84 human colon epithelia, *J.Nutr.,* 2005, 135(10), 2320-2325.
[15] De Sousa, R.R.; Queiroz, K.C.; Souza, A.C.; Gurgueira, S.A.; Augusto, A.C.; Miranda, M.A.; Peppelenbosch, M.P.; Ferreira, C.V.; Aoyama,H. Phosphoprotein levels, MAPK activities and NF kappaB expressions are effected by Fisetin, *J. Enzyme Inhib. Med. Chem.,* 2007, 22(4), 439-444.
[16] Cazarolli, L.H.; Alberton, E.H.; Figueiredo, M.S.; Folador, P.; Damazio, R.G.; Pizzolatti, M.G.; Silva, F.R. Flavonoids: Prospective drug candidates, *Mini Reviews in Medicinal Chemistry,* 2008, 8(13), 1429-1440.
[17] Cushnie, T.P.T.; Lamb, A.J.; Antimicrobial activity of flavonoids, *International Journal of Antimicrobial Agents,* 2005, 26(5), 343-356.
[18] Friedman, M. Overview of antibacterial, antitoxin, antiviral and antifungal activities of tea flavonoids and tea, *Molecular Nutrition and Food Research,* 2007, 51(1), 116-134.
[19] Cushnie, T.P.T.; Lamb, A.J. Recent advances in understanding the antibacterial properties of flavonoids, *International Journal of Antimicrobial Agents,* 2011, 38(2), *99-107.*
[20] Manner, S.; Skogman, M.; Goeres, D.; Vuorela, P.; Faliarero,A. Systematic exploration of natural and synthetic flavonoids for the inhibition of Staphylococcus Aureus biofilms, *International Journal of Molecular Sciences,* 2013, 14(10), 19434-19451.

[21] Skogman, M.E.; Vuorela, P.M.; Fallarero, A. Combining biofilm matrix measurements with biomass and viability assays in susceptibility assessments of antimicrobials against St. Aureus biofilms, *Journal of Antibiotics,* 2012, 65, 453-459.

[22] Dall'Agnol, R.; Ferrez, A.; Bemardi, A.P. Antimicrobial activity of some Hypericum species, *Phytomedicine,* 2003, 10, 511-516.

[23] El Abyad, M.S.; Morsi, N.M.; Zaki, D.A.; Shaaban, M.T. Priliminary screening of some Egyptian weeds for antimicrobial activity, *Microbios.,* 1990, 62, 47-57.

[24] Tsuchiya, H.; Sato, M.; Miyazaki, T. Comparative study on the antibacterial activity of phytochemical flavonones against Methicillin resistant Staphylococcus Aureus, *J. Ethnopharmacol.,* 1996, 50, 27-34.

[25] Yamamoto, Y.; Gaynor, G.; Richard, B. Therapeutic potential of inhibition of the NF-κB pathway in the treatment of inflammation and cancer, *Journal of Clinical Investigation,* 2001, 107(2), 135-42.

[26] Halliwell, B. Inhibition of mammalian 5-lipoxygenase and cyclo oxygenase by flavonooids and phenolic dietary additives relationship to antioxidant activity and to iron ion reducing ability, *Biochemical Pharmacology,* 1991, 42, 1673-1681.

[27] Scott, B.C.; Butler, J.; Halliwell, B.; Aruoma, O.I. Evalution of the antioxidant actions of ferulic acid and Catechins, *Free Radical Research Communications,* 1993, 19, 241-253.

[28] Hertog, M.; Feskens, E.; Hollman, P.; Katan, M.; Kromhout, D. Dietary antioxidant flavonoids and risk of coronary heart diseases, *The Zutphen Elderly Study,* 1993, 342, 1007-1011.

[29] Keli, S.; Hertog, M.; Feskens, E.; Kromhout, D. Flavonoids, antioxidant vitamins and risk of stroke, *Archives of Internal Medicines,* 1995, 154, 637-642.

[30] Bell, S.J.; Goodrick, G.K. A functional food product for the management of weight, *Critical review in food science and nutrition,* 2002, 42, 163-178.

[31] Mc Kay, D.L.; Blumberg, J.B. The role of tea in human health, an update, *Journal of The American College of Nutrition,* 2002, 21, 1-13.

[32] Esselen, M.; Fritz, J.; Hutter, M.; Marko, D. Delphinidin modulates the DNA-Damaging properties of TopoisomeraseII poisons, *Chemical Research in Toxicology,* 2009, 22(3*),* 554-564.

[33] Bandele, O.J.; Clawson, S.J.; Osheroff, N. Dietary polyphenols as topoisomerase II poisons. B-ring substituents determine the mechanism of enzyme-mediated DNA cleavage enhancement, *Chemical Research in Toxicology,* 2008, 21(6), 1253-1260.

[34] Barjesteh Van Waalwijk Van Doorn-Khosrovani, S.; Janssen, J.; Maas, L.M.; Godschalk, R.W.; Nijhuis, J.G.; Van Schooten, F.J. Dietary flavonoids induced MLL translocations in primary human CD34+cells, *Carcinogenesis,* 2007, 28(8), 1703-1709.

[35] Taylor, P.W.; Hamilton-Miller, J.M.T.; Stapleton, P.D. Antimicrobial properties of green tea Catechins, *Food Science and Technology bulletin,* 2005, 2(7), 71-81.

[36] Gonzalez-Segovia, R.; Quintanar, J.L.; Salinas, E.; Ceballos-Salazar, Aviles-Jimenez, F.; Torres-Lopez, J. Effect of the flavonoid Quercetin on inflammation and lipid peroxidation induced by Helicobacter Pylori in gastric mucosa of Guinea pig, *Journal of Gastroenterology,* 2008, 43(6), 441-447.

[37] Oh, D.R.; Kim, J.R.; Kim, Y.R. Genistein inhibits vibrio vulnificus adhesion and cytotoxicity to He La cells, *Archives of Pharmaceutical research,* 2010, 33(5), 787-792.

[38] Choi, O.; Yahiro, K.; Morinaga, N.; Miyazaki, M.; Noda, M. Inhibitory effects of various plant polyphenoles on the toxicity of Staphylococcal alpha-toxin, *Microbial Pathogenesis,* 2007, 432(5-6), 215-224.

[39] Bagchi, M.; Milnes, M.; Williams, C.; Ye, X.; Stohs, S.; Bagchi, D. Acute and chronic stress-induced oxidative gastrointestinal injury in rats and the protective ability of a novel grape seed Proanthocyanidin extract, *Nutrition Research,* 1999, 19(8), 1189-1199.

[40] Izzi, V.; Masuelli, L.; Tresoldi, I.; Sacchetti, P.; Modesti, A.; Galvano, F.; Bei, R. The effect of dietary flavonoids on the regulation of redox inflammatory networks, *Frontiers in Bioscience,* 2012, 17(7), 2396-2418.

[41] Gomes, A.; Couto, D.; Alves, A.; Dias, I.; Freites, M.; Proto,G.A.; Duarte, J.A.; Fernandes, E. Trihydroxyflavones with antioxidant and anti-inflammatory efficacy, *Bio Factors,* 2012, 38(5), 378-386.

[42] Ravishankar, D.; Rajora, A.K.; Greco, F.; Osborn, H.M.I. Flavonoids as prospective compounds for anti cancer therapy, *The International Journal of Biochemistry and Cell Biology,* 2013, 45(12), 2821-2831.

[43] Ferretti, G.; Bacchetti, T.; Masciangelo, S.; Saturni, L. Celiac disease,inflammation and oxidative damage-A nutrigenetic approach, *Nutrients,* 2012, 4(12), 243-257.

[44] Babu, P.V.A.; Liu, D.; Gilbert, E, R. Recent advances in understanding the anti diabetic actions of dietary flavonoids, *The Journal of Nutritional Biochemistry,* 2013, 24(11), 1777-1789.

[45] Manach, C.; Mazur, A.; Scalbert, A. Polyphenols and prevention of cardiovascular diseases, *Current Opinion in Lipidology,* 2005, 16(1), 77-84.

[46] Martinez-Micaelo, N.; Gonzalez-Abuin, N.; Ardevol, A.; Pinent, M.; Blay, M.T. Procinidins and inflammation, molecular targets and health implications, *Biofactors,* 2012, 38(4), 257-265.

[47] Romagnolo, D.F.; Selmin, O.I. Flavonoids and cancer prevention. A review of the evidence, *J. Nutr.,* 2012, 31(3), 206-238.

[48] Higdon, J.; Drake, V.; Frei, B. Non antioxidant roles for dietary flavonoids-Reviewing the relevance to cancer and cardiovascular diseases, *Nutraceuticals world, Rodman Media,* 2009.

[49] Van Dam, R.M.; Naidoo, N.; Landberg, R. Dietary flavonoids and the development of type 2 diabetes and cardiovascular diseases, *Current Opinion in Lipidology,* 2013, 24(1), 25-33.

[50] Tangney, C.C.; Rasmussen, H.E. Polyphenols inflammation and cardiovascular diseases, *Current Atherosclerosis Reports,* 2013, 15(5), 324.

[51] Siasos, G.; Tousoulis, D.; Tsigkou, V.; Kokkou, E.; Oikonomou, E.; Vavuranakis, M.; Basdra, E.K.; Papavassillou, A.G.; Stefanadis, C. Flavonoids in atherosclerosis:an overview of their mechanism of action, *Current Medicinal Chemistry,* 2013, 20(21), 2641-2660.

[52] Galeotti, F.; Barile, E.; Curir, P.; Dolci, M.; Lanzotti, V. Flavonoids from Carnation (Dianthus Caryophyllus) and their antifungal activity, *Phytochemistry Letters,* 2008, 1, 44.

[53] Delcour, Jan A.; Didier Janssens de Varebeke. A new colorimetric assay for flavonoids in Pilsner beers, *Journal of the Institute of Brewing,* 1985, 91(1), 37-40.

[54] Bello, I.A.; Ndukwe, G.I.; Audu, O.T.; Habila, J, D.A bioactive flavonoid from Pavetta Crassipes K. Schum,*Organic and Medicinal Chemistry Letters,* 2011, 1(1), 14.

[55] Yisa, J. Phytochemical analysis and antimicrobial activity of Scoparia Dulcis and Nymphaea Lotus, *Australian Journal of Basic and Applied Sciences,* 2009, 3(4), 3975-3979.

[56] Zhang, D.W.; Cheng, Y.; Wang, N.L.; Zhang, J.C.; Yang, M.S.; Yao, X.S. Effect of total flavonoids and flavonol glycosided from Epimedium Koreanum Nakai on the proliferation and differentiation of primary osteoblasts, *Phytomedicines,* 2008, 1591-20, 55-61.

[57] Liu, C.M., Zhao, X.L.; Liu, Z.Q.; Xing, J.P. Isolation and extraction of total flavonoids from Epimedium Koreanum Nakai by super critical fluid extraction, *Chemical Research in Chinese Universities,* 2004, 20, 708.

[58] Li, W.K.; Xiao, P.G.; Tu, G.Z.; Ma, L.B.; Zheng, R.Y. Flavonol glycosides from Epimedium Koreanum, *Phytochemistry,*1995, 38(1), 263-265.

[59] Li, W.K.; Xiao, P.G.; Liao, M.C. The structure of new flavonol glycoside from Epimedium Koreanum Nakai, *Chem. J. Chinese U.,* 1995, 16(10), 1575-1576.

[60] Guo, B.L.; Li, W.K.; Yu, J.G.; Xiao, P.G. Brevicornin, A flavonol from Epimedium Brevicornum, *Phytochemistry,* 1996, 2(43), 991-992.

[61] Li, W.K.; Zhang, R.Y.; Xiao, P.G. Flavonoids from Epimedium Wanshanense, *Phytochemistry,* 1996, 2(43), 527-530.

[62] Temidaya, A.R. Extraction and isolation of flavonoids present in the methanolic extract of leaves of Acanthospermum Hispidium DC, *Global Journal of Medicinal Plant Research,* 2013, 1(1), 111-123.

[63] Xiao, J, B.; Ren, F.; Xu, M. Flavones from Marchantia Convolute, *Journal of Pharmaceutical and Allied Sciences,* 2006, 3(1), 310-313.

[64] Moon, B.; Lee, Y.; Shin, C.; Lim, Y. Complete assignment of the ^{1}H and ^{13}C NMR data of flavones derivatives, *Bulletin of Korean Chemical Society,* 2005, 26(4).

[65] Fang, W.; Raun, J.; Wang, Z.; Zhao, Z.; Zou, J.; Zhao, D.; Cai,Y. Acetylated flavonone glycosides from rhizomes of Cyclosorusacuminatus, *Journal of Natural Products,* 2006, 69(11), 1641-1644.

[66] Gangwal, A. Extraction, estimation and thin layer chromatography of flavonoids:A review, *World Journal of Pharmacy and Pharmaceutical Sciences,* 2013, 2(3), 1099-1106.

[67] Gangwal, A.; Parmar, S.K.; Sheth, N.R. Triterpenoids, flavonoids and sterols from Lagenaria Siceraria fruits, *Der Pharmacia Lettre,* 2010, 2(1), 307-317.

[68] Bimakr, M.; Rahman, R.A.; Taip, F.S.; Ganjloo, A.; Salleh, L.M.; Selamat, J.; Hamid, A.; Zaidul, I.M.S. Comparison of different extraction methods for the extraction of major bioactive flavonoid compounds from Spearmint(Mentha Spicata L.) leaves, *Food and Bioproducts Processing,* 2011, 89, 67-72.

[69] Cavero, S.; Garcia-Risco, M.; Marin, F.; Jaime, L.; Santoyo, S.; Senorans, F.; Reglero, G.; Ibanez, E. Super critical fluid extraction of antioxidant compounds from Oregano chemical and functional characterization via LC-MS and in vitro assays, *Journal of Super Critical Fluids,* 2006, 38, 62-69.

[70] Yumiko, Y.S.; Ya-Pai, H.; Takeshi, S. Distribution of flavonoids and related compounds from seaweeds in Japan, *Journal of Tokyo University of Fisheries,* 2003, 89, 1-6.

[71] He, G.Q.; Xiong, P.H.; Chen, Q.H.; Ruan, H.; Wang, Z.Y.; Traore,L. Optimisation of conditions for super critical fluid extraction of flavonoids from Hops (Humulus Lupulus L.), *Journal of Zhejiang University Science,* 2005, 6B(10), 999-1004.

[72] Lu, K.L.; Ku, Y.R.; Wen, K.C.; Ho, L.K.; Cheng, Y.S. Analysis of flavonoids and Coumarins in Ixeris Laevigata Var. Oldhami by HPLC, *Journal of Liquid Chromatography and Related Technologies,* 2000, 16(23), 2573-2583.

[73] Lin, M.C.; Tsai, M.J.; Wen, K.C. Super critical fluid extraction of flavonoids from Scutellariae Radix, *Journal of Chromatography A,* 1999, 2(830), 387-395.

[74] He, J.Z.; Shao, P.; Liu, J.H.; Ru, Q.M. Super critical carbon dioxide extraction of flavonoids from Pomelo (Citrus Grandis L. Osbeck) peel and their anti-oxidant activity, *International Journal of Molecular Science,* 2012, 13(10), 13065-13078.

[75] Giamnuzzo, A.N.; Boggetti, H.J.; Nazareno, M.A.; Mishima, H.T. Super critical fluid extraction of Naringin from the peel of Citrus Paradise, *Phytochem. Anal.,* 2003, 14(40), 221-223.

[76] Cai, W.; Gu, X.; Tang, J. Extraction, purification and characterization of the flavonooids from Opuntia Milpa Alta skin, *Journal of Food Science,* 2010, 28(2), 108-116.

[77] Arcoleo, A.; Ruccia, M.; Cusmando, S. Flavonoid pigments from Opuntia I. Isorhamnetin from flowers of Opuntia Ficus Indica, *Annal Chimica (Rome),* 1961, 51, 81-82.

[78] Sakushima, A.; Hisada, S.; Ojihara, Y.; Nishibe, S. Studies on the constituents of Apocynaceae plants. Gas chromatography-Mass spectrometric determination of new flavonoid triglycosides from the leaves of Cerbera Manghas L., *Chemical and Pharmaceutical Bulletin,* 1980, 28(4), 1219-1223.

[79] Sakushima, A.; Nishibe, S.;Hisada, S.; Noro,Y., Hisada,Y. *Yakugaku Zasshi,* 1976, 96, 1046.

[80] Miller, J.M.; Bohm, B.A. Flavonol and dihydroflavonol glycosides of Echinocereus Triglochidiatus Var. Gurney, *Phytochemistry,* 1982, 21(4), 951-952.

[81] Markham, K.R. Techniques of flavonoid identification, *Academic Press, London,* 1982, 251-262.

[82] Zanutto, F,V.; Boldrin, P.K.; Varanda, S.F.; de Souza, S.F.; Sano, P.T.; Vilegas, W.; dos Santos, L.C. Characterization of flavonoids and naphthopyranones in methanol extracts of Paepalanthus Chiquitensis Herzog by HPLC-ESI-IT-MS and their mutagenic activity, *Molecules,* 2013, 18, 244-262.

[83] Marinova, D.; Ribarova, F.; Atanassova, M. Total phenolics and total flavonoids in Bulgarina fruits and vegetables, *Journal of the University of Chemical Technology and Metallurgy,* 2005, 40, 255-260.

[84] Jeong, S.J.; Jun, K.Y.; Kang, T.H. Flavonoids from the fruits of Opuntia Ficus Indica Var. Saboten, *Korea Journal of Pharmacognosy,* 1999, 30, 84-86.

[85] Ducrey, B.; Wolfender, J.L.; Marston, A.; Hostettmann, K. Analysis of flavonol glycosides of thirteen Epilobium species(Onagraceae) by LC-UV and thermospray LC-MS, *Phytochemistry,* 1995, 38(1), 129-137.

[86] Javanmardi, J.; Stushnoff, C.; Locke, E.; Vivanco, J.M. Antioxidant activity and total phenolic content of Iranian Ocimum accessions, *Food Chemistry,* 2003, 83, 547-550.

[87] Li, H.; Hao, Z.; Wang, X.; Huang, L.; Li, J. Antioxidant activities of extracts and fraction from Lysimachia Foenum-Graecum Hance, *Biosource Technology,* 2009, 100, 970-974.

[88] Properties and medical uses of flavonolignans [Silymarin] from Silybummarianum, *Phytotherapy Research (U.K.),* 1996, 10/Suppl.1, 525-526.

[89] Kumar, S. Antioxidant free radical scavenging potential of Citrullus Colocynthis (L.) Schrad. Methanolic fruit extract, *Acta Pharma,* 2008, 58, 215-220.

[90] Heim, K.E.; Taqliaferro, A.R.; Boliya, D.J. Flavonooid antioxidants: Chemistry, metabolism and structure activity relationship, *J. Nutr. Biochem.,* 2002, 13(10), 572-584.

[91] Babaei, H.; Sadeghpour, O.; Nahar, L.; Delazar, A.; Nazemiyeh, H.; Mansouri, M.R.; Poursaeid, N.; Asnaashari, S.; Moghadam, S.B.; Sedigheh, B.; Sarker, S.D. Antioxidant and vasorelaxent activities of flavonoids from Amygdddddalus Lycioides Var. Lorrida, *Turkish Journal of Biology,* 2008, 32(3), 203-208.

[92] Burda, S.; Oleszek, W. Antioxidant and anti-radical activities of flavonoids, *Journal of Agricultural and Food Chemistry,* 2001, 49(6), 2774-2779.

[93] Cao, G.; Sofic, E.; Prior, R.L. Antioxidant and prooxidant behaviour of flavonoids, structure-activity relationship, *Free Radic. Biol. Med.,* 1997, 22(5), 749-760.

[94] Jeong, J.M.; Choi, C.H.; Kang, S.K.; Lee, I.H.; Lee, J.Y.; Jung, H. Antioxidant and chemo sensitizing effects of flavonoids with hydroxyl and /or methoxy groups and structure activity relationship, *J. Pharm. Pharm. Sci.,* 2007, 10(4), 537-546.

[95] Jung, H.A.; Jung, M.J.; Kim, J.Y.; Khung, H.Y.; Choi, j.S. Inhibitory activity of flavonoids from Prunus Davidina and othe flavonoids on total ROS and hydroxyl radical generation, *Arc. Pharm. Res.,* 2003, 26(10), 809-815.

[96] Kandaswami, C.; Jr. Middleton,E. Free radical scavenging and antioxidant activity of plant flavonoids, *Adv. Exp. Med. Biol.,* 1994, 366, 351-376.

[97] Bergman, M.; Perelman, A.; Dubinsky, Z.; Grossman,S. Scavenging of reactive oxygen species by a novel glucurinated flavonoid antioxidant isolated and purified from Spinach, *Phytochemistry,* 2003, 62(5), 753-762.

[98] Robak, J.; Gryglewski, R.J. Bioactivity of flavonoids, *Pol. J. Pharmacol.,* 1996, 48(6), 555-564.

[99] Garcia-Lafuente, A.; Guillamon, E.; Villares, A.; Rostaqno, M.A.; Martinez, J.A. Flavonoids as anti-inflammatory agents-implications in cancer and cardiovascular disease, *Inflamm. Res.,* 2009, 58(9), 537-552.

[100] Kim, H.P.; Son, K.H.; Chang, H.W.; Kang, S.S. Anti-inflammatory plant flavonoids and cellular action mechanism, *J. Pharmacol. Sci.,* 2004, 96(3), 229-245.

[101] Ediavilla, M.V.; Sanchez-Campos, S.; Gonzalez-Gallego, J. Potential of flavonooids as anti-inflammatory agents. Modulation of pro inflammatory gene expression and signal transduction pathways, *Curr. Drug Metab.,* 2009, 10(3), 256-271.

[102] Rathee, P.; Chaudhary, H.; Rathee, S.; Rathee, D.; Kumar, V.; Kohli, K. Mechanism of action of flavonoids as anti-inflammatory agents: A review, *Inflamm. Allergy Drug Targets,* 2009, 8(3), 229-235.

[103] Manthey, J.A. Biological properties of flavonoids pertaining to inflammation, *Microcirculation,* 2000, 7, 29-34.

[104] Struckmann, J.R.; Niccolaides, A.N. Flavonoids: A review of the pharmacology and therapeutic efficacy of Daflon 500mg in patients with chronic venous insufficiency and related disorders, *Angiology,* 1994, 45(6), 419-428.

[105] Kumar, S.; Panddey, A.K. Chemistry and biological activities of flavonoids: An overview, *Scientific World Journal,* 2013, *29.*

[106] Yao, L.H.; Jiang, Y.M.; Shi, J.; Tomos-Barberan, F.A.; Datta,N.; Singanusong, R.; Chen, S.S. Flavonoids in food and their health benefits, *Plant Foods Hum. Nutr.,* 2004, 59(3), 113-122.

[107] Xiao, Z.P.; Peng, Z.Y.; Peng, M.J.; Yan, W.B.; Quyang, Y.Z.; Zhu, H.L. Flavonoids health benefits and their molecular mechanism, *Mini Rev. Med. Chem.,* 2011, 11(2), 169-177.

[108] Kandaswami, C.; Lee, L.T.; Lee, P.P.; Hwang, J.J.; Ke, F.C.; Huang, Y.T.; Lee, M.T. The anti tumor activities of flavonoids, *In Vivo,* 2005, 19(5), 895-909.

[109] Ren, W.; Qiao, Z.; Wang, H.; Zhu, L.; Zhang, L. Flavonoids: Promising anti-cancer agents, *Med. Res. Rev.,* 2003, 23(4), 519-534.

[110] Lopez-Lazaro, M. Distribution and biological activities of the flavonoid Luteolin, *Mini Rev. Med. Chem.,* 2009, 9(10), 31-59.

[111] Lin, Y.; Shi, R.; Wang, X.; Shen, H.M. Luteolin, a flavonoid with potential for cancer prevention and therapy, *Current Cancer Drug Targets,* 2008, 8(7), 634-646.

[112] Benavente-Garcia, O.; Castillo, J. Updates on uses and properties of Citrus flavonoids, new findings in anticancer, csrdiovascular and anti-inflammatory activity, *Journal of Agric. Food Chemistry,* 2008, 56(15), 6185-6205.

[113] Zou, Y.; Lu, Y.; Wei, D. Hypocholesterolemic effects of a flavonoid-rich extract of Hypericum Perforatum L. In rats fed a cholesterol-rich diet, *Journal of Agric. Food Chem.,* 2005, 53, 2462-2466.

[114] Rohman, A.; Riyanto, S.; Yuniarti, N.; Saputra, W.R.; Utami, R.; Mulatsih, W. Antioxidant activities, total phenolic and total flavonoid of extracts and fractions of red fruit (Pandanus Conoideus Lam), *International Food Research Journal,* 2010, 17, 97-106.

[115] Sindhu, S.; Manorama, S. Screening of Polycarpaea Corymbosa Lam.(Caryophylaceae) for its in vitro antioxidant activity,*Asian Journal of Pharmaceutical and Clinical Research,* 2012, 5(4), 175-178.

[116] Anand, S.P.; Mohan, G. Antioxidant activity, phenol and flavonoid contents of Acacia Sinuata(Lourr) Merr., *Bioscience Discovery,* 2014, 5(1), 24-27.

[117] Iraqui, P. Phytochemical analysis and evaluation of antioxidant property of leaf extract of Ficus Glomerata, *World Journal of Pharmacy and Pharmaceutical Science,* 2013, 2(4), 2044-2051.

[118] Aliyu, A.B.; Ibrahim, M.A.; Musa, A.M.; Lawal. A.Y.; Oshanimi, J.A.; Usman, M.; Abdulkadir, I.E.; Oyewale, A.O.; Amupitan, J.O. Free radical scavenging and total antioxidant capacity of methanol extract of Ethulia Conyzoides growing in Nigeria, *Romanian Biotechnological Letters,* 2012, 17(4), 7458-7465.

[119] Liu, B.; Yang, J.; Zhu, X.; Lv, G. Improvement of the antioxidant activity of Apigenin in linoleic acid system by O-prenylation, *Journal of Medicinal Plants Research,* 2012, 6(2), 181-186.

[120] Pekkarinen, S.S.; Heinonen, I.M.; Hopia, A.I. Flavonoid Quercetin, Myrcetin, Kaemferol and (+)-Catechin as antioxidants in methyl linoleate, *Journal of the Science of Food and Agriculture,* 1999, 79, 499-506.

In: Natural Heterocycles
Editors: K. L. Ameta and Saksit Chanthai
ISBN: 978-1-63463-424-3

Chapter 9

ANTHOCYANIN AND TOTAL PHENOLICS IN HOT CHILLI VARIETIES WITH THEIR POLYPHENOL OXIDASE AND PEROXIDASE ACTIVITIES

Prapha Arnnok and Saksit Chanthai*
Department of Chemistry, Faculty of Science, Khon Kaen University, Khon Kaen, Thailand

ABSTRACT

Anthocyanin was extracted from hot chilli pepper using 2%(v/v) HCl in methanol and then cleaned up using C_{18} solid-phase extraction prior to analysis by RP-HPLC-PDA. The separation method was optimized for determination of anthocyanin in twenty varieties of the chilli samples. Total phenolics (TPCs) in the samples were extracted using methanol:0.05% aqueous HCl (90/10, v/v) and estimated by Folin-Ciocalteu reagent. Both contents of anthocyanin and TPCs in these chilli fruits were ranged of 0.796-4.70 mg CGE kg^{-1} FW (62.9-70.3 mg CGE kg^{-1} DW) and 0.782-4.52 g GAE kg^{-1} FW (81.8-90.2 g GAE kg^{-1} DW), respectively. Their enzymatic activities of polyphenol oxidase (PPO) and peroxidase (POD) were measured spectrophotometrically using catechol and guaiacol as substrates, respectively. Under the optimum conditions, both PPO and POD activities in the samples were 0.292-0.929 and 0.229-15.8 $AUmin^{-1}g^{-1}$ FW, respectively. The relationship between the enzyme activities and the contents of anthocyanin or TPCs was evaluated. It is demonstrated that both enzyme activities and their contents of anthocyanin and TPCs in almost chilli fruits show similar trends in a potential related effect on the enzymatic browning.

1. INTRODUCTION

Phenolic compounds are a group of chemical substances in plants, which play an important role during enzymatic browning because they are substrates for the browning enzymes. The phenolics are normally complex organic substances, which contain more than

* Corresponding author: e-mail: sakcha2@kku.ac.th.

one phenolic group. Polyphenolics can be divided into many different subcategories, such as flavonoids and non-flavinoid components. Anthocyanins are a subgroup of plant constituents known as flavonoids [1]. They occur in all higher plants, mostly flowers and fruits but also in leaves, stems and roots. Anthocyanins are the plant pigments, which are responsible for red, blue and purple colors of many fruits, vegetables and flowers. These polyphenolic substances are glycosides of polyhydroxy- and polymethoxy derivatives of 2-phenylbenzopyrylium or flavylium ion [2] as shown in Figure 1(a).

There are seven different side groups on the flavylium ion. These side groups can be hydrogen atom, hydroxide or methoxy group. The color of anthocyanidin varies according to these substitutions. The main groups of anthocyanins, which have different combination of side groups and their name, are shown in Table 1 [2]. In nature, anthocyanidins are always found attached to one or more sugar and/or acylated sugars. Four substituted monosides are particularly common: glucose, galactose, rhamnose and arabinose. When anthocyanidins are coupled to sugars, anthocyanins are formed. Thus, an anthocyanin molecule normally consists of two parts: a light-absorbent (chromophore) termed an anthocyanidin, and a glycoside attachment, i.e. cyanidin-3-glucoside (Figure 1(b)). As sugars can be coupled at different places and many different sugars are present in plants, it is clear that a very large range of anthocyanins can be formed [3]. Nevertheless, the most common anthocyanin structure in plants is cyanidin-3-glucoside.

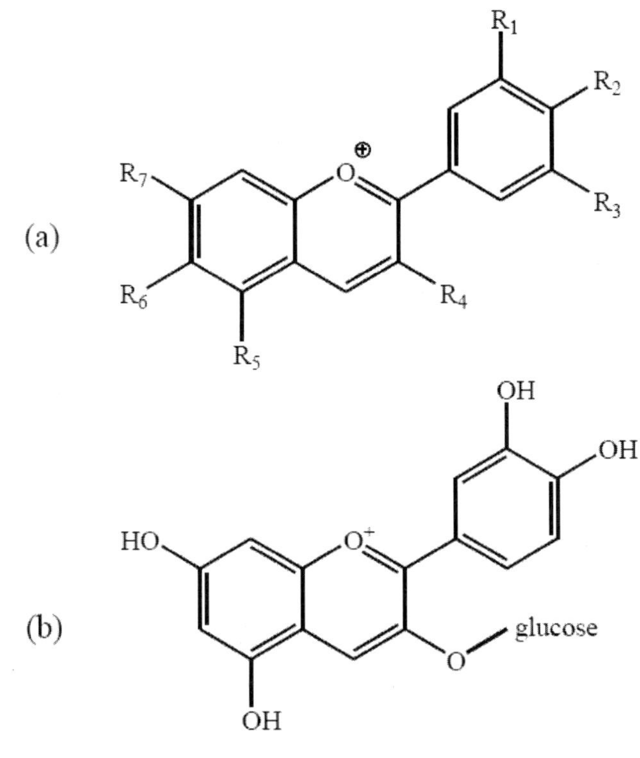

Figure 1. (a) Flavylium ion, the basic structure of anthocyanins and (b) a common anthocyanin, cyanidin-3-glucoside [2].

Table 1. The main groups of anthocyanidins showing R_1-R_7 for the side groups of each colored compound [2]

Anthocyanidin	R_1	R_2	R_3	R_4	R_5	R_6	R_7	Main color
Apigeninidin	-H	-OH	-H	-H	-OH	-H	-OH	orange
Aurantinidin	-H	-OH	-H	-OH	-OH	-OH	-OH	orange
Capensinidin	$-OCH_3$	-OH	$-OCH_3$	-OH	$-OCH_3$	-H	-OH	bluish-red
Cyanidin	-OH	-OH	-H	-OH	-OH	-H	-OH	magenta
Delphinidin	-OH	-OH	-OH	-OH	-OH	-H	-OH	purple, blue
Europinidin	$-OCH_3$	-OH	-OH	-OH	$-OCH_3$	-H	-OH	bluish red
Hirsutidin	$-OCH_3$	-OH	$-OCH_3$	-OH	-OH	-H	$-OCH_3$	bluish-red
Luteolinidin	-OH	-OH	-H	-H	-OH	-H	-OH	orange
Pelargonidin	-H	-OH	-H	-OH	-OH	-H	-OH	orange, salmon
Malvidin	$-OCH_3$	-OH	$-OCH_3$	-OH	-OH	-H	-OH	purple
Peonidin	$-OCH_3$	-OH	-H	-OH	-OH	-H	-OH	magenta
Petunidin	-OH	-OH	$-OCH_3$	-OH	-OH	-H	-OH	purple
Pulchellidin	-OH	-OH	-OH	-OH	$-OCH_3$	-H	-OH	bluish-red
Rosinidin	$-OCH_3$	-OH	-H	-OH	-OH	-H	$-OCH_3$	red
Triacetidin	-OH	-OH	-OH	-H	-OH	-H	-OH	red

The color of anthocyanins depends on structure and pH. The anthocyanidin system undergoes a variety of molecular transformations as the pH changes. In strongly acidic environments, the flavylium cation form predominates, which has a red color. The benzopyrilium cation is the most stable species. Increasing the pH leads to a decrease in the color intensity and the concentration of the flavylium cation which undergoes hydration to produce a colorless carbinol pseudobase at around pH 4-5. The highly conjugated benzopyrilium structure is disrupted due to a nucleophilic attack of water at the 2-position of the anthocyanidin skeleton. A rapid proton loss of the flavylium cation takes place as the pH shifts higher. Now the equilibrium is shifted toward quinoidal base at pH 6-7 and it has purple-blue color. When pH increases further the carbinol form yields, through opening of the central pyran ring, the light yellow chalcone [4]. Anthocyanins are becoming increasing important as antioxidant properties and health benefits, including anti-cancer, anti-inflammatory and vasoprotective effects, preventing coronary heart diseases and improving visual acuity [5-13]. They are important compounds not only because they have health related properties, but also because they are widely used as colorants in food industry [5,12-14]. However, anthocyanins are unstable during processing and storage of fruits and vegetables. The browning of fruits and vegetables is generally thought to be a rapid degradation of anthocyanins caused by polyphenol oxidase and peroxidase, producing the browning-colored by-products [15-17].

Polyphenol oxidase (PPO) is a copper-containing enzyme which is probably present in all plant. It is also known as tyrosinase, diphenol oxidase, *o*-phenolase, catechol oxidase, catecholase, phenolase and lactase. They appear to reside in the plastids and chloroplasts of plants, although freely existing in the cytoplasm of senescing or ripening plants [3]. Polyphenol oxidase is thought to play an important role in the resistance of plants to microbial and viral infections and to adverse climatic conditions. It also occurs in animals and is thought to increase disease resistance in insects and crustaceans. In the presence of oxygen, the enzyme catalyzes the oxidation of phenolic compounds to form the corresponding quinone intermediates which polymerize to form undesirable pigments. This enzyme catalyzes two types of the oxidative reaction: the hydroxylation of monophenols to *o*-diphenols and the oxidation of *o*-diphenols to *o*-quinones [17-19] which lead to production of the brown pigment, melanins.

Peroxidase (POD) is an oxidoreductase enzyme that is directly involved in many plant functions such as hormone regulation, defense mechanisms, indolacetic degradation and lignin biosynthesis [20]. It is a heme protein catalyzing a redox reaction in which hydrogen peroxide acts as the acceptor and another compound acts as the donor of hydrogen atom [19,20]. POD can reduce H_2O_2 to water while its oxidation of a variety of substrates is also present. The overall reaction is as follows: the POD that involving in an enzymatic browning since diphenols may function as reducing substrates in this reaction. The involvement of POD in the browning effect is limited by the availability of electron acceptor compounds such as superoxide radicals, hydrogen peroxide and lipid peroxidase [20].

Enzymatic browning is a chemical process involving PPO and/or other enzymes that create a brown pigment. It is a significant problem in a number of fruits such as apple, pear, banana, grape, longan and litchi [15,21-25]; vegetables such as potato, mushroom and lettuce [19,23,26] and sea foods such as shrimp, lobster and crab [18]. It usually impairs the sensory properties of the products due to associated changes in the color flavor and texture besides nutritional properties [25]. However, some enzymatic browning reactions are very beneficial to some commodities such as resins, prune, tea, coffee and cocoa [18]. The discoloration in fruits, vegetables and sea foods by an enzymatic browning resulted from the conversion of phenolic compounds to *o*-quinones which subsequently polymerize to the brown pigments. The main enzymes responsible for the enzymatic browning effect are both of PPO and POD [14,15,19,21].

Chilli fruits (*Capsicum annuum L.*) are popular vegetables because of the combination of color, test and nutrition. They are used as foods and spice. Moreover, the red chilli fruit has been used for many years as a source of pigments to add or change the color of foodstuffs. Fresh chilli peppers are a good source of vitamin C and E as well as provitamin A and carotenoids with well known antioxidant properties [20,27-33]. The carotenoid pigments in fresh chilli peppers have been widely studied to improve color retention during processing and storage [27,28,33]. However, the anthocyanin content has been limited in the relation with the browning effect in the chilli fruits. Therefore, this study is aimed at the relationship of enzyme activity, anthocyanin content and total phenolics in hot chilli samples. HPLC has been used to determine the anthocyanin contents. Both PPO and POD activities have been evaluated using spectrophotometric procedure. In addition, total phenolic compounds were measured by Folin-Ciocalteu assay.

2. Extraction and Analytical Methods

Chemicals and reagents: Polyvinylpyrrolidone was obtained from Fluka (UK). Di-sodium hydrogen phosphate was purchased from Fisher Chemical (UK). Potassium dihydrogen phosphate was purchased from Scharlau Chemie S.A. (Spain). Tri-sodium citrate was obtained from Ajax Chemical (Australia). Citric acid and phosphoric acid were AR grade, Carlo Erba (Italy). All reagents used were at least analytical reagent (AR) grade such as methanol, ethyl acetate, acetic acid, citric acid, hydrochloric acid, hydrogen peroxide (30%, m/m) and Folin-Ciocalteu's reagent were obtained from Carlo Erba (Italy). Methanol was of HPLC grade that obtained from Lab Scan (Thailand). Gallic acid, guaiacol and catechol were 99% purity obtained from Across (USA). Cyanin-3-glucoside used as standard anthocyanin was HPLC grade obtained from Fluka (USA). Aqueous solutions were prepared with de-ionized water obtained from RiOs^{TM} type 1 Simplicity 185 (Millipore Waters, USA) throughout the experiments. All standards and samples were kept cool at 4°C.

Instruments: The experiments were carried out on a Waters liquid chromatograph (Waters, USA). It consists of a Waters 600E multi-solvent delivery system, a Waters in-line degasser AF, a Rheodyne injector with sample loop of 20 μL and a Waters 2996 photodiode array detector. Empower software was used for data acquisition. Vertisep AQS C18 column (250 × 4.6 mm i.d., 5 μm) and guard column (10 × 4.6 mm i.d., 5 μm) were used. A spectrophotometric measurement was performed using Agilent 8453 UV-Visible spectrophotometer and JENWAY 6400 visible spectrophotometer. The 1000 series centifugal (Labquip, England), Model R-114 rotary evaporator (Buchi, Switzerland), VS-202D Shaker (Scientific Co. Ltd., Korea), HTS-1003 hotplate stirrer (LMS laboratory & material supplies, Japan) and Vortex (Model G-560 Vortex-2 Genie, Scientific Industries, USA) were used in the sample preparation step. The DSC-18 cartridge (20 mL/5 g, 55 μm) was obtained from Supelco (USA). Pipetman (Pipetman, France) micropipette and filter membrane, 0.45 μm (Whatman International Ltd., UK) were used.

Sample materials: Chilli pepper fruits used in this study were obtained from local markets in Khon Kaen, Thailand. Various kinds of the chilli fruits (red whole/flesh chilli, RP/RF or orange chilli, OP; green whole/flesh chilli, GP/GF) were washed several times with tap water and the seeds and peduncle were removed. The pericarp was homogenized by using homogenizer (Moulinex Optiblend 2000, France) for 2 min and stored at 4°C for further experiments. The type of samples includes chilli fruits, chilli fleshes, dried chilli (D), chilli with fish sauce and chilli with vinegar. In case of the pericarp sample, the seeds and peduncle were removed and the pericarps were homogenized. For fish sauce with chilli (FC) and vinegar with chilli (VC), fish sauce and vinegar were removed by rinsing several times with tap water. The chilli pieces were air dried and homogenized in the same manner. The homogenized samples were kept storing at 4°C for further experiments.

Determination of Anthocyanin

The chromatographic conditions used were optimized with the aim of obtaining the separation of analyte compound from other compounds within a short analysis time. The stationary phase was a reversed phase column. Several mobile phases have been described in the literatures for the analysis of anthocynin [6,34,35]. In this study, the organic solvent selected for the preliminary experiment was methanol due to the solubility of anthocyanin. The mobile phases containing various percentages of methanol in aqueous acetic acid were investigated. Using the optimal isocratic conditions, anthocyanin was separated within 5 min, as shown in Figure 2(a). The retention time of the studied compound was 4.2 min.

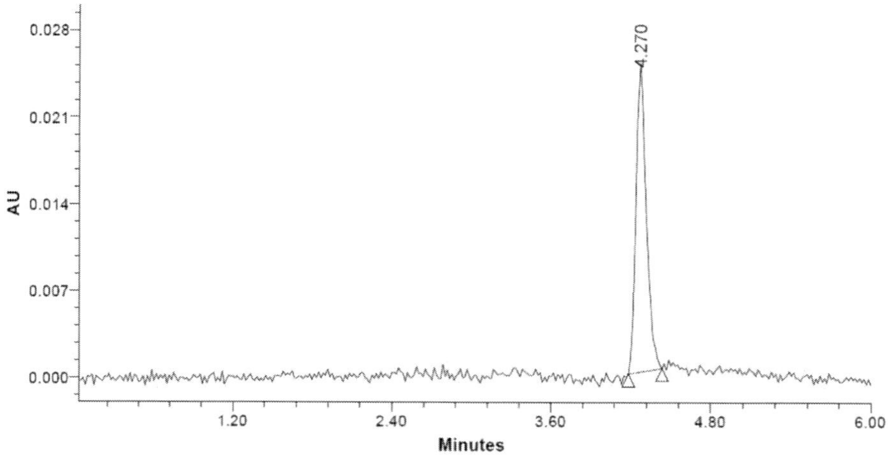

(a)

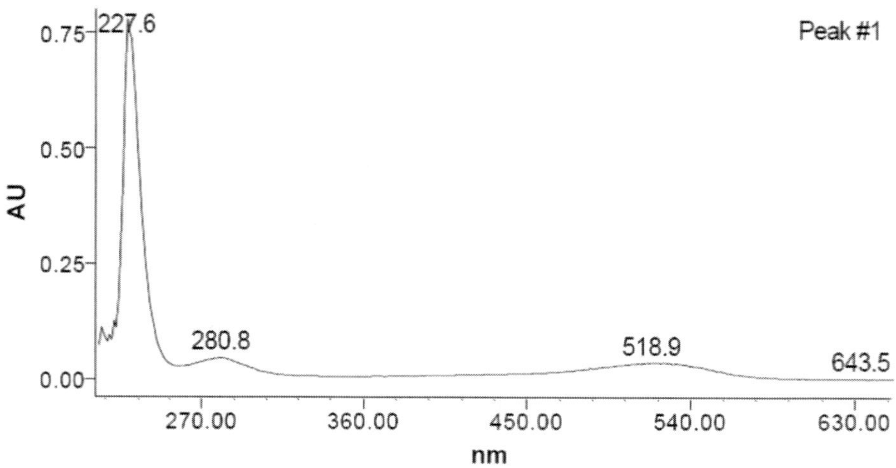

(b)

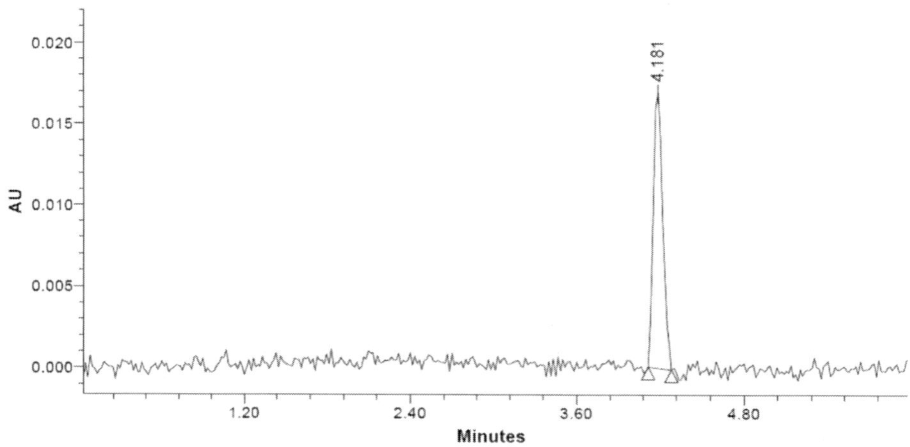

(c)

Figure 2. (Continued).

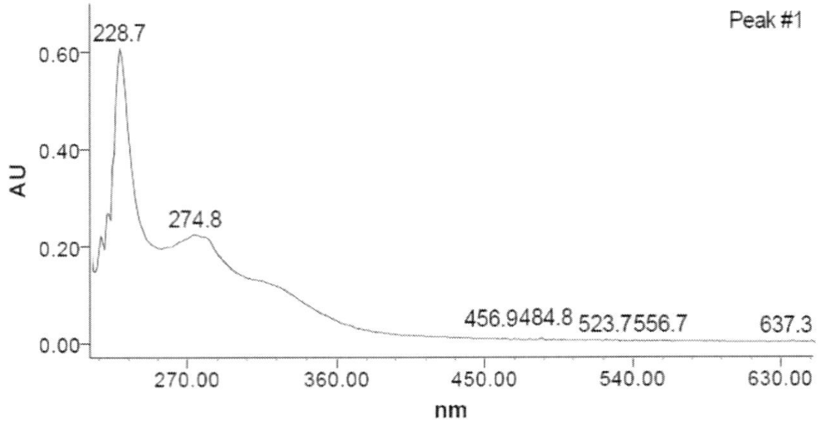

(d)

Figure 2. Chromatogram of (a) cyanidin-3-glucoside 10 mgL^{-1} with (b) its UV-Visible spectrum, and (c) chromatogram of anthocyanin from the chilli extract with (d) its UV-Visible spectrum.

In this study, photodiode array (PDA) detector is used. PDA detection allows the identification of the compound using the absorption spectrum and peak purity. Using PDA detector at maximum wavelength of 520 nm, the analyte compound is detected and its UV-Visible spectrum is shown in Figure 2(b). Identification of the compound was achieved by comparing retention time and absorption spectrum of standard and sample. The chromatogram of the chilli extract and its visible spectrum are shown in Figure 2(c) and 2(d), respectively.

Method Validation

In order to evaluate the potential of the present method for quantitative uses, linearity, LOD, LOQ, precision and recovery were investigated. According to the proposed method, the calibration curve for determination of anthocyanin was constructed under the optimum conditions using cyanin-3-glucoside as standard. The linear range for cyanin-3-glucoside was 5.0 to 45.0 mg L^{-1} and the linear regression equation was as follows: $y = 13{,}595x - 36{,}907$, with a correlation coefficient (r^2) greater than 0.998. The anthocyanin content was determined by comparing the peak area obtained with that of the standard curve of cyanin-3-glucoside. The results were expressed as μg cyanin-3-glucoside equivalents (CGE) per kg sample.

LOD and LOQ of the present method were deduced based on the concentration of the analyte which produced a signal to noise ratio of three times and ten times, were 0.7 and 5.0 mgL^{-1}, respectively. Those HPLC chromatograms are shown in Figure 3(a) and 3(b).

The repeatability and reproducibility were investigated in terms of relative standard deviation (RSD) of retention time and peak area. The repeatability was deduced from ten replicates in one day (intra-day precision, n = 10) and reproducibility was calculated from the experiment in five consecutive days (inter-day precision, n = 5×3). The repeatability and reproducibility of the retention times for anthocyanin were in good precision which their RSDs were lower than 0.4% while those of their peak area were lower than 8%.

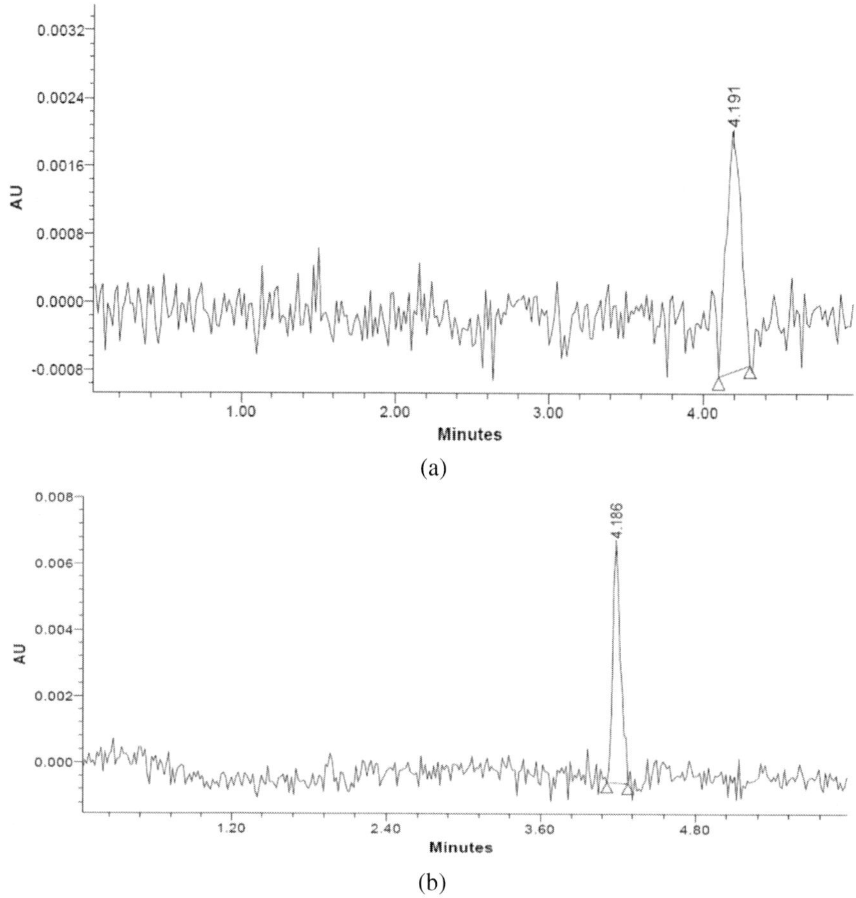

Figure 3. Chromatograms of cyanin-3-glucoside at LOD (a) and LOQ (b) levels.

Sample Preparation

In order to evaluate the optimum conditions for anthocyanin extraction from hot chilli pepper, one sample (RP2) was chosen to optimize the extraction conditions. The parameters affecting the yield of anthocyanin were investigated in details. There were the amounts of sample, the sample-to-solvent ratio, HCl concentration and extraction time.

Effect of amount of sample: Since it had no reports about the anthocyanin contents in Thai chilli peppers, the preliminary study was done giving a little amount of anthocyanin in the chilli pepper samples. In order to get the amounts of anthocyanin in the linear range of calibration curve and avoid the use of a large amount of the sample size, it was necessary to evaluate the optimal sample size which was used for the anthocyanin extraction.

Various amounts of the homogenized sample (2, 5 and 10 g) were extracted by magnetic stirring method, using 1%HCl in methanol as an extract solvent for 15 min. The volume (mL) of solvent was used at ten times of the sample weight (g). As expected, an increase in the amounts of sample resulted in an increase in the peak area of anthocyanin. At the sample used of 10 g gave the highest yield of anthocyanin (Figure 4 (a)) and its content was in the range of the calibration curve. Nevertheless, the sample size less than 10 g gave the peak area of anthocyanin under limit of the linear range of the calibration curve. Thus, the sample size of 10 g was chosen for the following steps.

Effect of sample-to-solvent ratio: Different sample-to-solvent ratios were tried out. The anthocyanin contents varied in response to the sample-to-solvent ratio used. Ten grams of the homogenized sample were extracted with various volumes of the extract solvent (30, 50, 70 and 100 mL). The sample was extracted using 1% HCl in methanol as an extraction solvent for 15 min. The extraction yield of anthocyanin increased with an increasing of the volume of the solvent as shown in Figure 4(b). However, the yield was not significantly different when the sample-to-solvent ratio less than the ratio of 1:5 by volume. Therefore, the sample-to-solvent ratio of 1:5 was chosen.

Effect of HCl concentration: Acidified methanol was used to extract anthocyanin from hot chilli pepper. Addition of acid into the extraction solvent may increase the stability of anthocyanin [45]. It is known that the use of acid is necessary to obtain the flavylium cation form, which is red and stable in an acid medium, and the maximal anthocyanin stability is obtained at pH 1.8 [1]. But at low pH levels or high concentration of HCl can make any partial hydrolysis and decomposition of anthocyanin. From the results as shown in Figure 4 (c), the yield of anthocyanin increased with the increasing of HCl concentration until the concentration of 2% HCl at which the yield was kept constant. Thus, 2%(v/v) HCl in methanol was chosen for the evaluation of other parameters.

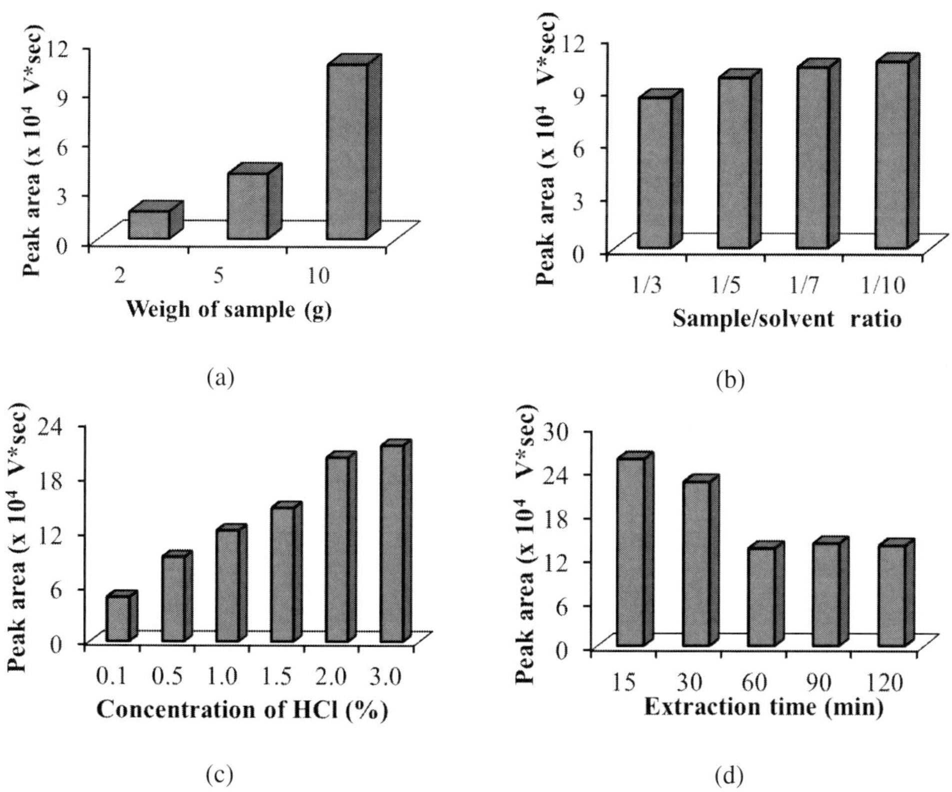

Figure 4. Effects of (a) sample size, (b) sample-to-solvent ratio, (c) HCl concentration, and (d) extraction time on the extraction of anthocyanin.

Effect of extraction time: The effect of an extraction time was studied by using 10 g of sample with 50 mL of the acidified methanol as the extraction solvent. Anthocyanin content was the highest in terms of the peak area when the extraction time was 15 min and it decreased at higher extraction times, probably due to the decomposition of anthocyanin (Figure 4 (d)). To avoid this possible loss of anthocyanin during the extraction, a two-step extraction of 15 min in association with the sample-to-solvent ratio of 1:5 was used.

Clean up using SPE: Solid-phase extraction (SPE) used for clean up the sample was necessary to remove matrices from the analyte prior to analysis. The C18 cartridge was used. It was conditioned with 10 mL methanol, followed by 10 mL of 0.01% HCl in de-ionized water. Five mL of the extract solution were diluted with 5 mL of deionized water and introduced into the conditioned cartridge. The water soluble and other phenolic compounds were removed by washing the mini-column with 10 mL of 0.01% aqueous HCl followed by 20 mL of ethyl acetate. The anthocyanin was successfully eluted with 10 mL of methanol containing 0.01% HCl. The purified sample from this process gave a clear chromatogram when injected into HPLC. It was an indication of the efficiency of SPE for the clean-up of anthocyanin. The optimum conditions for anthocyanin preparation in hot chilli peppers are summarized as follows:

Sample size:	10 g
Sample-to-solvent ratio:	1/5
Extract solvent:	2%(v/v) HCl in methanol
Extraction time:	15 min
Extraction method:	magnetic stirring

Efficiency of the magnetic stirring method: The extraction efficiency of anthocyanin from the samples was studied. The method was applied for the determination of anthocyanin in red chilli fruit (RP2) and green chilli fruit (GP2). The chilli pepper fleshes were extracted using magnetic stirring and cleaned up using a C18 cartridge. The recovery was studied by spiking 10 μg and 20 μg of standard anthocyanin into the samples before extraction, clean up and analysis by HPLC. The recoveries of anthocyanin for the two types of the samples were found in the range of 75.77 - 95.18% as shown in Table 2.

Determination of Total Phenolic Compounds

To optimize the extraction of total phenolics, the effects of experimental parameters including sample-to-solvent ratio, methanol-to-water ratio, HCl concentration, extraction method and extraction time were investigated in details.

Table 2. Percentage recoveries of anthocyanin

Sample	Anthocyanin found (μg)				
	Real sample	Added 10 μg	% Recovery	Added 20 μg	% Recovery
RP2	18.27	27.79	95.18	35.96	88.45
GP2	17.38	26.49	91.12	32.54	75.77

Effect of sample-to-solvent ratio: The yield of phenolic compounds as a function of the solvent-to-sample ratio was evaluated. As methanol is the most frequently used in the phenolics extraction, it is chosen as a solvent for the following step. The homogenized sample was shaken in 20 mL of methanol for 30 min. The results are shown in Figure 5(a). The extraction yield of the phenolic compounds was not significantly different when the solvent-to-sample ratio increased from 4 to 40 and the solvent-to-sample ratio of 60 gave the highest yield. However, the solvent-to-sample ratio of 10 was chosen to reduce the amounts of the organic waste.

Effect of methanol-to-water ratio: The use of water in combination with organic solvents contributes to the creation of a moderate polar medium that ensures the extraction of phenolic compounds [36,37]. However, the use of water as only the solvent yields an extraction with a high content of impurities such as organic acids, sugars and soluble proteins, which can interfere in the phenolic identification and quantification. The effect of the solvent-to-water ratio on the extraction yield of phenolic compounds was investigated. Then, the extraction in solvents contained various percents of methanol in 0.01% aqueous HCl. As the results shown in Figure 5(b), the proportion of methanol in the extraction medium had a significant effect on the extraction yield of the phenolic compounds. The highest values were obtained when the methanol-to-water ratio was 90/10 (v/v). It is suggested that the water proportions that equal or are higher than 50% do not ensure the recovery of the phenolic compounds from the chilli sample. Moreover, a minimum of 70% methanol has been reported to be inactivating of PPO, which are widely distributed in plants, and to allow high recovery of phenolic compounds [38]. Due to the results obtained, it was decided to continue working with 0.01%HCl in 90% methanol for the evaluation of other parameters.

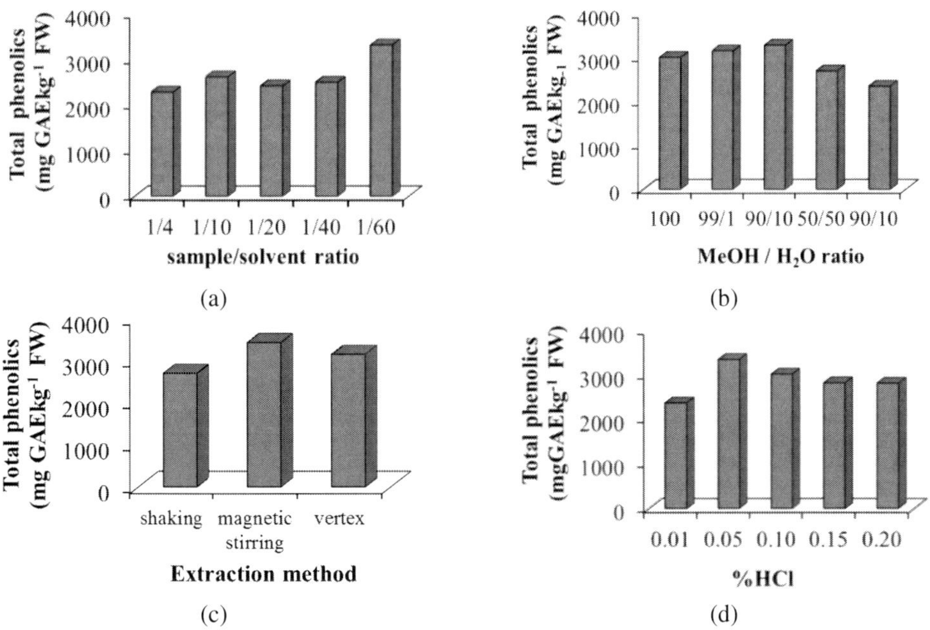

Figure 5. Effects of (a) sample-to-solvent ratio, (b) methanol-to-water ratio, (c) HCl concentration, and (d) extraction method on the extraction of total phenolics from chilli sample.

Effect of HCl concentration: The effect of acidified solvents on the extraction of total phenolics was evaluated using aqueous HCl in 90% methanol as the solvent (Figure 5(c). A small quantity of water (10%, v/v) was added to the methanol in order to create a moderate polar medium to facilitate the extraction of phenolic glycosides which are more water soluble. The suitably acidified conditions for the extraction of phenolic compounds were 0.05% HCl. At the concentrations of HCl higher than 0.05%, the extraction yields of total phenolics decreased properly due to an acid hydrolysis and decomposition [39]. Thus, 0.05% aqueous HCl in 90% methanol was chosen for further experiments.

Effect of the extraction method: Three extraction methods were simply compared for the extraction of total phenolics from the chilli sample. There are shaking, magnetic stirring and vortex methods. The results are shown in Figure 5(d). From these data, high extraction efficiency was obtained using magnetic stirring method. Therefore, the magnetic stirring method was used to extract the total phenolics from the sample.

Effect of the extraction time: The effect of the extraction time was also studied with the methanol:0.05% aqueous HCl (90/10,v/v) by magnetic stirring method. From the results, no significant differences were found when the extraction time increased from 15 to 120 min. Therefore, the extraction time of 15 min could be chosen for a routine analysis in order to reduce the analysis time.

The optimum conditions for total phenolics extraction from the chilli pepper samples are summarized as follows:

Sample size:	2 g
Sample-to-solvent ratio:	1/10
Extract solvent:	methanol:0.05% aqueous HCl (90:10,v/v)
Extraction time:	15 min
Extraction method:	magnetic stirring

It has been reported that the phenolic compounds of plant materials are subject to be a substrate of the enzymes [17,19,21,23,40]. Thus, it is an important to consider the relation between the total phenolics and the enzyme activities of hot chilli peppers. The total phenolic contents of the methanolic extracts are determined using Folin-Ciocalteu method. Briefly, the appropriate extract dilution is oxidized with the Folin-Ciocalteu reagent and the reaction is neutralized with sodium carbonate. The absorbance of the resulting blue color is measured at 745 nm after 90 min using an UV-Visible spectrophotometer. The total phenolics are determined by a comparison of the values obtained with standard curve of gallic acid. Figure 6 shows the spectrum of the mixed solution of gallic acid, Folin-Ciocalteu reagent and sodium carbonate. Its maximum wavelength is 745 nm and the absorbance at 745 nm increases with the increase in an amount of gallic acid. A five-point calibration curve of gallic acid in the range of 20 - 100 mgL^{-1} with r^2 of 0.9993 was constructed. The total phenolic contents were expressed as GAE in mg per kg sample.

The proposed methods were then applied for the determination of anthocyanin and total phenolic contents in twenty samples used consisting of sixteen hot chilli pepper fruits, two kinds of the chilli products, and two samples of dried and ground chilli peppers.

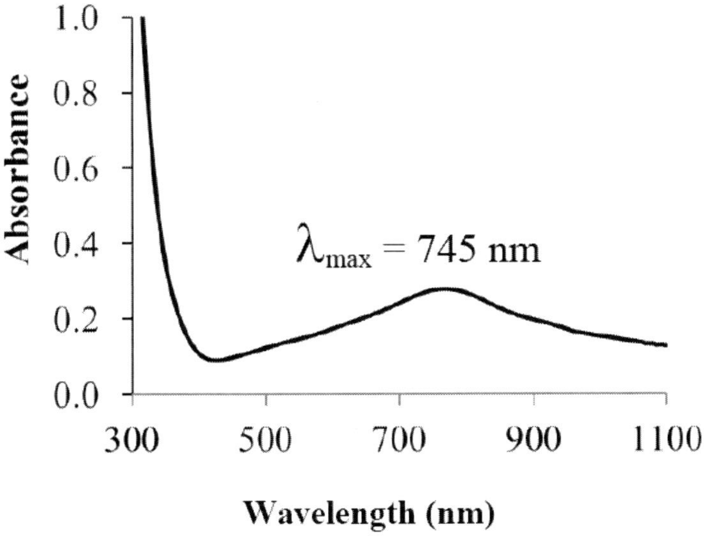

Figure 6. UV-Visible spectrum of gallic acid solution in the presence of Folin-Ciocalteu reagent and sodium carbonate.

Table 3. Effect of amount of the enzyme extract, substrate concentration, pH and temperature on PPO and POD activity

Optimized parameter	PPO activity (AU/min/g FW)[a]	POD activity (AU/min/g FW)[a]
Amount of the enzyme extract (μL)[b]		
20	5.40 ± 0.09	-
25	-	27.2 ± 1.07
50	6.51 ± 0.70	60.2 ± 2.12
75	-	81.0 ± 2.95
100	7.65 ± 0.46	101 ± 17.92
150	9.04 ± 0.52	-
200	10.7 ± 0.17	-
Substrate concentration (mM)		
Catechol		
16.7	0.115 ± 0.009	-
23.3	0.290 ± 0.028	-
33.3	0.539 ± 0.030	-
50.0	0.559 ± 0.052	-
66.7	0.431 ± 0.061	-
Guaiacol		
2.0	-	0.027 ± 0.004
6.0	-	0.111 ± 0.032
12.0	-	0.245 ± 0.034

Table 3. (Continued)

Optimized parameter	PPO activity (AU/min/g FW)[a]	POD activity (AU/min/g FW)[a]
18.0	-	0.349 ± 0.057
24.0	-	0.351 ± 0.043
H_2O_2		
0.0	-	0.011 ± 0.007
0.7	-	0.228 ± 0.015
1.6	-	0.381 ± 0.074
3.3	-	0.542 ± 0.030
4.9	-	0.592 ± 0.061
6.6	-	0.614 ± 0.063
9.9	-	0.538 ± 0.032
pH		
3	0.077 ± 0.015	0.056 ± 0.005
4	0.032 ± 0.008	0.310 ± 0.026
5	0.158 ± 0.031	0.427 ± 0.054
6	0.347 ± 0.027	0.696 ± 0.070
7	0.444 ± 0.028	0.714 ± 0.068
8	0.228 ± 0.041	0.354 ± 0.031
Temperature (°C)		
20	0.218 ± 0.018	0.711 ± 0.038
30	0.278 ± 0.018	0.862 ± 0.088
40	0.263 ± 0.050	0.947 ± 0.051
50	0.274 ± 0.063	0.840 ± 0.126
60	0.233 ± 0.060	0.862 ± 0.056

[a] activity expressed as $AUmin^{-1}g^{-1}FW$ (mean ± SD, $n = 3$)
[b] activity expressed as the rate of reaction, $AUmin^{-1}$ (mean ± SD, $n = 3$)
- *No data*

Optimization of Enzyme Activity Measurement

To determine the optimum conditions for both PPO and POD activity measurements, the following parameters including an amount of the enzyme extract, substrate concentration, pH and temperature were investigated. The results obtained are summarized in Table 3.

Effect of amount of the enzyme extract: PPO and POD are oxidative enzymes which catalyze the oxidation of phenolic substrates mainly due to enzymatic browning [15]. They catalyze the oxidation of phenolic compounds to *o*-quinone which polymerize to form undesirable pigments [19]. In this study, catechol and guaiacol were used as the substrates for PPO and POD, respectively. The enzymatic oxidation of catechol by PPO/O_2 transforms the substrate into yellow products with a maximum absorbance at 410 nm. While the enzymatic oxidation of guaiacol by POD/H_2O_2 direct the substrate into orange-pink products with a maximum absorbance at 470 nm (Figure 7).

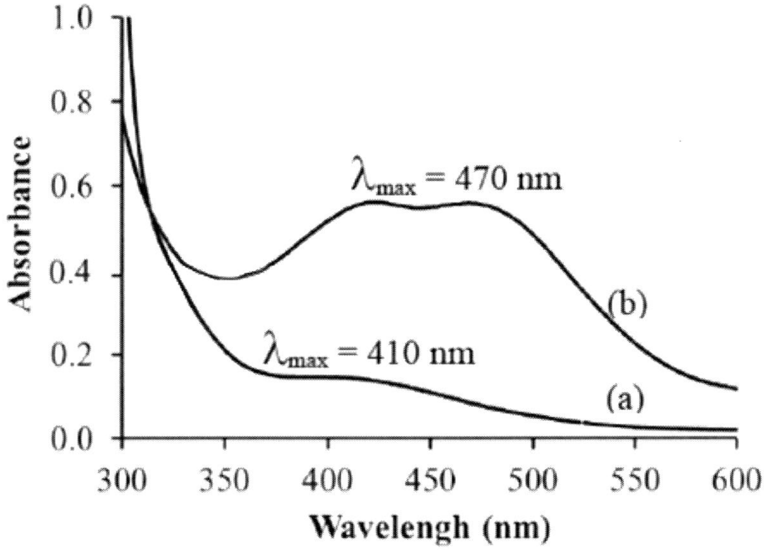

Figure 7. Visible absorption spectra (at 25°C) obtained from (a) the oxidation of catechol by PPO activity and (b) the oxidation of guaiacol by POD activity.

The effect of various amounts of the enzyme extract on PPO and POD activities was studied resulting in the rate of substrate oxidation by the enzymes. The substrate oxidation was found to be dependent on the amounts of the enzyme extract. An increase in the enzyme concentrations resulted in the corresponding linearly increasing in both enzyme activities, in the enzyme concentration range assayed (20 to 200 μL for PPO; r^2 = 0.9959, and 25 to 100 μL for POD; r^2 = 0.9852).

Effect of substrate concentration: As shown in Figure 7, the oxidation of catechol by PPO of the chilli pepper generated the products which showed the absorbance maximum at 410 nm. Thus, the optimal catechol concentration was determined by measuring the increase in the absorbance at 410 nm, using different amounts of the substrate (Figure 8(a)). As expected, an increase in the substrate concentration resulted in an increase in the pigment formation. The rate of which stayed practically constant at saturating catechol concentration. Therefore, the concentration of 33.3 mM catechol was routinely chosen because at higher concentrations of the substrate did not significantly affect the formation of the *o*-quinone intermediate.

When the oxidation of guaiacol by POD was carried out in the presence of H_2O_2, the quinone intermediate formed gave the absorbance maximum at 470 nm (Figure 7). The optimal guaiacol concentration was determined. Their results are shown in Figure 9(b). The optimal concentration of guaiacol was found to be 18.0 mM. In addition, when the H_2O_2 concentration increased, at a fixed saturating concentration of guaiacol, the POD exhibited the highest activity at 4.9 mM of H_2O_2.

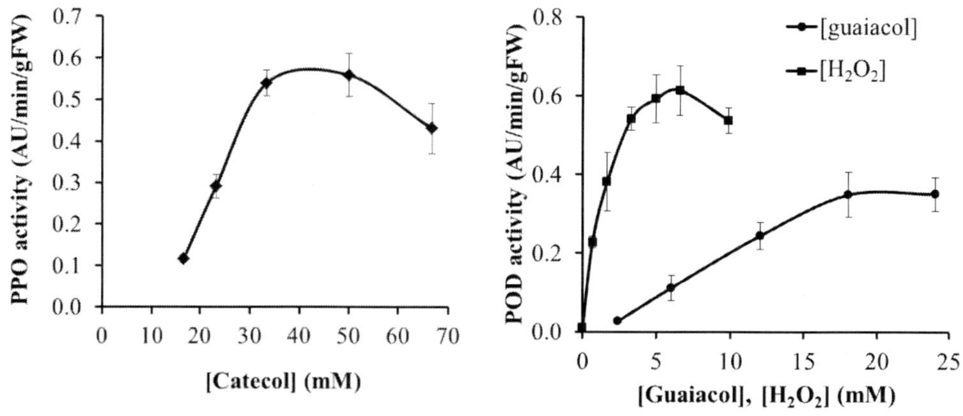

(a) (b)

Figure 8. (a) Effect of catechol concentration on the PPO activity, and (b) effect of guaiacol and hydrogen peroxide concentrations on the POD activity.

Effect of pH: The activity of PPO and POD was measured at different pH values using catechol and guaiacol as substrates, respectively. As shown in Figure 9, the optimum pH around 7 of both enzymes was obtained. It is known that the optimum pH for any enzymes depends on plant materials and substrate in the activity assay. In general, most plants show the maximum enzyme activity at or near neutral pH. Different optimum pH values for both enzymes obtained from various sources and substrates used have been reported. The optimum pH values are 6.8 and 5.5 for butter lettuce PPO using 4-methycatechol and catechol as substrates, respectively [26], pH 6.5 for longan fruit PPO using 4-methycatechol as substrate [46], pH 6.0-8.5 for kiwi fruit POD using *p*-phenyllenediamine as substrate, and pH 6.0 for spring cabbage POD using guaiacol as substrate [47].

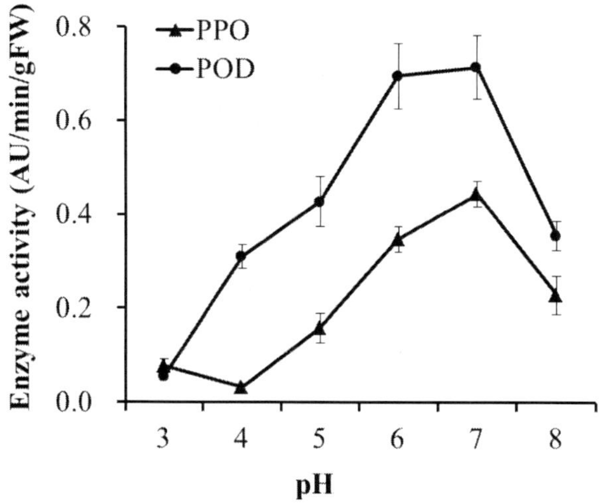

Figure 9. Effect of the pH solution on the chilli PPO and POD activity.

Effect of temperature: The optimum temperature for enzyme activity usually depends on experimental conditions. Generally, the reaction rate decreases because of thermal denaturation when the temperature increases. This situation is similar for most enzymes. Temperature dependence in the enzyme activities is presented in Figure 10. It was found that the highest activities of PPO and POD were obtained at 30°C and 40°C, respectively. PPO showed the highest activity at 30°C, and its activity decreased slightly between 40 and 60°C. The POD activity increased when the temperature increased from 20 to 40°C, and then decreased probably due to denaturation of the enzyme at higher temperatures. From previous studied, the temperature at which PPO showed the highest activity was in the range of 25-30°C, and then decreased at temperature above 40°C [48]. In case of POD, the enzyme was highly active up to 40°C and lost its activity at higher temperatures [47,49,50]. From the obtained results, the optimum temperature of the both enzymes was found between 30°C and 40°C. Thus, we determined the enzyme activities of the chilli samples at ambient temperature (30±3°C).

The optimum conditions for both PPO and POD activities measurement are summarized in Table 4.

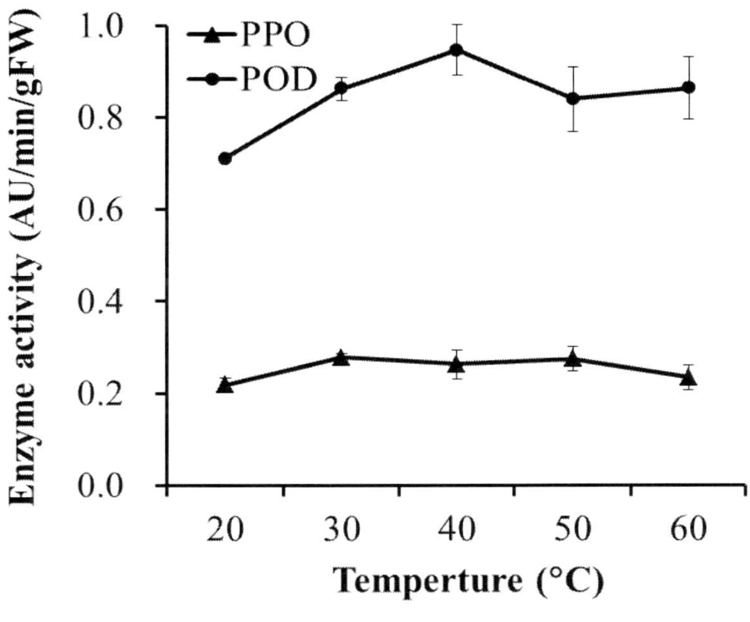

Figure 10. Effect of temperature on the chilli PPO and POD activity.

Table 4. The optimum conditions for the PPO and POD measurement

Enzyme	Enzyme extract	Substrate concentration	pH	Temperature
PPO	50 μL	33.3 mM catechol	7	ambient
POD	40 μL	18.0 mM guaiacol	7	ambient
		4.9 mM H_2O_2		

Table 5(a). Enzyme activities, anthocyanin and total phenolic contents in chilli samples based on fresh weight basis

Sample	PPO activity (ΔA/min/gFW)	POD activity (ΔA/min/gFW)	Anthocyanin (mg kg^{-1}FW)	Total phenolics (mg kg^{-1}FW)
RP1	0.730 ± 0.030	1.13 ± 0.063	0.795 ± 0.008	1993.4 ± 35.7
RP2	0.402 ± 0.078	5.29 ± 0.363	1.82 ± 0.019	3304.5 ± 180.9
RP3	0.429 ± 0.088	1.89 ± 0.181	2.78 ± 0.132	2728.7 ± 269.3
RP4	0.696 ± 0.040	2.54 ± 0.189	0.842 ± 0.014	1581.7 ± 38.6
RP5	0.612 ± 0.021	1.11 ± 0.348	1.35 ± 0.074	4516.2 ± 135.2
RF5	0.622 ± 0.037	1.54 ± 0.058	1.80 ± 0.096	4297.1 ± 185.2
OP1	0.728 ± 0.015	3.60 ± 0.201	0.968 ± 0.015	2227.7 ± 13.6
GP1	0.846 ± 0.094	3.06 ± 0.149	0.894 ± 0.01	897.49 ± 41.2
GP2	0.292 ± 0.085	0.229 ± 0.028	1.732 ± 0.102	1590.0 ± 33.1
GP3	0.609 ± 0.068	1.33 ± 0.148	1.860 ± 0.029	2370.6 ± 66.0
GP4	0.787 ± 0.087	5.57 ± 0.137	0.892 ± 0.022	1244.2 ± 49.8
GP5	0.848 ± 0.022	8.08 ± 0.451	1.01 ± 0.024	1255.9 ± 63.8
GP6	0.929± 0.007	14.4 ± 0.224	1.28 ± 0.012	782.07 ± 19.7
GP7	0.839 ± 0.015	8.69 ± 0.171	2.28 ± 0.178	2898.6 ± 171.7
GF8	0.617 ± 0.057	15.8 ± 0.760	4.70 ± 0.458	3094.3 ± 128.8
FC	-	0.899 ± 0.063	1.84 ± 0.022	678.85 ± 27.5
VC	-	-	0.803± 0.013	132.203 ± 4.1

- No data.

Table 5(b). Enzyme activities, anthocyanin and total phenolic contents in chilli samples based on dry weight basis

Sample	PPO activity (ΔA/min/gDW)	POD activity (ΔA/min/gDW)	Anthocyanin (mg kg^{-1}DW)	Total phenolics (mg kg^{-1}DW)
RP1	6.07 ± 0.25	9.41 ± 0.53	6.62 ± 0.06	16596 ± 296.8
RP2	1.95 ± 0.38	25.6 ± 1.76	8.84 ± 0.11	16021 ± 877.1
RP3	3.09 ± 0.64	13.6 ± 1.3	19.9 ± 0.83	19642 ± 1938.2
RP4	6.68 ± 0.38	24.4 ± 1.8	8.03 ± 0.05	15195 ± 371.0
RP5	3.62 ± 0.12	6.57 ± 0.16	8.18 ± 0.48	26671 ± 798.4
RF5	2.39 ± 0.14	5.89 ± 0.22	6.87 ± 0.33	16493 ± 710.7
OP1	5.94 ± 0.13	29.4 ± 1.64	7.98 ± 0.12	18176 ± 110.8
GP1	9.08 ± 1.00	32.9 ± 1.60	9.60 ± 0.10	9628.7 ± 441.8
GP2	2.41 ± 0.68	1.87 ± 0.23	14.3 ±0.85	13125 ± 273.3
GP3	4.82 ± 0.54	10.6 ± 1.17	14.8 ± 0.26	18766 ± 522.4
GP4	16.1 ± 1.79	114 ± 2.80	18.3 ± 0.49	25486 ± 1019.6
GP5	9.79 ± 0.25	93.3 ± 5.21	11.7 ± 0.32	14500 ± 736.6
GP6	11.6 ± 0.10	180 ± 2.80	15.9 ± 0.18	9782.8 ± 246.4
GP7	5.66 ± 0.10	58.7 ± 1.15	15.1 ± 0.87	19558 ± 1158.7
GF7	3.61 ± 0.39	49.1 ± 2.21	17.2 ± 2.11	17484 ± 1105.1
GF8	4.08 ± 0.37	104 ± 5.02	31.0 ± 3.02	20452 ± 851.2
FC	-	3.57 ± 0.25	42.7 ± 0.22	2723.4 ± 110.3
VC	-	-	3.20 ± 0.03	2199.6 ± 69.0
D1	-	-	70.3 ± 4.39	9016.0 ± 510.3
D2	-	-	62.9 ± 2.77	8183.9± 694.1

- No data.

Enzyme activities and the contents of both anthocyanin and total phenolics of these chilli samples were determined. The calculated results are based on both fresh weight and dry weight as shown in Table 5 (a) and 5 (b), respectively. Both enzyme activities were found in the chilli pepper fruits (RP1-GP8). While that was not found in the chilli products and the dried chilli samples, except in the fish sauce sample, of which the POD activity was found to be 0.899 AU $min^{-1}g^{-1}$ fresh weight. It was probably due to enzymatic process that could be present in the fish sauce. Fish sauce contained many chemical compositions such as free amino acid, organic acids, nucleosides and nucleic acid bases, creatine and creatinine which could be changed by the enzymatic process [41]. The PPO activity in hot chilli fruits was found to be 0.292–0.929 $AUmin^{-1}g^{-1}$ fresh weight, whereas the POD activity was ranged of 0.229–15.8 $AUmin^{-1}g^{-1}$ fresh weight. The highest activities of PPO and POD were found in the sample GP6 and GF8, respectively. And the lowest activity of the both enzymes was found in the chilli sample GP2. From literatures, it has reported that the activity of the enzymes depends on stage of maturity, genotypes, condition during growth, post harvest handing and shelf life [42].

The Relationship between PPO and POD Activities in Hot Chilli Samples

Figure 11 shows the relationship between PPO and POD activities in hot chilli samples. It is demonstated that both enzyme activties in almost fruit samples have a similar tendency. The samples with high PPO activity also have high POD activity excepted the sample RP1 and RP2. Generally, the oxidation of phenolics in plant materials is mainly catalyzed by PPO and POD in which may also contribute in this reaction. It has been reported that the oxidation of phenolic compounds by PPO can generate H_2O_2 and the POD keeps the generated H_2O_2 to get further oxidation of the phenolic compounds [21]. Thus, both enzyme activities would function the same trends in the plant materials.

HPLC coupled with PDA detection was used for the quantification of anthocyanin in hot chilli peppers. As cyanidin-3-glucoside is the most common anthocyanin in plants, it was chosen as the anthocyanin standard. Quantification was done based on the basis of standard curve of the anthocyanin standard. The results were then expressed as mg CGE per kg of fresh weight and dry weight. As shown in Table 5(a), the anthocyanin contents in hot chilli fruits (RP1–GF8) were ranged of 0.796–4.70 mg $CGEkg^{-1}$ fresh weight. The highest anthocyanin contents in the chilli fruits based on fresh weight was found in the sample GF8, while the lowest of this was found in the sample RP1. On the other hand, the anthocyanin contents in the sample FC and VC were 1.84 and 0.803 mg $CGEkg^{-1}$ fresh weight, respectively. When the results were expressed as dry weight basis (Table 5(b)), it was suggested that the anthocyanin contents in the ground chilli samples (D1 and D2) were higher than that of the chilli fruits. It has been reported that the browning index of the ground sample increased with an increasing of storage time from 1 to 10 days [43]. From the results obtained, it could be possibly implied that the dark color of the ground chilli resulted from the formation of the anthocyanin pigment. In case of flower, the accumulation of anthocyanin in blotches of the petals caused the color of the blotches darker than that of the non-blotches [44].

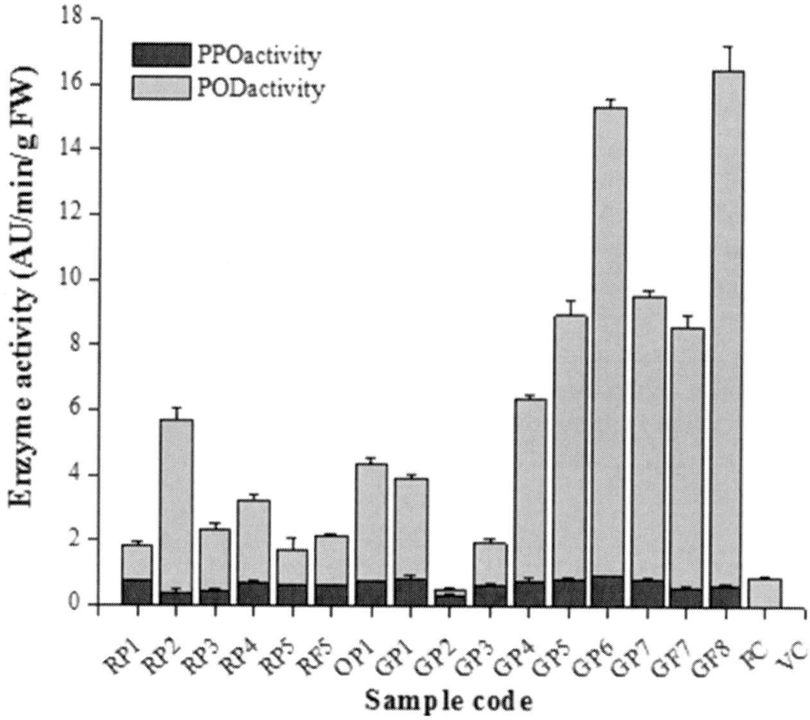

Figure 11. PPO and POD activities found in the fresh chilli samples.

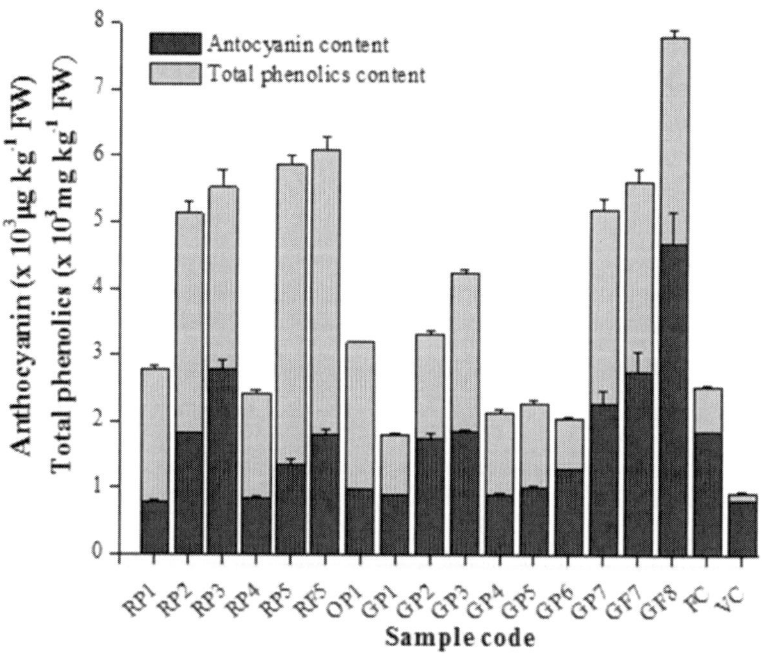

Figure 12. Comparison of anthocyanin and total phenolic contents in the chilli samples.

Total phenolics of the methanolic extracts of the chilli samples were determined using Folin-Ciocalteu method. The results were expressed as mg GAE per kg sample. The total phenolic contents based on fresh weight are shown in Table 5(a). The total phenolic contents in the chilli fruits (RP1-GF8) were found to be 782.07–4516.2 mg GAEkg^{-1} fresh weight. The highest content of total phenolics based on both fresh weight and dry weight was found in the RP5 sample and the lowest one was found in the VC sample. In addition, it can be suggested that the total phenolic contents in the ground chilli peppers were lower than that of the chilli fruits that was calculated based on dry weight. The phenolic compounds are susceptible towards temperature, oxygen and UV light. Temperature may cause the loss of phenolics due to decomposition. Light may also have a similar effect. Oxygen may destroy phenolic compounds, as do other oxidizing reagents [45].

Therefore, the phenolic compounds in the ground chilli could be lost during processing and storage which exposed to high temperature, UV light and oxygen. The relationship between anthocyanin and total phenolic contents in the chilli samples is illustrated in Figure 12. Since anthocyanin is a subgroup of the phenolic compounds in plants, a similar tendency of the anthocyanin content and total phenolic contents was observed for the chilli samples. In addition, the total phenolic contents were found to be 1000-folds of the anthocyanin contents in these chilli samples.

CONCLUSION

Anthocyanin was extracted from hot chilli peppers using 2%(v/v) HCl in methanol by magnetic stirring for 15 min and then cleaned up using C18 solid-phase extraction The separation method was applied for the determination of anthocyanin in twenty kinds of hot chilli samples. The total phenolics were also extracted by the same procedure using methanol : 0.05% aqueous HCl (90/10, v/v) and estimated by Folin-Ciocalteu reagent. The anthocyanin and total phenolics in the chilli fruits were found in the ranges of 0.796-4.70 mg CGE kg^{-1} fresh weight and 0.782-4.52 g GAE kg^{-1} fresh weight, respectively. Based on dry basis, these were 6.62-31.1 mg CGE kg^{-1} and 13.13-26.7 g GAE kg^{-1}, respectively. The anthocyanin and total phenolic contents in ground chilli samples were also determined and found to be 62.9-70.3 mg CGE kg^{-1} dry weight and 81.8-90.2 g GAE kg^{-1} dry weight, respectively.

The activities of polyphenol oxidase (PPO) and peroxidase (POD) and the contents of anthocyanin and total phenolics in hot chilli (*Capsicum annuum* L.) were evaluated. Their activities were measured spectrophotometrically using catechol and guaiacol as substrates, respectively. The optimum conditions for the measurement of PPO activity were composed of 50 μL of the enzyme extract and 33 mM catechol. While the POD exhibited the highest activity when using 40 μL of the enzyme extract, 18 mM guaiacol and 4.9 mM H_2O_2. The optimum pH for both enzymes was around 7. Furthermore, it was found that both PPO and POD lost their activities due to thermal denaturation at temperatures above 30°C and 40°C, respectively. The PPO and POD activities in hot chilli fruits were 0.292-0.929 and 0.229-15.8 AUmin^{-1}g^{-1} fresh weight, respectively. Furthermore, the relationship between the enzyme activities and both contents of anthocyanin and total phenolics was evaluated. It was clearly demonstated that both enzyme activties and their correlations between the anthocyanin and the total phenolic contents in almost chilli fruits had similar trends involving the plant metabolic process.

ACKNOWLEDGMENTS

The Centre of Excellence for Innovation in Chemistry (PERCH-CIC), Commission on Higher Education, Ministry of Education, Thailand and the Hitachi Scholarship Foundation, Tokyo, Japan are gratefully acknowledged for research financial support.

REFERENCES

[1] Markakis, P. (1982). *Anthocyanins as Food Colors*. New York: Academic Press.
[2] Kong, J-M., Chia, L-S., Goh, N-G., Chai, T-F. & Brouillard, R. (2003). Analysis and biological of anthocyanins. *Phytochemistry*, *64*, 923-933.
[3] Delgado-Vargas, F. & Paredes, O (Eds) (2003). *Natural Colorants for Food and Nutraceutical Uses*. Washington, D.C.: CRC Press.
[4] Strack, D. (1997). *Plant Biochemistry*. London: Academic Press.
[5] Giusti, M. M. & Wrolstad, R. E. (2001). *Current Protocols in Food Analytical Chemistry*. New York: John Wiley & Sons.
[6] Stintzing, F. C., Stintzing, A. S., Carle, B. F. & Wrolstad, R. E. (2002). Color and antioxidant properties of cyanidin-based anthocyanin pigments. *Journal of Agricultural and Food Chemistry*, *50*, 6172-6181.
[7] Chaovanalikit, A., Dougherty, M. P., Camire, M. E., Darokar & Briggs, J. (2003). Ascorbic acid fortification reduces anthocyanins in extruded blueberry-corn cereals. *Journal of Food Science*, *68*, 2136-2140.
[8] Joniec, C. M., Mes, P. & Myers, J. R. (2003). Characterization and inheritance of the anthocyanin fruit (Aft) Tomato. *Journal of Heredity*, 94, 449-456.
[9] Kähkönen, M. P., Heinämäki, J., Olliainen, V. & Heinoneh, M. (2003). Berry anthocyanins: isolation, identification and antioxidant activities. *Journal of the Science of Food and Agriculture*, *83*, 1403-1411.
[10] Kallithraka, S., Mohdaly, A. A. A., Makris, D. P. & Kefalas, P. (2005). Determination of major anthocyanin pigments in Hellenic native grape varieties (*Vitis vinifera* sp.): associated with antiradical activity, *Journal of Food Composition and Analysis*, *58*, 375-386.
[11] Katsube, N., Iwashita, K., Tsushida, T., Yamaki, K. & Kobori, M. (2003). Induction of apoptosis in cancer cells by bilberry (*Vaccinium myrtillus*) and the anthocyanins . *Journal of Agricultural and Food Chemistry*, *51*, 68-75.
[12] Gómez-Plaza, E., Miñano, A. & Lóprz-Roca. (2006). Comparison of chromatic properties, stability and antioxidant capacity of anthocyanin-based aqueous extracts from grape pomace obtained from different vinification methods. *Food Chemistry*, *97*, 87-94.
[13] Prata, R. B. A. E. & Oliveira L. S. (2007). Fresh coffee husks as potential sources of anthocyanins. *LWT-Food Science and Technology*, *40*, 1555-1560.

[14] Muñoz, J., Garcia-Molina, F., Varen, R., Rodriguez-Lopez, J. N., García-Ruíz, P. A., Garcia-Cánovas & Tudela, J. (2007). Kinetic characterization of the oxidation of chlorogenic acid by polyphenol oxidase and peroxidase, characteristic of *o*-quinone. *Journal of Agricultural and Food Chemistry*, *55*, 920-928.

[15] Jiang, Y., Duan, X., Joyce, D., Zhang, Z. & Li, J. (2004). Advances in understanding of enzymatic browning in harvest litchi fruit. *Food Chemistry*, *88*, 443-446.

[16] Jiménez-Atiénzar, M. Escribano, J., Cabanes, J., Gandía-Herrero, F. & García-Carmona, F. (2005). The flavonoid eriodityol as substrate of peach polyphenol oxidase. *Journal of Food Science*, *7*, c540-c543.

[17] Zhang, Z., Pangi, X., Xuewu, D., Zuoliang, J. & Jiang, Y. (2005). Role of peroxidase in anthocyanin degradation in litchi fruit pericarp. *Food Chemistry*, *90*, 47-52.

[18] Lee, C. Y. & Whitaker, J. R. (1995). *Enzymatic Browning and Its Prevention*. Washington, D.C.: American Chemical Society.

[19] Chisari, M., Barbagallo, R. N. & Spagna, G. (2007). Characterization of polyphenol oxidase and peroxidase and influence on browning of cold stored strawberry. *Journal of Agricultural and Food Chemistry*, *55*, 3469-3479.

[20] Serrano-Martínez, A., Fortea, F. M., Del Amor, F. M. & Núñez-Delicado, E. (2008). Kinetic characterization and thermal inactivation study of partially purified red pepper (*Capsicum annuum* L.) peroxidase. *Food Chemistry*, *107*, 193-199.

[21] Richard-Forget, F. C. & Gauillard, F. A. (1997). Oxidation of chlorogenic acid catechins and 4-methylcatechol in model solution by combination of pear (*Pyrus communis* Cv. Williams) polyphenol oxidase and peroxidase: a possible involvement of peroxide in enzymatic browning. *Journal of Agricultural and Food Chemistry*, *45*, 2475-2476.

[22] Muñoz, O., Sepúlveda, M. & Schwaerz, M. (2004). Effects of enzymatic treatment on anthocyanin pigments from grapes skin from Chilean wine. *Food Chemistry*, *87*, 487- 490.

[23] Duan, X., Su, X., You, Y., Qu, H., Li, Y. & Jiang, Y. (2007). Effect of nitric oxide on pericarp browning of harvested longan fruit in relation to phenolic metabolism. *Food Chemistry*, *104*, 571-576.

[24] Lee, M-K. & Park, I. (2007). Studies on inhibition of enzymatic browning in some foods by Du-Zhong (*Eucommia uimoides* Oliver) leaf extract. *Food Chemistry*, *1114*, 154-163.

[25] Ye, S., Yo-Xin, Y., Heng, Z., Yuan-Peng, D., Feng, C. & Shu-Wei, W. (2007). Polyphenolic compound and the degree of browning in processing apple varieties. *Agricultural Sciences in China*, *6*, 607-612.

[26] Gawlik-Dziki, U., Złotek, U. & Świeca, M. (2007). Characterization of polyphenol oxidase from butter lettuce (*Luctuca sativa var. capitata* L.). *Food Chemistry*, *107*, 129-135.

[27] Markus, F., Daood, H. G., Kapitany, J. & Biass, P. A. (1999). Change in the carotenoid and antioxidant content of spice red pepper (paprika) as a function of ripening and some technological factors. *Journal of Agricultural and Food Chemistry*, *47*, 100-107.

[28] Krinsky, N. I. (2001). Carotenoids as antioxidation. *Nutrition*, *17*, 815-817.

[29] Perucka, I. & Masterska, M. (2001). Phenylalanine ammonia-lyase and antioxidant activities of lipophilic fraction of pepper fruits *Capsicum annuum* L. *Innovative Food Science and Emerging Technologies*, *2*, 189-192.

[30] Navarro, J. M., Flores, P., Garrido, C. & Martinez, V. (2006). Changes in the contents of antioxidant compounds in pepper fruits at different ripening stages, as affected by salinity. *Food Chemistry*, *96*, 66-73.

[31] Chatterjee, S., Niaz, Z., Gautan, S., Adhikari, S., Pasad, S. V. & Sharma, A. (2007). Antioxidant activity of some phenolic constituents from green pepper (*Piper nigum* L.) and fresh nutmeg mace (*Myristica fragrans*). *Food Chemistry*, *102*, 515-523.

[32] Conforti, F., Statti, G. A. & Menichini, F. (2007). Chemical and biological variability of hot pepper fruits (*Capsicum annuum var. acuminatum* L.) in relation to maturity stage. *Food Chemistry*, *102*, 1094-1104.

[33] Deepa, N, Kaur C., George, B., Singh, B. & Kapoor, H. C. (2007). Antioxidant constituents in some sweet pepper (*Capsicum annuum* L.). *LWT-Food Science and Technology*, *40*, 212-219.

[34] Da Costa, C. T., Nelson, B. C., Margolis, S. A. & Horton, D. (1998). Separation of blackcurrent anthocyanins by capillary zone electrophoresis. *Journal of Chromatography A*, *799*, 321-327.

[35] Longo, L., Scardino, A. & Vasapollo, G. (2007). Identification and quanitification of anthocyanins in the berries of *Pistacia lentiscus* L., *Phillyrea latifolia* L. and *Rubia peregrine* L. *Innovative Food Science and Emerging Technologies*, *8*, 360-364.

[36] Lapornic, B., Prošek, M. & Wonder, A. G. (2005). Comparison of extract prepared from plant by-products using different solvents and extraction time. *Journal of Food Engineering*, *71*, 214-222.

[37] Liyana-Pathirana, C. & Shahidi, F. (2006). Optimization of extraction of phenolics from wheat using response surface methodology. *Food Chemistry*, *93*, 45-56.

[38] Robards, K., Prenzler, P. D., Ticker, G., Swatsitang, P. & Glover, W. (1999). Phenolic compounds and their role in oxidative process in fruits. *Food Chemistry*, *66*, 401-436.

[39] Revilla, E., Ryan, J. M. & Martín-Oetega, G. (1998). Comparison of several procedures used for the extraction of anthocyanins from red grapes. *Journal of Agricultural and Food Chemistry*, *46*, 4592-4597.

[40] Mohapatra, D., Frias, J. M., Oloveira, Z. M., Bira, Z. M. & Kerry, J. (2008). Development and validation of a model to predict enzymatic activity during storage of cultivated mushrooms (*Agaricus bisporus* spp.). *Journal of Food Engineering*, *86*, 39-48.

[41] Park, J-N., Fukumato, Y., Fujita, E., Tanaka, T., Washio, T., Otsuka, S., Shimazu, T., Watanabe, K. & Abe, H. (2001). Chemical composition of fish sauces produced in southeast and East Asian countries. *Journal of Food Composition and Analysis*, *14*, 113-125.

[42] Boonsiri, K., Ketsa, S. & Van Doorn, W. G. (2007). Seed browning of hot peppers during low temperature storage. *Posthavest Biochemistry and Technology*, *45*, 358-365.

[43] Göğüş, F. & Eren, S. (1998). Effect of temperature and pH on nonenzymic in minced dried pepper during storage. *Turkish Journal of Engineering Environmental Science*, *22*, 33-38.

[44] Zhang, J., Wang, L., Shu, Q., Liu, Z., Li, C., Zhang, J., Wei, X. & Tian, D. (2007). Comparison of anthocyanins in non-blothes and blotches of the petals of Xibei tree peony. *Scientia Horticulturae*, *114*, 104-111.

[45] Escribano-Bailón, M. T. & Santos-Buelga, C. (2003). *Method in Polyphenol Analysis*. United Kingdom: The Royal Society of Chemistry.

[46] Jiang, Y-M. (1999). Purification and some properties of polyphenol oxidase of longan fruit, *Food Chemistry*, *66*, 75-79.

[47] Belcarz, A., Ginalska, G., Kowalewska, B. & Kulesza, P. (2008). Spring cabbage peroxidase – Potential tool in biocatalysis and bioelectrocatalysis. *Phytochemistry*, *69*, 627-636.

[48] Doğan, S. & Doğan, M. (2004). Determination of kinetic properties of polyphenol oxidase from Thymus (*Thymus logicaulis* subsp. *chaubardii* var. *chaubardii*). *Food Chemistry*, *88*, 69-77.

[49] Fatima, A. & Husain, Q. (2007). A role of glycosyl moieties in the stabilization of bitter gourd (*Momordica charantia*) peroxidase. *International Journal of Biological Macromolecules*, *41*, 56-63.

[50] Saraiva, J. A., Nunes, C. S. & Coimbra, M. A. (2007). Purification ad characterization of olive (*Olea europaea* L.) peroxidase – Evidence for the occurrence of pectin binding peroxidase. *Food Chemistry*, *101*, 1571-1579.

In: Natural Heterocycles
Editors: K. L. Ameta and Saksit Chanthai
ISBN: 978-1-63463-424-3

Chapter 10

CARBAZOLE AND AZACARBAZOLE WITH CYTOTOXICITY AND ANTI MITOCHONDRIAL POTENTIAL

Sudesh Kumar[1,•], Priyanka Jain[1], K. L. Ameta[2] and Dharma Kishore[1]
[1]Department of Chemistry, Banasthali University,
Banasthali, India
[2]Department of Chemistry, FASC,
Mody Institute of Technology and Science,
Lakshmangarh, Rajasthan, India

ABSTRACT

Carbazole alkaloids which occur in *Murrayakoenigii* (Rutaceae) affect on the growth of the humanleukaemia cell line HL-60. Carbazoles also exhibit biological activities like, anti-inflammatory, antibiotic, antimicrobial, antifungal, antitumor, inhibit mitosis and are antimalaria. Azacarbazoles include carboline series and pyridoindoles which are useful as anticancer, antimicrobial as exemplified as harman derivative, eudistomins, lavendomycin and picrasmajavanica alkaloids.

Keywords: Carbazoles, Azacarbazoles, mukanol and Indole

INTRODUCTION

Heterocyclic compounds have been found to occur widely in nature and have proved to be of immense significance to life. These are pharmacologically active and are in regular clinical use. Carbazole was isolated first from coaltar in 1872 by Graebe and Glazer [1].

• E-mail: sudeshneyol@gmail.com.

In 1965, Chakraborty et al. isolated carbazole alkaloids from the stems of *Murrayakoenigii* (Rutaceae) which has effect on the growth of the humanleukaemia cell line HL-60. Besides this, other carbazole alkaloids, such as mahanine, pyrafoline and murrafoline showed significant cytotoxicity against HL-60 cells and induced the loss of mitochondrial membrane potential [2]. Curry leaf tree (*Murrayakoenigii* L., Rutaceae) has various important uses in the traditional system of medicine in Eastern Asia [3]. Carbazoles are among exclusive types of N-containing aromatic heterocycles possessing desirable electronic and charge-transport properties, as well as large pi-conjugated system. Carbazole nucleus has always stimulated the endeavors to find new pathways for its synthesis and for its novel derivatives because of their extensive photophysical, photochemical and biological properties especially their amazing pharmacological profile and their role as drug molecules. Carbazole and its derivatives are privileged natural product motifs that are endowed with enormous properties and exclusive biological profile.

Structure of Carbazole (Nomenclature)

Carbazole contains a benzene nucleus fused to the indole ring at its 'a' face. The numbering in carbazole is followed as shown in (Figure 1). Tautomer's other than 9H-carbazoles are unknown except as reduced derivatives. Therefore, the term carbazole is generally taken as equivalent to 9H-carbazole. In case, any substituent is present on carbazole nitrogen, it is named by either using prefix 9 or the prefix N.

Biological Aspect of Carbazoles

Carbazoles nucleus is present in a series of natural products which are widely distributed in higher plants. Carbazoles exhibit a broad range of biological activities like, anti-inflammatory [4], antibiotic [5], antimicrobial [3], antifungal [6], antitumor [7] as well as they inhibit mitosis and are active against malaria. [8]

Due to their polycyclic, planar and aromatic structure for intercalation into DNA, they remained in the mainstay as important molecules to the main targets for cytotoxic activities. Carbazoles cytotoxicity is related to DNA dependent enzyme inhibition such as topoisomerase I / II and telomerase [9]. Other targets such as cyclin – dependent kinases and estrogen receptors have also emerged actively.

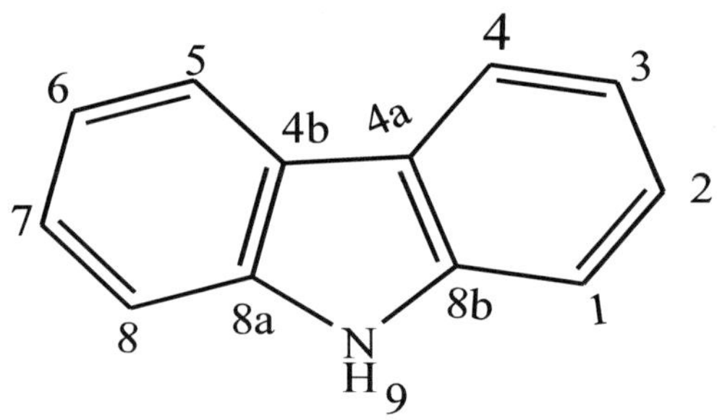

Figure 1: Structure of carbazole.

In view of the broad spectrum of biological activities of carbazole derivatives, it is considered of interest to present a review on the important pharmacological properties of these molecules. For the sake of brevity, it is more appropriate to present the important features of these molecules into two main parts:

1. Carbazole alkaloids
2. Azacarbazoles

1. CarbazoleAlkaloids

Carbazole nucleus exhibits impressive pharmacological properties due to the planarity of the system and consequently to its DNA-chain intercalating ability which makes them suitable for antineoplastic or mutagenic applications [10]. Recently, new synthetic methods of carbazole alkaloids viz; carbazomycin B (a), rimcazole (b) ellipticine (c), and olivacine (d) have been developed (figure 2) [11].

Furthermore, the anti-inflammatory properties of carbazomycin B [12], antipsychotic properties of rimcazole [13], and the antitumor properties of ellipticine, olivacine [14, 15] have been extensively studied. Carbazoles have also been widely used as building blocks for potential organic semiconductors [16], organic light-emitting diodes [17] and electro-luminescent materials.

Anti-thrombocytopenia activity of carbazole (KT6352 & KF41399) [18], anti-HIV properties of (CA-0, CA-1&CA-2) [19], antimalarial properties of carbazole derivatives (cathine-6-one and apidocarpine) [8] and antibacterial properties of novel pyridazinocarbazole [6] a-l (figure 2) have been reported.

2. Azacarbazoles

Azacarbazoles (pyridoindole or carboline derivatives) have attracted considerable interest due to the recent discoveries of several naturally occurring compounds containing this skeleton.

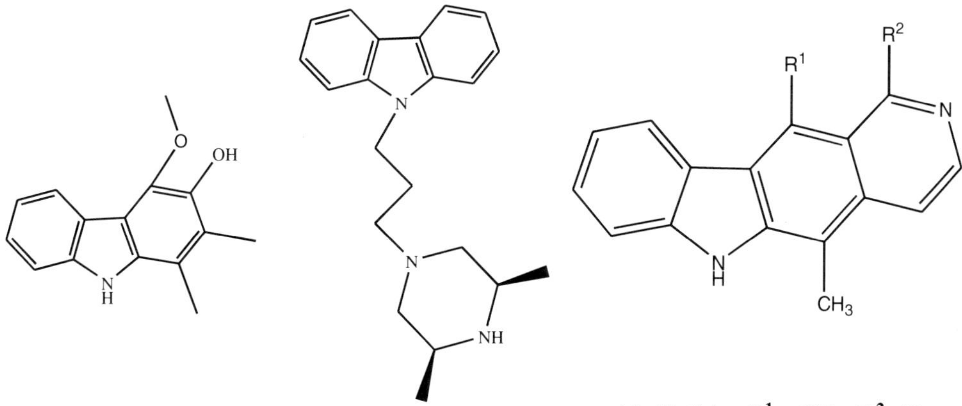

Carbazomycin B
(a)

Rimcazole
(b)

(c) elipticine $R^1 = CH_3$, R^2 =H
(d) olivacine R^1 =H, R^2= CH_3

Figure 2. (Continued).

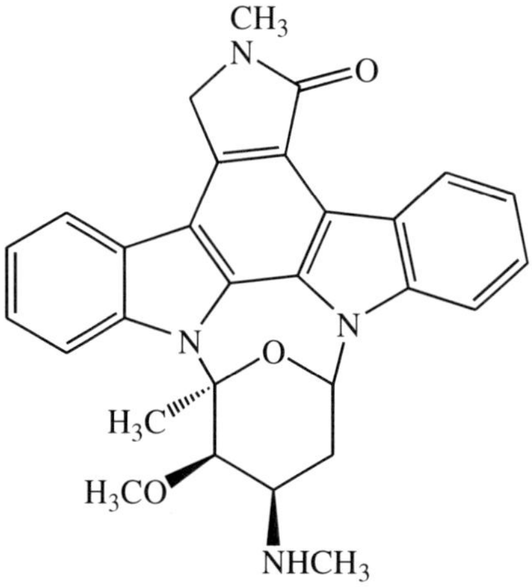

KT6352
(e)

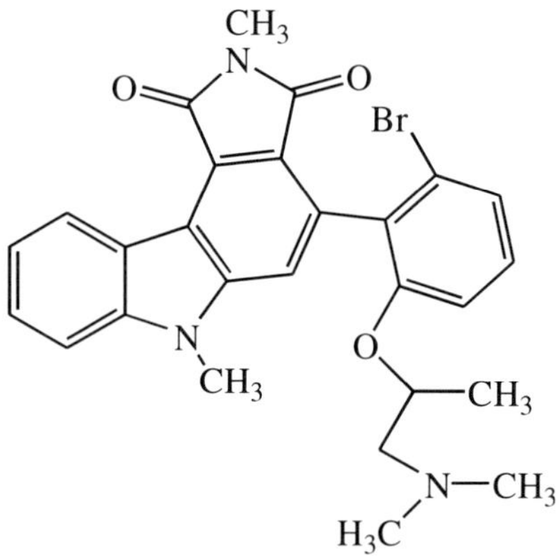

KF41399
(f)

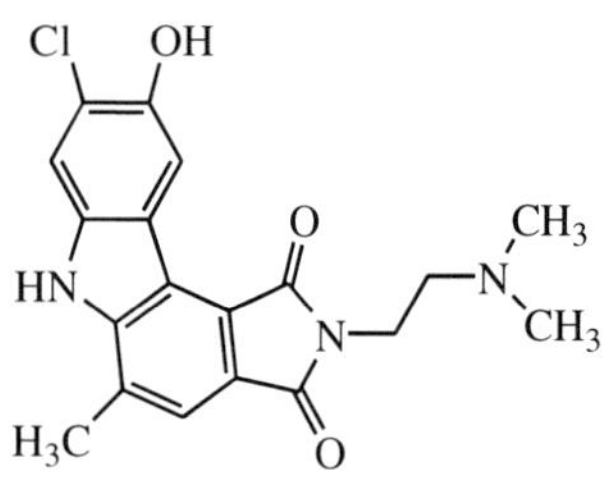

CA-0
(g)

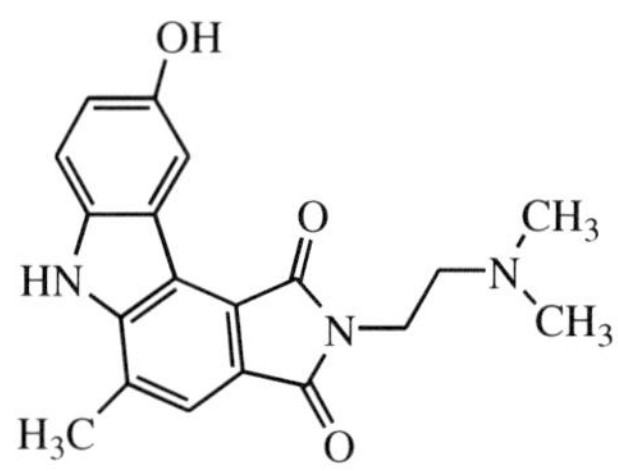

CA-1
(h)

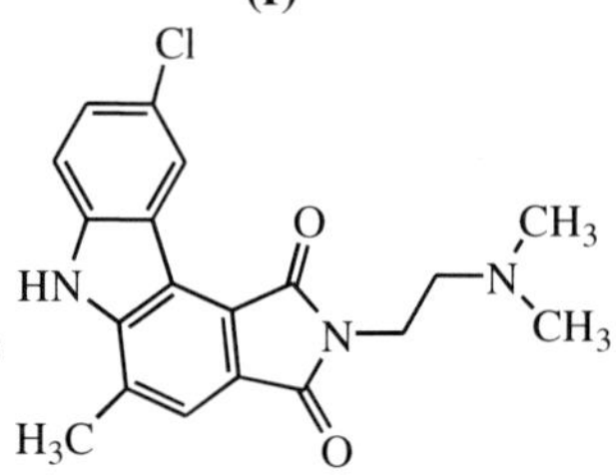

CA-2
(i)

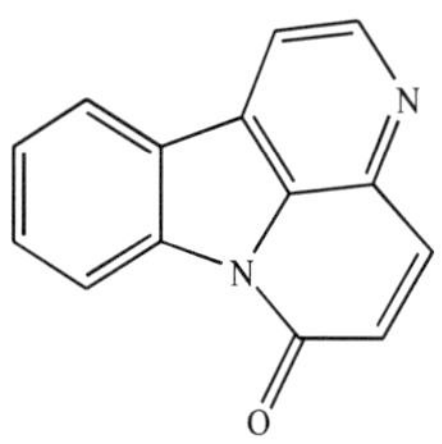

cathin-6-one
(j)

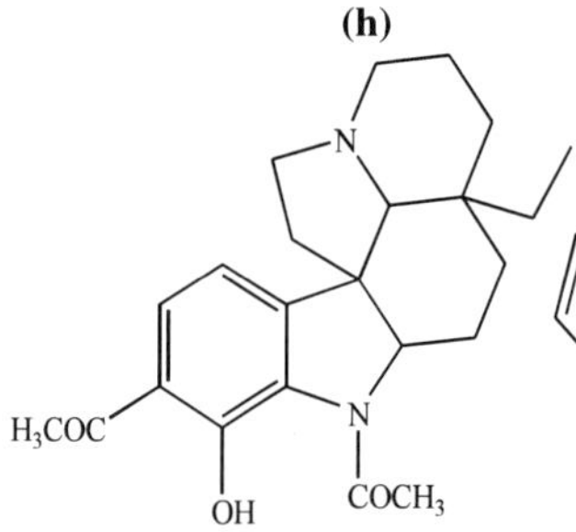

aspidocarpine
(k)

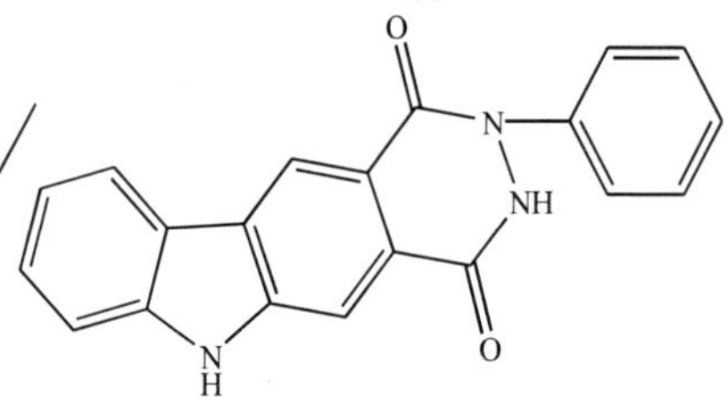

Novel pyridazino carbazole
(l)

Figure 2: Physiologically active carbazoles and azacarbazoles alkaloids (a-l).

These compounds have displayed a wide range of important biological activities [10]. Pyridoindole (figure 3, Azacarbazole derivatives) stobadine is a drug with low toxicity and is effective against atherosclerosis, stroke and neurodegenerative disease.

Azacarbazoles which include carboline series and pyridoindoles are useful as anticancer [20] and antimicrobial agents [21]. Numerous natural alkaloids belonging to the carboline series have displayed interesting biological properties such as harman derivative [22], eudistomins [23], lavendomycin and picrasmajavanica alkaloids. α-Carbolines (pyrido (2,3-b) indole derivatives) (figure 4) have gained considerable attention in the recent discoveries of several naturally occurring compounds containing this skeleton.

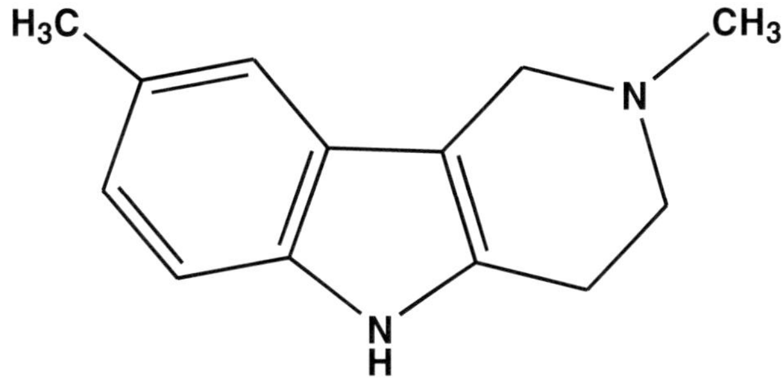

Pyridoindole

Figure 3: Structure of pyridoindole (azacarbazole derivative).

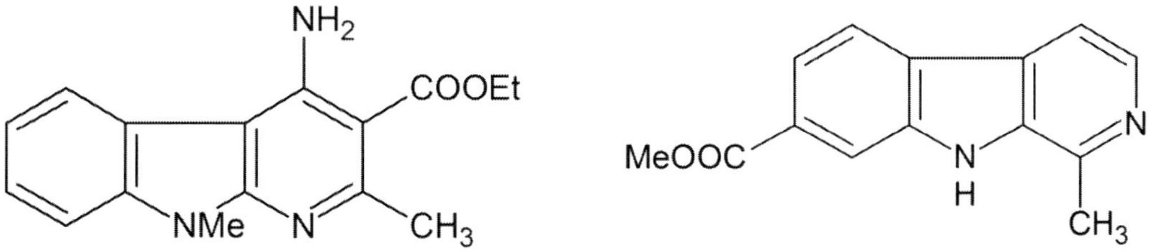

(a) **(b)**

Figure 4: Structure of α-carbolines.

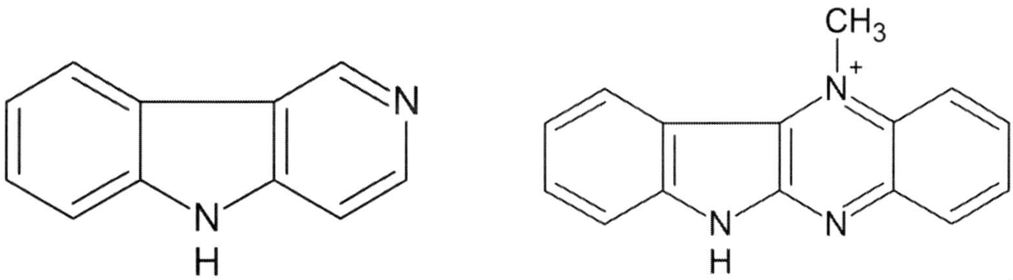

(a) **(b)**

Figure 5: Structure of δ-carbolines.

These compounds show a wide range of important biological activities. The marine alkaloid grossularine have displayed marked cytotoxicity towards murine and human tumor cells [29], whereas mescengricin (isolated from Streptomyces griseoflavus) is found to protect the nervous system against L-glutamate induced excitotoxicity [30]. Pyrido (2,3-b) indole derivatives (harmine) have showed marked activity as potential anxiolytic agents [31] (figure 4).

δ- Carbolines (9H- pyrido[3,4-b]indole) contain isoquinoline nitrogen in the ring which offer a rich source of biologically important molecules. Benz-δ-cabolines are also found to have wide scope, their derivative cryptolepine is found in the living systems (figure 5).

1,2-Dimethyl-1,2,3,4-tetrahydro-β-Carboline, leptocladin has been recognized as a constituent of *Arthrophytum leptocladum,* where it is found together with N-β-methyl trytamine (dipterine), which is a probable biogenic precursor [32] (figure 6).

A variety of β-carbolines have been found to occur in natural products which show various types of pharmaceutical activities. β-Carboline (harmalol) occurs in *Peganumharmala.* Harmo, harmine, harman and norharman are β-carbolines found in plants (figure 7).

Most of the 1-substituted carbazolines are biogically active molecules, namely pavettine, nitramarine, fascaplysin, eudistomin U and eudistomin T (figure 8).

Extensive studies have been carried out in the search for compounds derived from indole, which show pharmacodynamic actions. The pyridoindolestobadine showed antioxidant and pharmacodynamic properties.

Stobadine [33] displayed cardioprotection, neuroprotection, antimutagenic effects, antidisrhythmic action, antihistaminic properties and behavioural effects. Stobadine is found to protect sulphur containing amino acid by preventing oxidation of SH groups of amino acids.

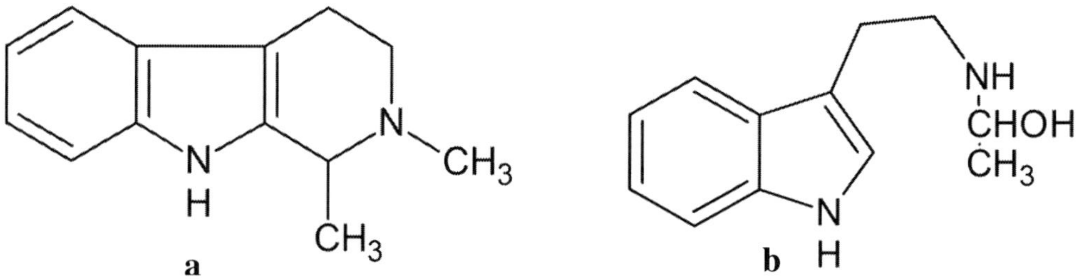

Figure 6: Structure of active constitiutes of leptocladin (a) and dipterine (b) isolated from Arthrophytum leptocladum.

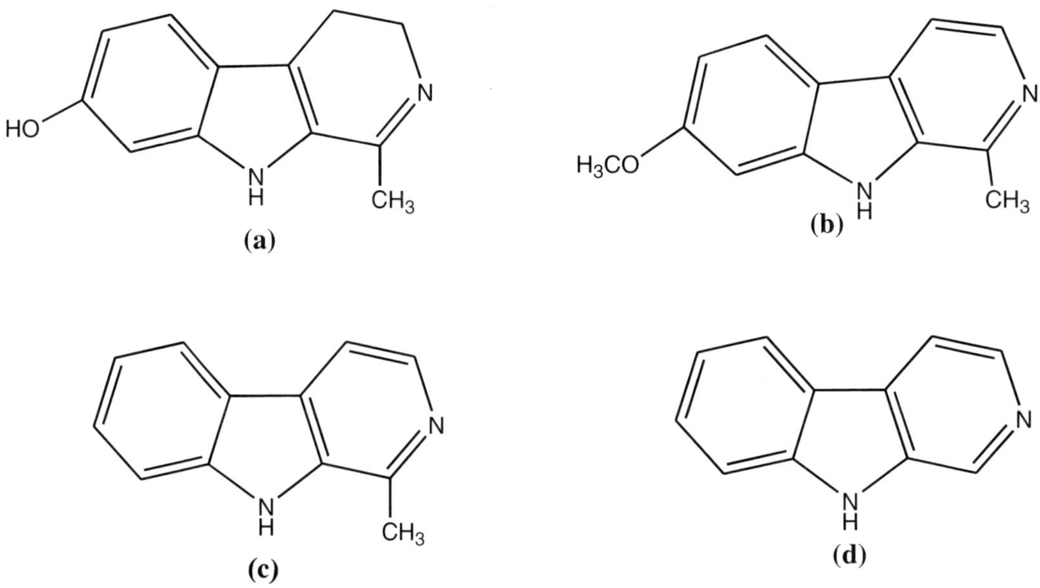

Figure7: Naturally occurring β-carboline alkaloids.

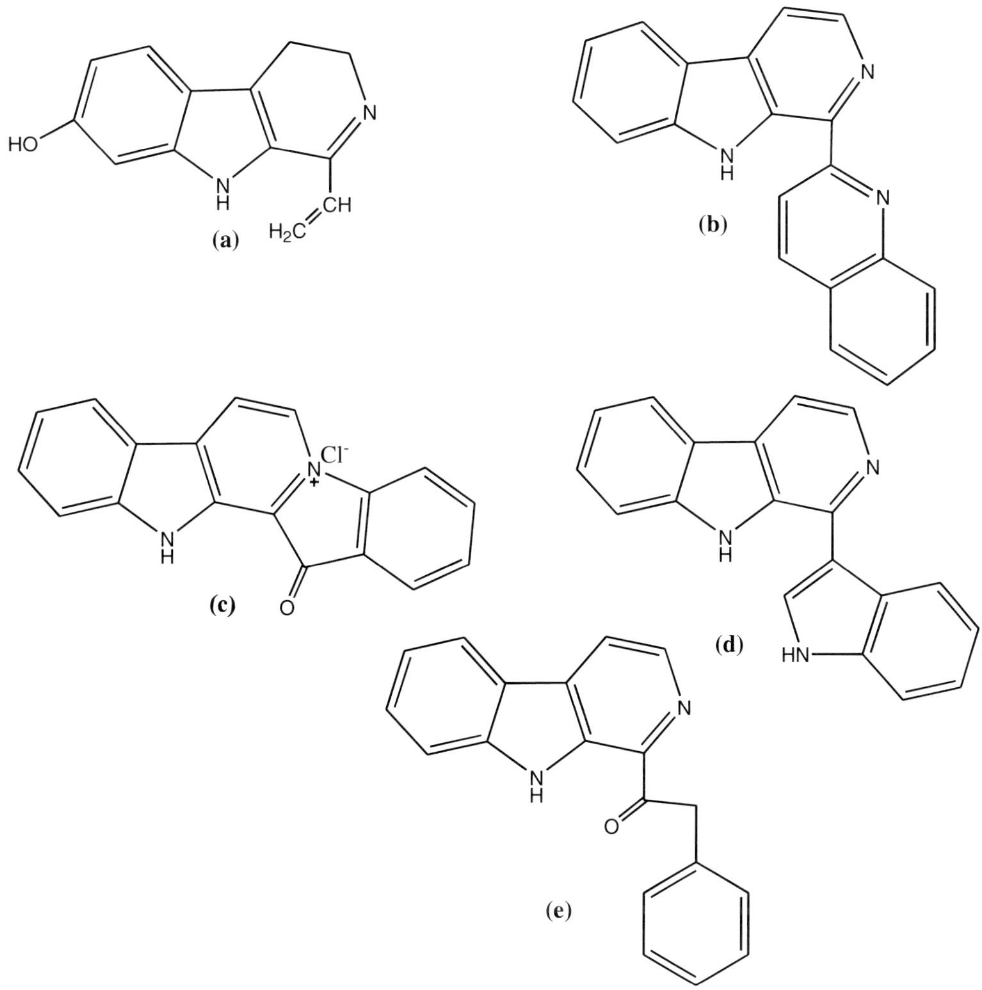

Figure 8: Physiologically active 1-substituted carboline derivatives.

Synthetic Aspects

Carbazole have considerable importance as pharmacotherapeutics. The broad spectrum of medicinal properties of carbazoles has triggered the development of a variety of methods for their synthesis and has led to an impressive armoury of synthetic strategies to be devised in the literature for this class of compounds.

Synthesis of Carbazoles

The synthesis of carbazoles has attracted much interest due to its unique biological activities.

Although many synthetic efforts have been reported, there are a few examples of the efficient construction of carbazole under mild conditions [34].

(a) Pyridine hydrochloride causes an interesting effective loss of water from various Ar-alkyl accompanied Ar-alkoxyl-substituted-1-oxo-1,2,3,4-tetrahydrocarbazole. The formation of the corresponding carbazole is accompanied by demethylation of phenolic methoxyl group to give hydroxycarbazole **2** along with the demethylated ketones **3** after a short time and the hydroxycarbazole **(2)** only after a long time [35]. (Scheme 1)
(b) Tsang and co-workers [36] proposed a method which involves the formation of N-acetyl carbazole (2) from 2-acetaminodiphenyl (1) using palladium catalyst and a reoxidant combination. Preliminary results show that the use of combination of 5% $Pd(OAc)_2$ at 120°C under an atmosphere of air or oxygen provides a near quantitative yield of (2) in toluene (Scheme-2).
(c) Iodoaniline **(1)** first reacts with silylaryltriflate **(2)** to afford the N-arylated-o-iodoaniline **(3)**, which subsequently undergoes intramolecular palladium-catalyzed arylation to give **4**. These can also alternatively undergo reaction in the same pot to produce the carbazole derivatives **(4)** in good yield [37] (Scheme-3).
(d) The deoxygenation [38] of o-nitrobiphenyls**(1)** followed by the insertion of nitrene
(e) into the o-carbon – hydrogen bond is a useful and versatile method for the synthesis of carbazoles [39]**(2)**. Refluxing triethylphosphite is the usual reagent with triethyl phosphate being the by product (Scheme-4).
The oxidative closure of irontricarbonyl complex **(2)** (R=H and Me), prepared by the reaction of an arylamine with the tricarbonyl (cyclohexadienyl) iron cation **(1),** has emerged as potential method for the synthesis of carbazole**3** (Scheme-5).

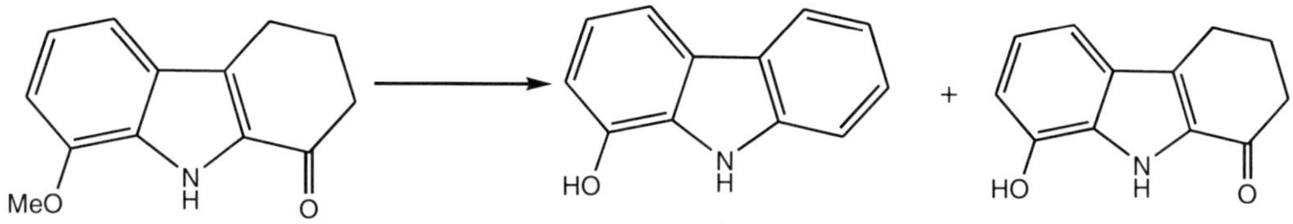

1 **2** **3**

Scheme 1: Pyridine hydrochloride induced hydrogenation and demethylation.

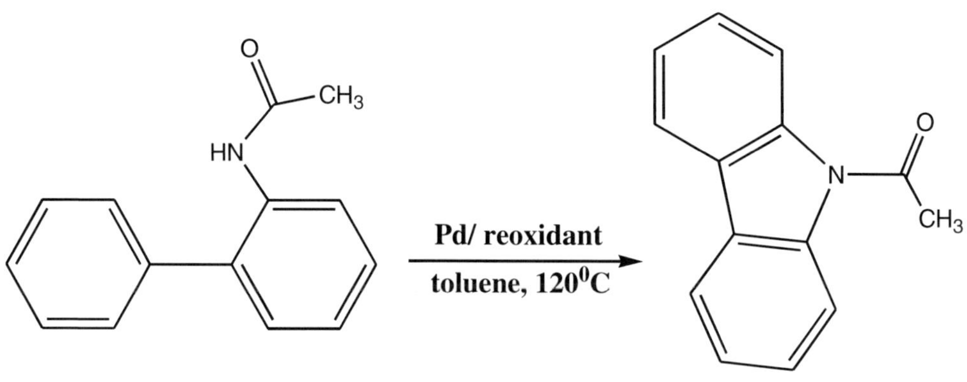

1 **2**

Scheme 2: Formation of N-acetyl carbazole by palladium catalyzed insertion of nitrene product from 2-acetaminodiphenyl.

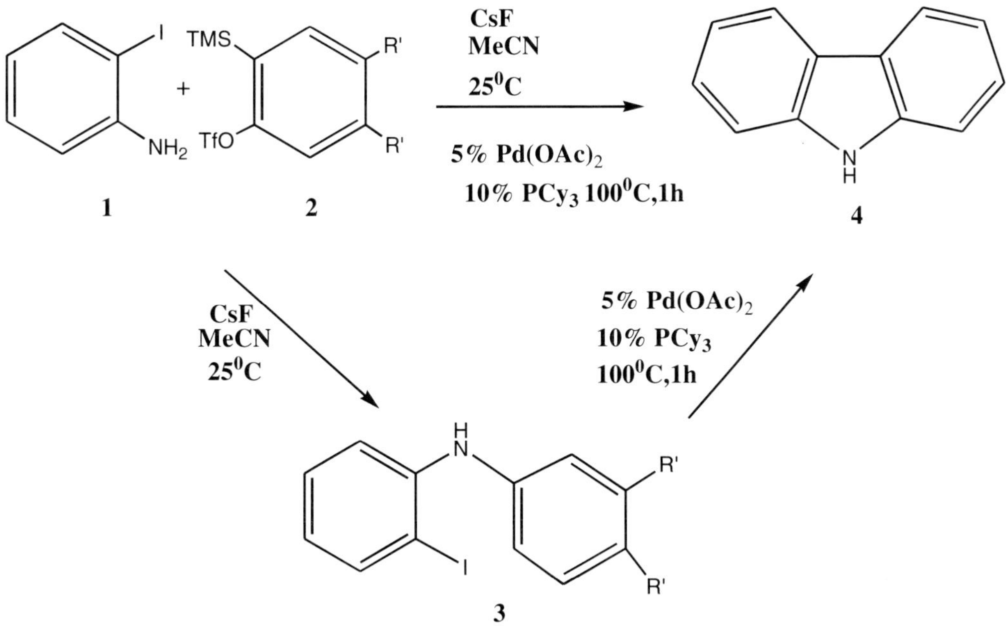

Scheme 3: Palladium catalyzed cyclocondensation of o-iodoaniline + o-trimethylsilyl aryl triflate or N-arylated-o-iodo aniline.

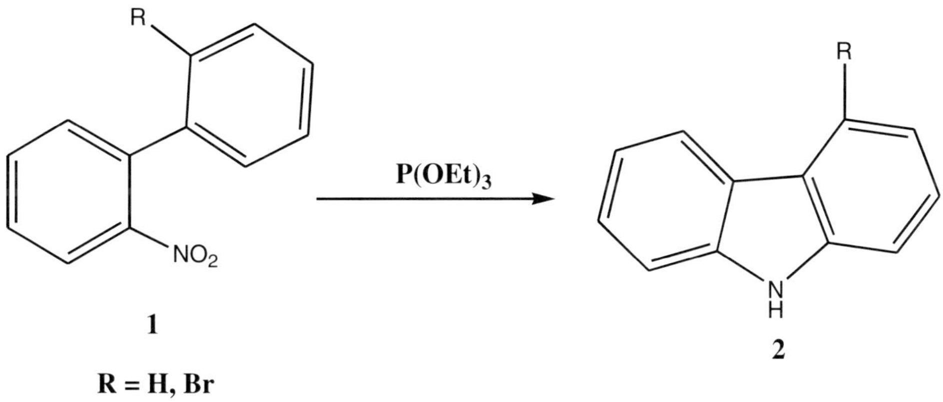

Scheme 4: Triethyl phosphate mediated deoxygenation of o-nitrobiphenyl followed by insertion of the formed nitrene to the o-carbon-hydrogen bond.

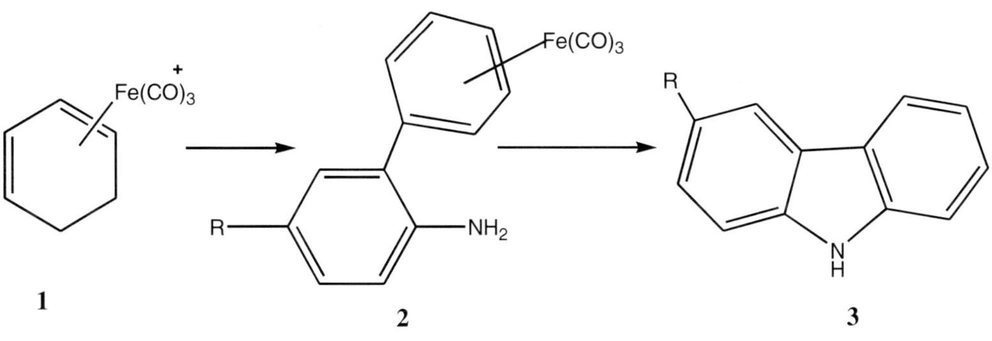

Scheme 5: Oxidative ring closure of arylamine tricarbonyl complex.

(f) An important route to the formation of carbazole ring system involves the thermal or photochemical decomposition of o- azidophenyls(**1**) prepared from the o-amino biphenyls via the diazonium salt. Nitrogen is evolved and a nitrene is produced that effectively inserts into the ortho C-H bond of the second ring to form **2** (Scheme-6).

(g) The direct photolysis of o- isocyanatobiphenyl(**1**) is yet another means for the generation of nitrene, which leads via cyclization to the carbazole(**2**). However, 6(5H)-phenanthridone(**3**) are also produced and are fomed exclusively on sensitized irradiation [40] (Scheme-7).

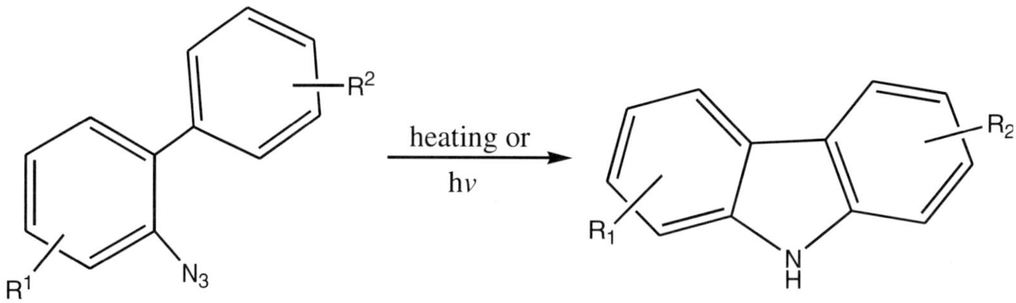

1 **2**

Scheme 6: Formation of carbazole ring system by the thermal or photochemical decomposition or azido biphenyls.

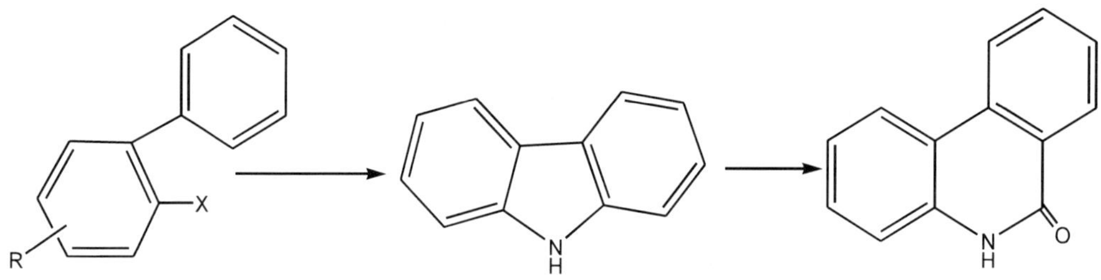

1 **2** **3**

Scheme 7: Formation of carbazole ring by the insertion of nitrene produced by the direct photolysis of o-isocyanato biphenyls.

Synthesis of Azacarbazoles

Synthesis of azacarbazoles (pyridoindoles, carbolines series) has attracted considerable interest due to the recent discoveries of several naturally occurring compounds containing this skeleton. The two best known synthetic approaches for this class of compounds involve construction of either pyridine ring from indole derivatives or synthesis of pyrrole ring via cross coupling between an appropriately substituted pyridine and aniline derivatives.

(a) The palladium catalyzed cross-coupling reaction between phenylboronic acid **(1)** and 3-fluoro-4-iodopyridine **(2)** using the Suzuki conditions gives the corresponding heterobiaryls **(3)** in good yields. The phenylpyridines were cyclized by treatment with boiling pyridinium chloride to give hydroxylcarbolines **(4)** [41, 42] (Scheme-8).

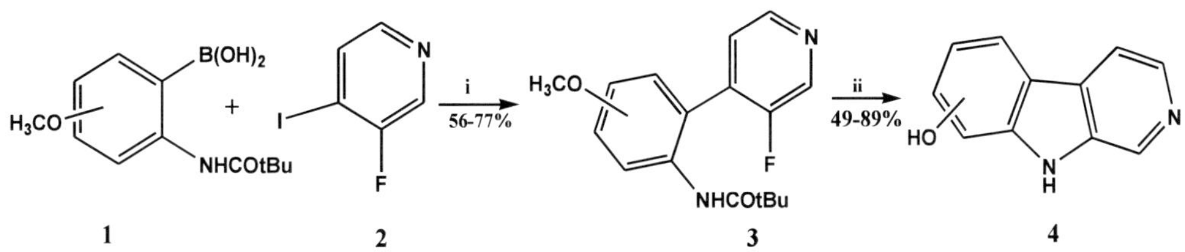

i) Pd(PPh$_3$)$_4$, 2M K$_2$CO$_3$,toluene, reflux 48 h.
ii) 1) Pyridinium chloride, 215^{0}C, 15 min.
2) NH$_4$OH, ice

Scheme 8: Formation of hydroxyl carbolines by the pyridinium chloride induced cyclocondensation of phenyl pyridines.

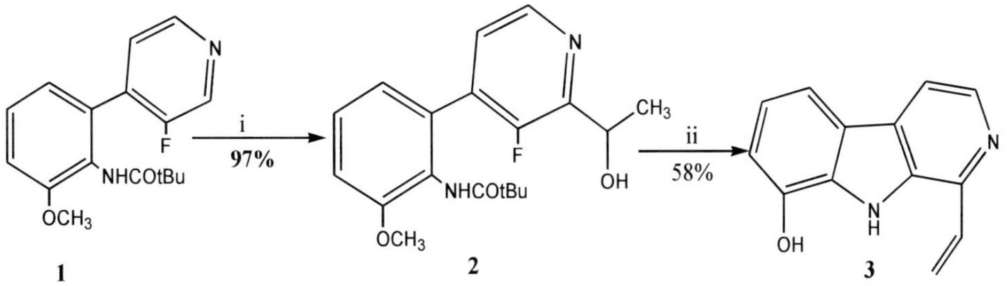

Scheme-9

(i) a) n- BuLi, THF, -75^{0}C, 1h 30 min. **b) CH$_3$CHO, 1h** **c) hydrolysis**
(ii) a) Pyridinium chloride, 215^{0}C, 15 min **b) NH$_4$OH, ice**

Scheme 9: Synthesis of cytotoxic agents produced by Cibricellina cribaria from 2-hydroxy ethyl substituted phenyl pyridines.

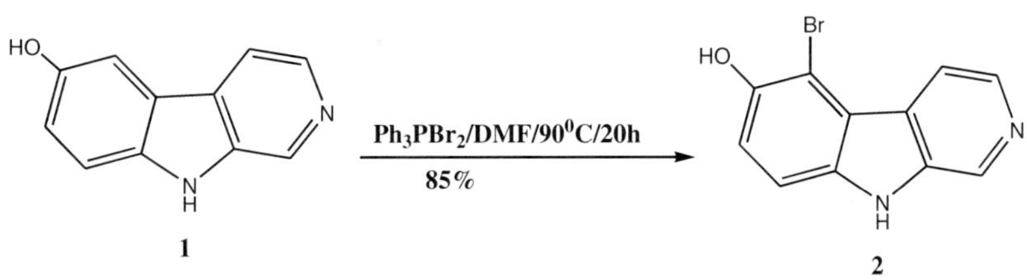

Scheme 10: Formation of 5-bromo-6-hydroxy carboline by triphenyl phosphine di-bromide induced bromination of 6-hydroxy carboline.

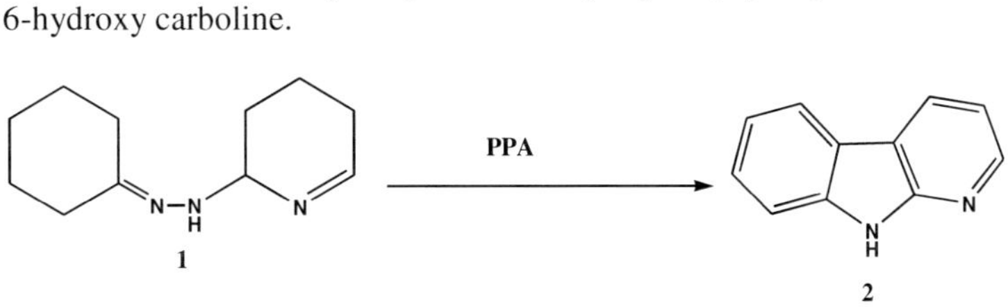

Scheme-11

PPA= Polyphosphoric acid

Scheme 11: Formation of 9H-pyrido [2,3-b] indole from the cyclisation of cyclohexanone-2-pyridyl hydrazine.

(b) The strategy used in *scheme-9* has been successfully extended to the total synthesis of major cytotoxic agents of "Cibricellinacribaria "1-vinyl-8-hydroxycarboline **(3)** from **1** (Scheme-9).

(c) 6-Hydroxycarboline **(1)** when subjected to bromination by triphenyl-phosphinedibromide gives 5-bromo -6-hydroxycarboline **(2)** (eudistomin D) [33] (Scheme-10).

(d) The cyclization of cyclohexanone – 2 – pyridylhydrazone **(1)** to 9*H*-Pyrido [2,3-b] indole **(2)** in 53% yield using polyphophoric acid*(Scheme-11)*is according toOkuda and Robinson cyclization [34].

Chemical Reactions of Carbazole Derivatives

*Formation of substituted N-amide from 1,4-dimethyl carbazole-2,3-diester:*Formation of N-substituted carbazole-2,3-dicarboximide **2** from carbazole derivative **1** takes place in presence of formic acid [35] (Scheme-1).

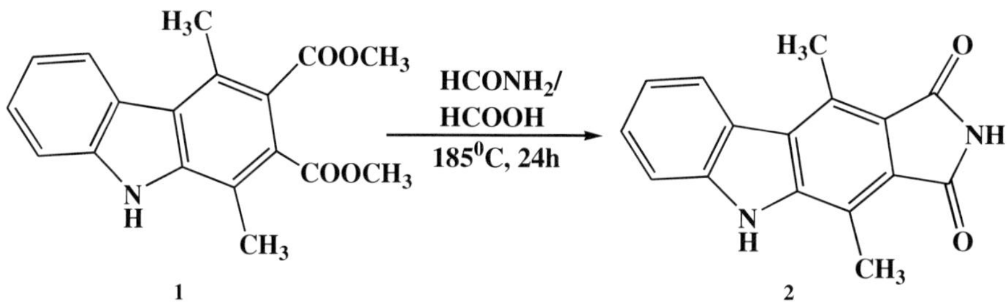

Scheme 12: Formamide/formic acid mediated cyclisation of 1 to 2.

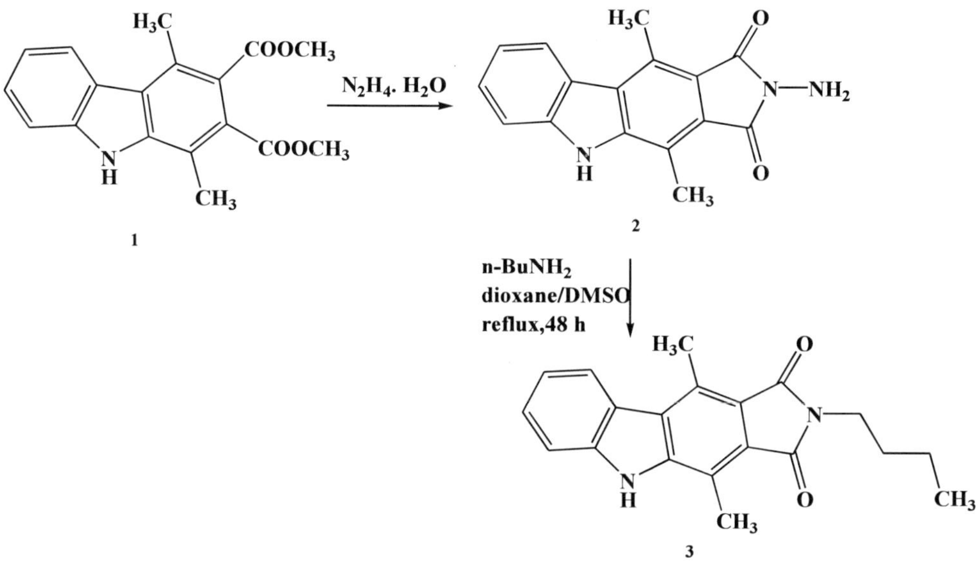

Scheme 13: Formation of N-butyl substituted amide (3) from 1,4-dimethyl carbazole-2,3-diester (1).

Formation of N-butyl substituted amide from 1,4-dimethyl carbazole-2,3-diester: The employment of a reactive intermediate is necessary [36] for preparation of the *N*-butyl substituted imide **3** from **1.** For this purpose, the N-aminoimide**2** [37] which is easily formed on treatment of **1** with hydrazine hydrate can be conveniently used as an "anhydride equivalent", thus giving **3** in good yield upon refluxing in excess n-butyl amine (Scheme-2).

Formation of substituted N-benzyl carbazole from 1,4-dimethyl carbazole-2,3-diester: Carbazole-2,3-diester derivative **1** can be transformed into a N-benzyl carbazole derivative **2** simply by heating with an excess of a high-boiling amine such as benzylamine (Scheme-3).

Formation of diethylaminopropyl-substituted compound 4a and 4b from 1a and 1b: **1a** [38] gives - the diethylaminopropyl-substituted target compound **4a** in high yield. In an analogous fashion, the mono-methyl substituted carbazole-2,3-diester **1b** [39] (lacking the methyl group at position 4) was smoothly cyclized into the imide **4b** (Scheme-4).

Formation of carbazoloazepinone from amino substituted carbazole: 3-Amino-9-ethyl carbazole react with ethyl succinyl chloride to undergo acylation followed by Friedel Crafts reaction to form the product [40] (Scheme-5).

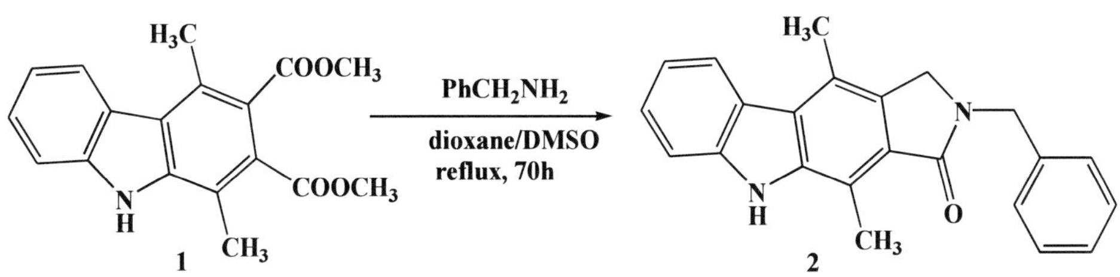

Scheme 14: Formation of N-benzyl carbazole derivatives (2) from 1,4-dimethyl carbazole-2,3-diester (1).

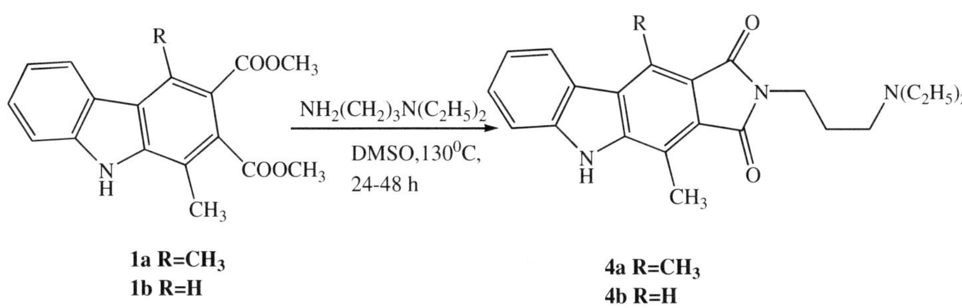

Scheme 15: Formation of diethyl amino propyl substituted derivative 4a and 4b from 1a and 1b.

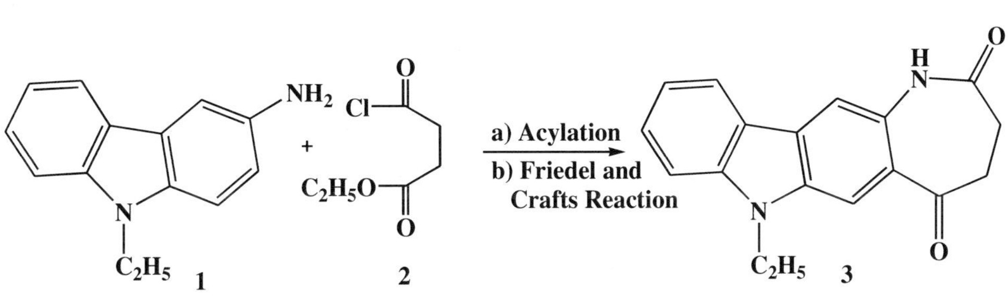

Scheme 16: Formation of carbazolo azepinone from amino substituted carbazole.

*Formation of carbazolo condensed pyrazoloazepinone from oxoketenedithioac*etal. *derivative of carbazole:* **1** reacted with amine hydrochloride and hydrazine hydrate to form the pyrazole and isoxazole derivative of carbazole **2** and **3** respectively [40]. (Scheme-6)

Formation of 4-nitro carbazole from carbazole: This scheme is according with the work of Lindeman [41] who found that treatment of a solution of carbazole in glacial acetic acid with sodium nitrite followed by nitric acid yielded principally 3-nitroso-carbazole which on heating with KOH or glacial acetic acid gave 3-nitro carbazole (Scheme-7).

*Formation of 1-nitro carbazole from carbazole:*1-Nitro carbazole was prepared by nitrating 3,6-carbazoledicarboxylic acid and subsequently decarboxylating through the Greaebe-Ullmann reaction [42] (Scheme-8).

Formation dihydroxycarbazole derivative from carbazole derivative: Bromination of the N-substituted tetrahydrocarbazoles**1** leads to the formation of unstable 10,11 dibromo derivatives 2 which are easily hydrolyzed to give the corresponding dihydroxy derivative **3** [43](Scheme-9).

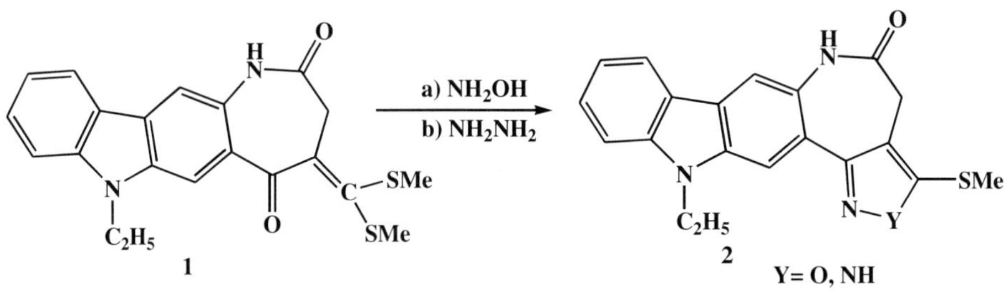

Scheme 17: Formation of carbazolo condensed isooxazolo azepinones 2a and 2b respectively from corresponding isoketene dithioacetyl derivatives (1).

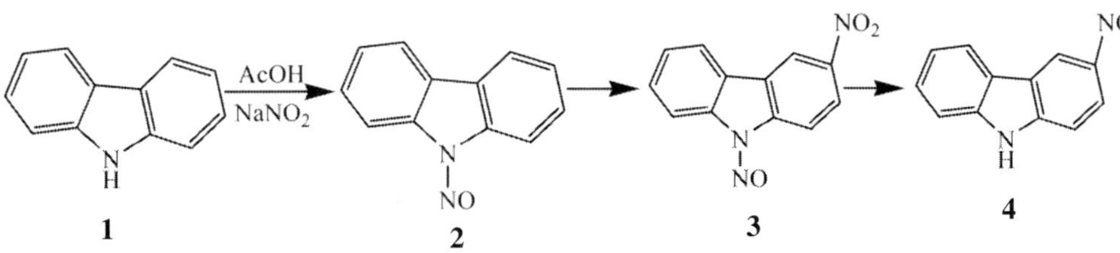

Scheme 18: Preparation of 4-nitrocarbazole (4) from carbazole (1).

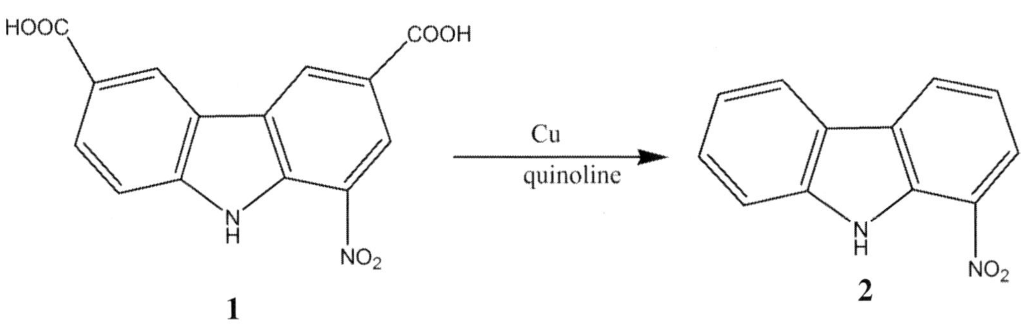

Scheme 19: Preparation of 1-nitro carbazole (2) from the corresponding acid (1).

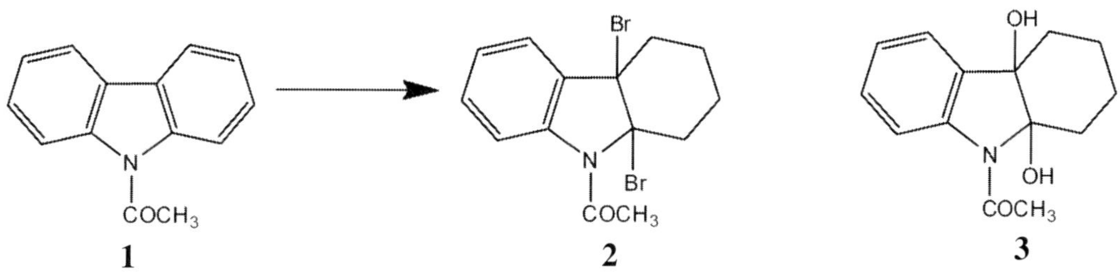

Scheme 20: Preparation of dihydroxycarbazole derivative (3) from the corresponding 1-acetyl carbazole (1).

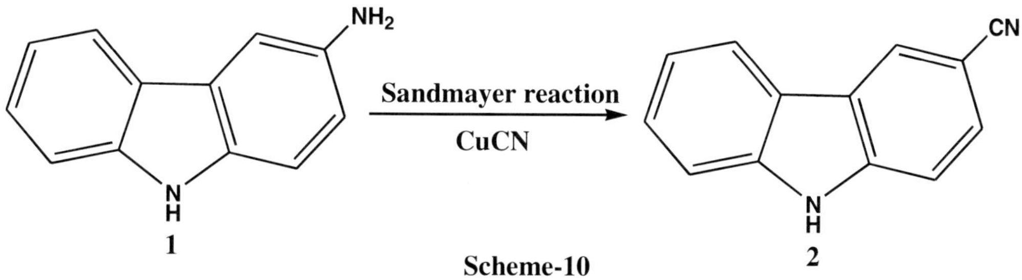

Scheme 21: Preparation of 3-cyanocarbazole (2) from the corresponding 3-amino carbazole derivatives (1).

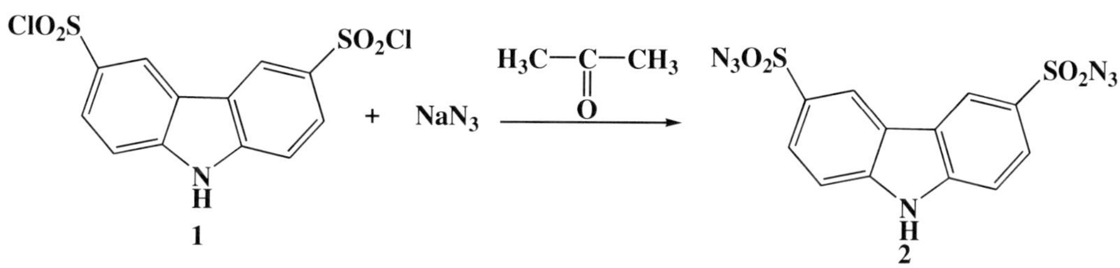

Scheme 22: Preparation of 3,6-disulfonylazide-9H-carbazole (2) from the corresponding sulphonylchloride derivative (1).

Formation of cyanocarbazole from carbazole derivative: 3-Amino carbazole 1 undergo Sandmeyer reaction to give product (2) [43] (Scheme-10).

Preparation of 3, 6-Disulfonylazide-9H-Carbazole

The compound **(1)** reacts with sodiumazide in acetone to give (3,6-disulfonylazide-9H-carbazole) **(2)** [44] (Scheme-11).

Preparation of N-Alkyl Carbazole

Carbazole reacts smoothly with alkyl halides under microwave irradiation. The reaction was carried out by mixing of carbazole with 50% excess of an alkyl halide and a catalytic amount of tetrabutyl- ammoniumbromide (TBAB).

The mixtures were adsorbed on potassium carbonate and irradiated in an open vessel in a domestic microwave oven for 5-10 min. This procedure was a fast synthetic method reported for the formation of N-alkyl carbazoles in dry media under microwave irradiation [45] (Scheme-12).

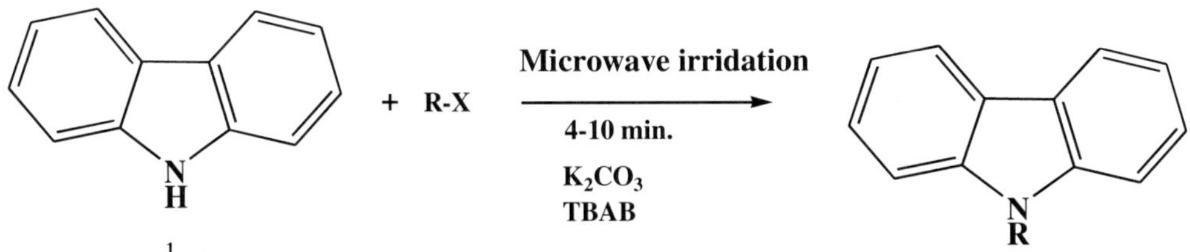

1 2

Scheme 23: Preparation of N-alkyl carbazole 2 from the carbazole (1).

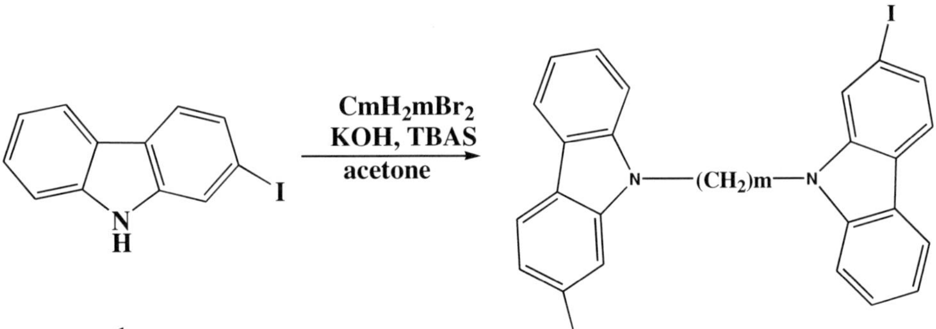

1 2

Scheme 24: Conversion of iodocarbazole to corresponding bis-carbazole derivative (2).

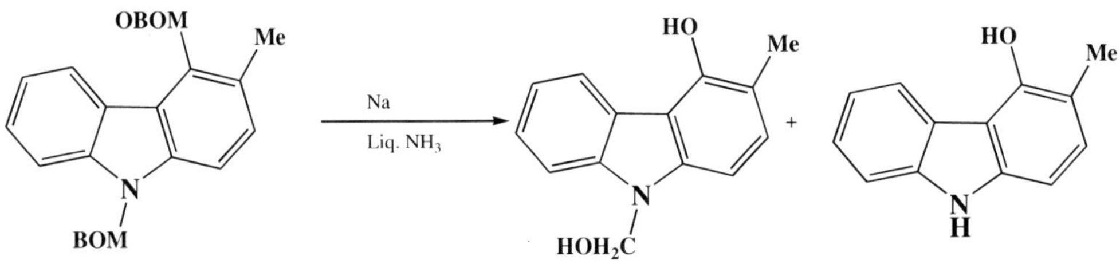

1 2 3

Scheme 25: Preparation of N-hydroxymethyl-4-hydroxy-3-methyl carbazole (2) and 4-hydroxy-3-methylcarbazole (3) by deprotonation of 1 under Birch reduction condition.

Preparation of 1,6-di-(3-iodo-9-carbazolyl)hexane (1) and 1-12-di(3-iodo-9-carbazolyl)dodecane (2)

2 have been prepared by alkylation of 3-iodo-9H-carbazole with dibromoalkanes [46] (Scheme-13).

Preparation of N-Hydroxymethyl-4-Hydroxy-3-Methylcarbazole and 3-Methyl-4-Hydroxycarbazole

Deprotection of compound **1** under Birch reaction conditions gave a mixture of N-hydroxymethyl-4-hydroxy-3-methylcarbazole **(2)** and 4-hydroxy-3-methyl carbazole**(3)** in 75% and 22 % yield, respectively. However, 2 was transformed to 3 with Triton B in a yield of 71 % [46] (Scheme-14).

Preparation of (E)-but-1-en-3-yne and (E,E)-buta-1,3-diene

One-pot synthesis of (E)-9-(2-iodovinyl)-9H-carbazole **(2)** via sequential ruthenium-catalysedsilylative coupling of N-vinylcarbazole with vinyltrimethylsilane and by subsequent iododesilylation. It is used as a new building block in the palladium-catalysedSonagashira and Suzuki-Miyaura coupling reactions to yield new carbazole-containing (E)-but-1-en-3-yne and (E,E)-buta-1,3-diene has been demonstrated [47] (Scheme-15).

Preparation of 3 by Suzuki-Miyaura Cross-CouplingReaction

Palladium catalysed cross-coupling reaction is highly useful for extending the π-system of bromide **2**. For example, the Suzuki-Miyaura cross-coupling of **2** with 4-methoxyphenylboronic acid gives **3** [48] (Scheme-16).

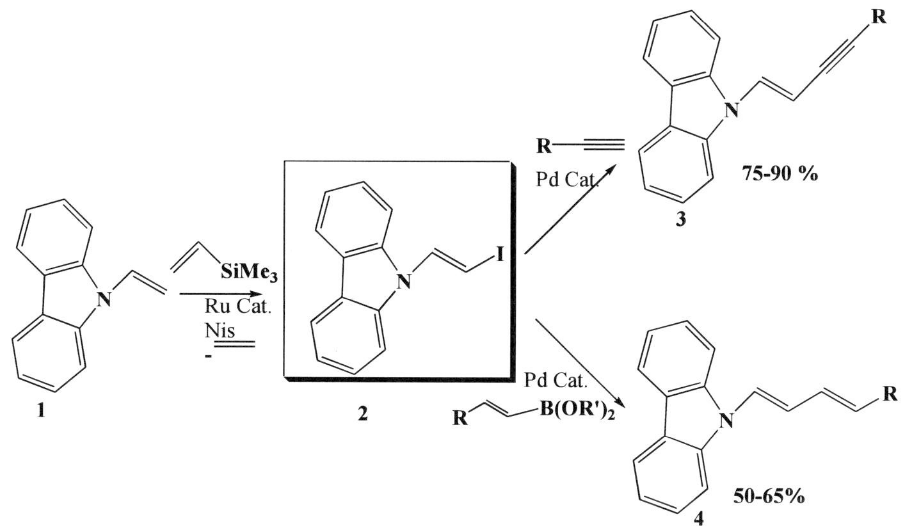

Scheme 26: One pot preparation of [E]-9-(2-iodovinyl)-9H-carbazole (2) via sequential ruthenium catalyzed silylation coupling of N-vinyl carbazole(1) with vinyl trimethylsilane and subsequent iodo silylation.

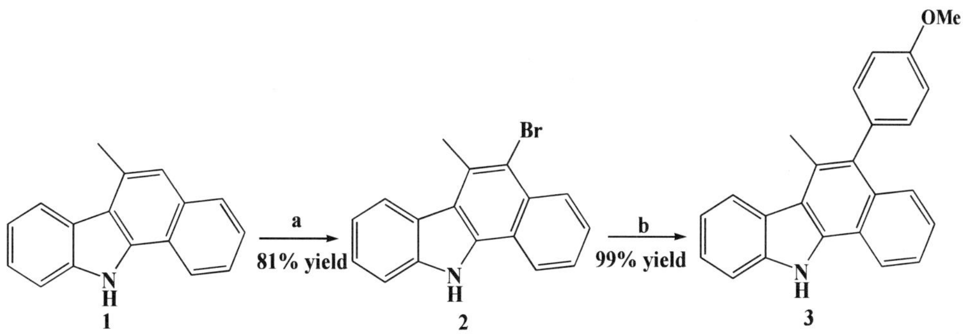

Reagents and conditions: (a) NBS (1.0 equiv), $FeCl_3$ (30 mol %), CH_2Cl_2, 0 °C, 30 min; (b) 4-MeO–C_6H_4–$B(OH)_2$ (2.0 equiv), $Pd(PPh_3)_4$ (20 mol %), K_3PO_4(3.0 equiv), DMF, 100°C, 24 h.

Scheme 27: Preparation of 3 by palladium catalyzed Suzuki-Miyaura cross coupling of 2 with 4-methoxy phenylboronic acid

Preparation of 11H-benzo[a] carbazole

Compound **1** has been demonstrated toundergo carbonylation in presence of SeO_2 oxidation followed by decarbonylation with rhodium-catalyst (Wilkinson's catalyst) which gives **2** and **3** [48] (Scheme-17).

Preparation of 6-methyl-benzo[a]carbazole-11-carboxylic acidtert-butyl ester

Compound **1** has been reported to react with di-tert-butyl dicarbonate [$(Boc)_2O$] with the 4-(*N,N*-dimethylamino)pyridine (DMAP) to give **2** in high yield [48] (Scheme-18).

Preparation of 6-methyl-11-pyrimidin-2-yl-1H-benzo[a]carbazole

Na complex of **1** prepared in situ from **1** and NaH reacted with 2-chloropyrimidine through S_NAr process to give **2** [48] (Scheme-19).

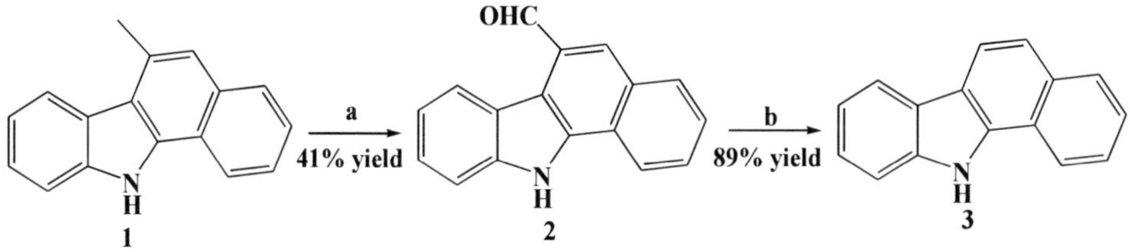

Reagents and conditions: (a) SeO_2 (1.5 equiv), K_2CO_3 (0.5 equiv), pyridine, 115 °C, 19h; (b) $RhCl(CO)(PPh_3)_2$ (10 mol %), $Ph_2P(CH_2)_3PPh_2$(25 mol %), xylenes, 140 °C, 24 h.

Scheme 28: Formation of 3 from 1 by oxidation with SeO2 folowed by decaroxylation with Wilkinson's catalyst.

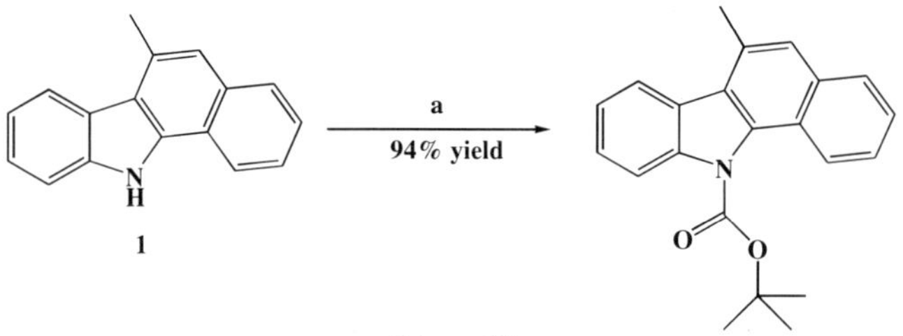

Scheme-18

Reagents and conditions: (a) $(Boc)_2O$ (2.0 equiv), DMAP (1.0 equiv), MeCN, rt, 5 h.

Scheme 29: Preparation of 6-methyl-benzo (a) carbazole-11-carboxylic acid t-butyl ester (2) by alkylation of (1) with t-butyric anhydride.

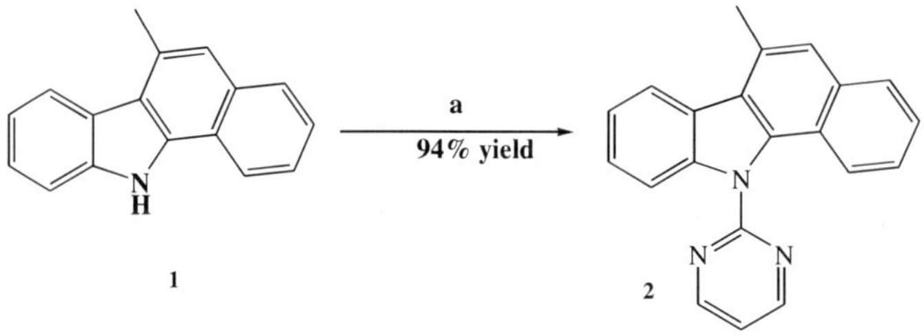

Reagents and conditions: (a)2-Cl–pyrimidine (1.8 equiv), NaH (2.2 equiv), DMF, 130°C, 25 h.

Scheme 30: Preparation of (2) from (1) by alkylation of later with 2-chloro-pyrimidine.

Preparation of [4-(6-methyl-benzo[a]carbazol-11-yl)-phenyl-diphenyl-amine

Compound **1** undergoes Buchwalk*N*-arylation reaction to give [4-(6-methyl-benzo[a]carbazol-11-yl)-phenyl-diphenyl-amine**2** [48] (Scheme-20).

Preparation of compound 3 fromcarbazolederivative 1

Aromatization of compound 1 was carried out using either excess of activated MnO_2 or 2-5 equivalent of DDQ to give compound **2**. This was further demethylated to give **3**, using either pyridine hydrochloride at 200°C or 2 -10 equivalent of BBr_3 in CH_2Cl_2 at room temperature [49] (Scheme-21).

Preparation of 4-iodo-9-methoxy-6H-pyrrolo[3,4-c]carbazole-1,3-dione (5a) or 4-iodo-9-methoxy-6-methyl-6H-pyrrolo[3,4-c]carbazole-1,3-dione (5b)

Compounds **5a** and **5b** were formed through the following sequence of reactions in presence of given reagents [49] (Scheme-22)

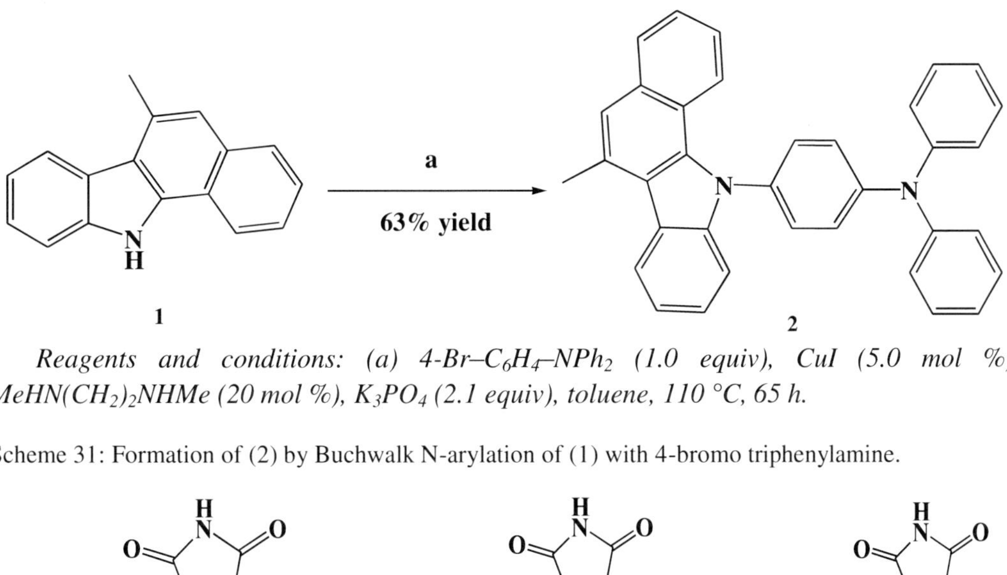

Reagents and conditions: (a) 4-Br–C_6H_4–NPh_2 (1.0 equiv), CuI (5.0 mol %), MeHN(CH_2)$_2$NHMe (20 mol %), K_3PO_4 (2.1 equiv), toluene, 110 °C, 65 h.

Scheme 31: Formation of (2) by Buchwalk N-arylation of (1) with 4-bromo triphenylamine.

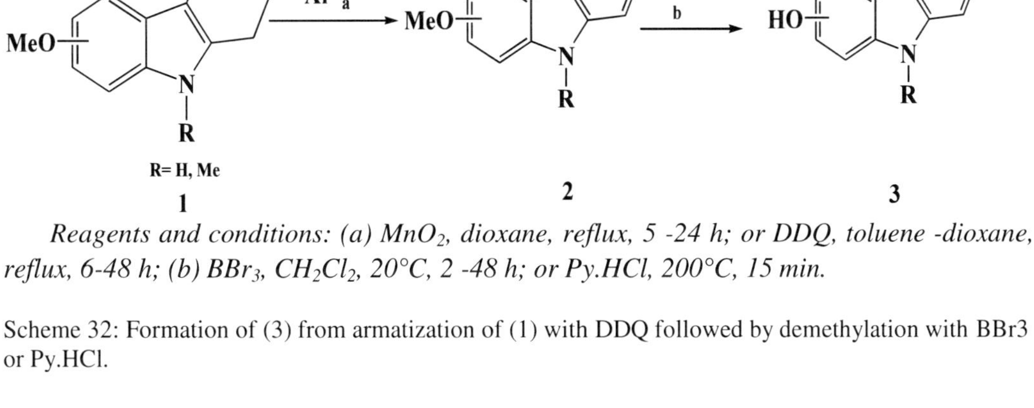

Reagents and conditions: (a) MnO_2, dioxane, reflux, 5 -24 h; or DDQ, toluene -dioxane, reflux, 6-48 h; (b) BBr_3, CH_2Cl_2, 20°C, 2 -48 h; or Py.HCl, 200°C, 15 min.

Scheme 32: Formation of (3) from armatization of (1) with DDQ followed by demethylation with BBr3 or Py.HCl.

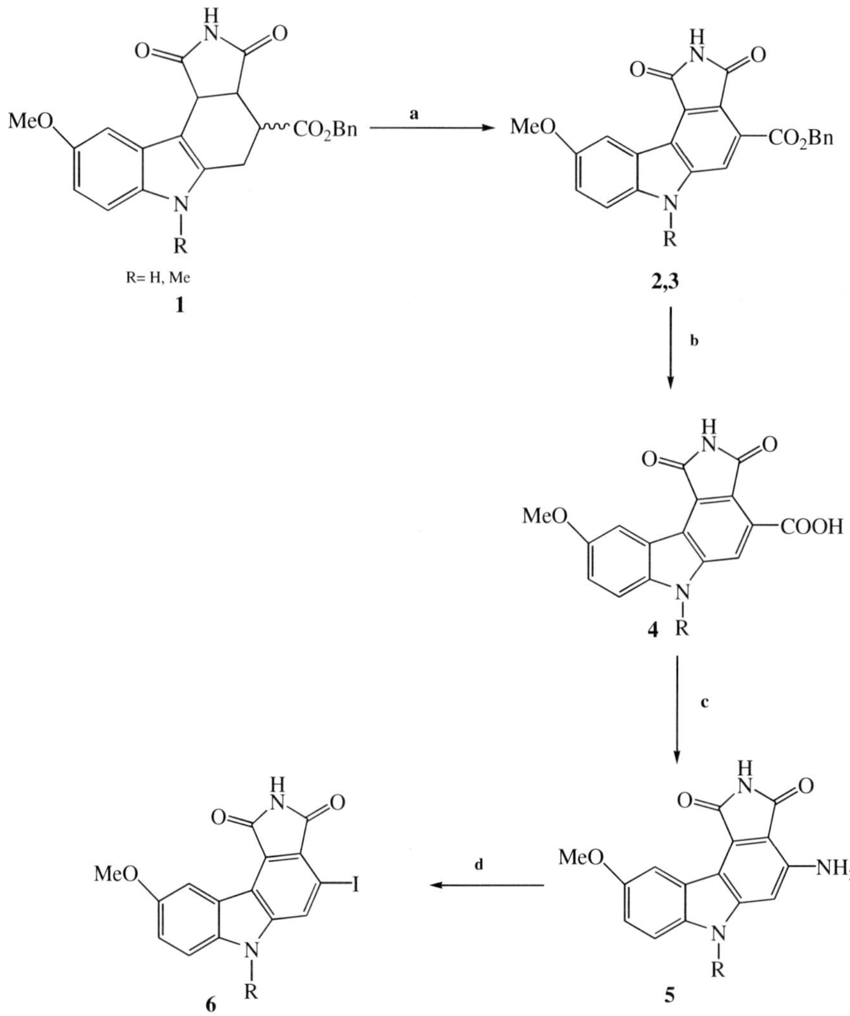

(a) MnO_2, dioxane, reflux, 2 h; (b) H_2/Pd-C, DMF-MeOH, 60 psi, 2 h; (c) DPPA, Et_3N, t-BuOH, reflux 16 h; then TFA, DCM, 20 °C, 1 h; (d) aqueous H_2SO_4, $NaNO_2$, 3°C, 3 min;then KI, CuI, 70 °C, 1 h.

Scheme 33: The sequence of reactions depicted in scheme 22 shows the formation of (6) from (1).

Preparation of Mukanol

2-Methoxy-3-methylcarbazole led to further 2-oxygenated carbazole alkaloids such as O-methylmukonal(**2**), and mukonal(**3**) by simple transformation. The oxidation of compound **1** with DDQ afforded O-methylmukonal (**2**) which on demethylation with boron tribromide form mukanol **(3)** [3] (Scheme-23).

Preparation of 2-methyl-1-phenyl-9H-carbazol-3-ylamine

The carboxylic group was transformed to an amino group via the intermediate isocyanate 2 in the presence of reagents H_2O_2 and NaOH [3] (Scheme-24).

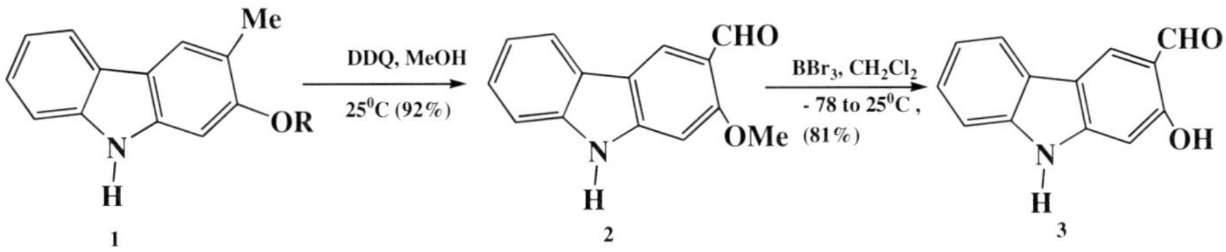

Scheme 34: Oxidation of (1) with DDQ (based on scheme-17) and demethylation with BBr3 (based on scheme-21) were applied on (1) in its conversion to (3).

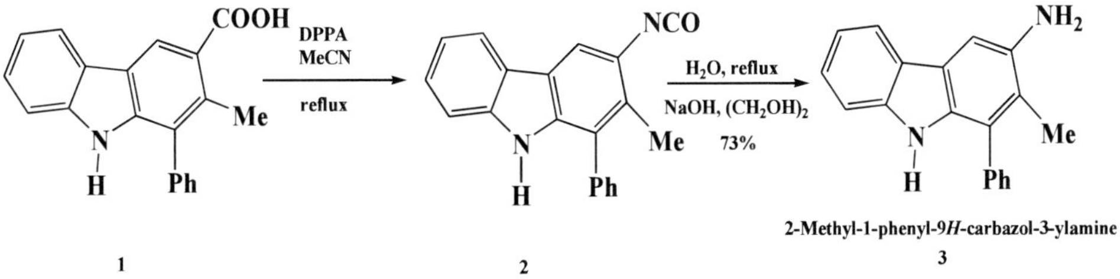

Scheme 35: Preparation of 2-methyl-1-phenyl-9H-carbazol-3-ylamine (3) from (1) by its conversion to isocyanate (2) followed by hydrolysis.

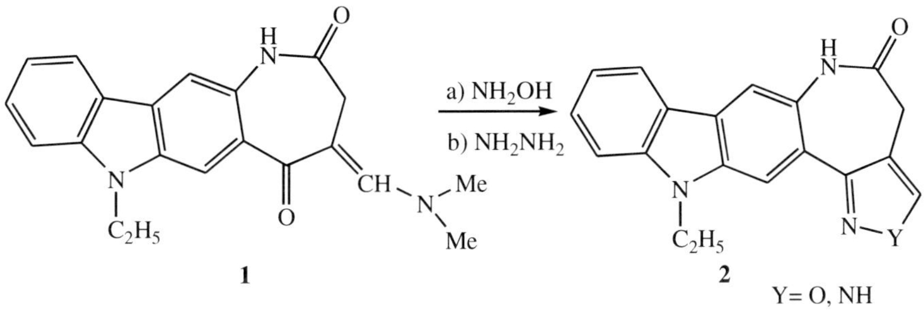

Scheme 36: Formation of the isoxazole and pyrazole derivative from the reaction of corresponding 2-dimethylaminomethylene ketone derivative (1) with NH2OH-HCl and H2NNH2 respectively.

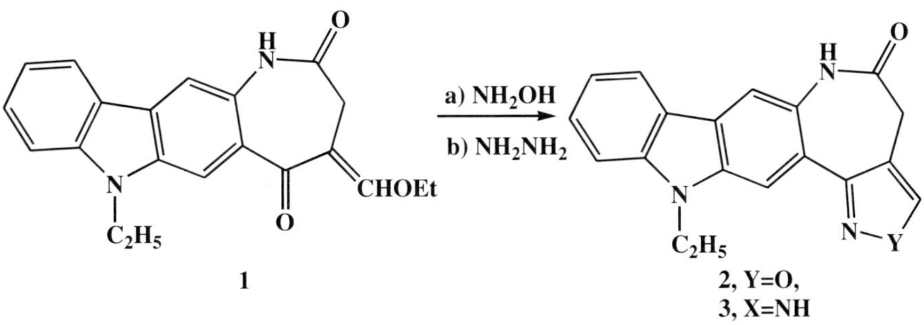

Scheme 37: Preparation of 2a and 2b from the enolic ether derivative (1).

Formation of carbazolo condensed pyrazolo/isoxazoloazepinonefrom the corresponding 2-(dimethylaminomethylene) ketone derivative

Hydroxyl amine hydrochloride and hydrazine hydrate reacted with **1** to give pyrazole and isoxazole derivative of carbazole **2** from **1** [40] (Scheme-25).

Formation of carbazolo condensed pyrazoloazepinone from the corresponding enolic ether derivative: Compound 1 reacted with hydroxyl amine hydrochloride and hydrazine hydrate to afford the corresponding pyrazole and isoxazoles derivative 2 [40] (Scheme-26).

REFERENCES

[1] Ito, C., Itoigawa, M., Nakao, K., Murata, T.,Tsuboi, M., Kaneda, N., Furukawa, H. (2006). Induction of apoptosis by carbazole alkaloids isolated from Murrayakoenigii. *Phytomedicine,* 13, 359-365. doi:10.1016/J.phymcd.2005.03.010.

[2] Ningappa, M. B., Srinivas, L. (2008). Purification and characterization of approximately 35 k Da antioxidant protein from curry leaves (Murrayakoenigii L.). *Toxicology in Vitro*, 22, 699-709. doi:10.1016/j.foodchem.2009.05.059.

[3] Knolker, H. J., Reddy, K. R. (2002). Isolation and synthesis of biologically active carbazole alkaloids. *Chem. Rev.,* 102, 4303. doi:10.1021/cr020059j.

[4] Sakano, K., Ishimaru, K., Nakamura, S. (1980). New Antibiotics, Carbazomycins A and B. *J. Antibio.*, 33, 683. doi:10.1111/j.1574-6968.2011.02335.x.

[5] Grammel, H., Wolf, H., Gilles, E. D., Huth, F., Laatsch, H. (1998). Carbazole antibiotics synthesis in a streptomycestendae bald mutant, created by acriflavine treatment. *Journal of Bioscience*, 53, 325-30. PMID: 9679324.

[6] Yaqub, G., Hannan, A., Akbar, E., Muhammad, U., Almas, H., Sadiq, Z., Iqbal, M. (2013). Synthesis, Antibacterial, and Antifungal Activities of Novel Pyridazino Carbazoles. *Journal of chemistry*, 7. http://dx.doi.org/10.1155/2013/818739.

[7] Kuoa, P. L., Hsub, Y. L., Changc, C. H., Linb, C. C. (2005). The mechanism of ellipticine-induced apoptosis and cell cycle arrest in human breast MCF-7 cancer cells. *Cancer Letters*, 223, 293–301. doi:10.1016/j.canlet.2004.09.046.

[8] Kaur, K., Jain, M., Kaur, T., Jain, R. (2009). Antimalarials from nature. *Bioorganic and Medicinal Chemistry*. doi:10.1016/j.bmc.2009.02.050.

[9] Asche C., Demeunynck M. (2007). Antitumor carbazoles. *Anticancer Agents Med. Chem.*, 7, 247-67. doi: 10.2174/187152007780058678.

[10] Kirsch, G. H. (2001). Heterocyclic Analogues of Carbazole Alkaloids. *Current Organic Chemistry,* 5, 507-518.doi: http://dx.doi.org/10.2174/1385272013375409.

[11] Crich, D., Rumthao, S. (2004). *Synthesis of carbazomycin B by radical arylation of benzene. Tetrahedron,* 60, 1513–1516. doi.org/10.1016/j.tet.2003.12.015.

[12] Hook, D. J., Yacobucci, J. J. O'Connor S., Lee, M., Kerns, E., Krishan, B., Matson, J., Hesler, G. (1990). Identification of the inhibitory activity of carbazomycins B and C against 5-lipoxygenase, a new activity for these compounds. *The Journal of Antibiotics,*43 (10), 1347-1348. doi.org/10.7164/antibiotics.43.1347.

[13] Yang, S., Alkayed, N. J., Hurn, P. D., Kirsch, J. R. (2009). Cyclic Adenosine Monophosphate Response Element-Binding Protein Phosphorylation and Neuroprotection by 4-Phenyl-1-(4-Phenylbutyl) Piperidine (PPBP), *Anesth. Analg.,* 108,964–970.

[14] Haider, N. (2006). Diazine analogues of the pyridocarbazole alkaloids. *Curr. Org. Chem.,* 10, 363-375.

[15] Haider, N., Käferböck, J. (2004). Intramolecular [4+2] cycloaddition reactions of indolylalkylpyridazines: synthesis of annulated carbazoles. *Tetrahedron*, 60, 6495-6507.

[16] Mutkins, K. ,Chen, S. S. Y., Pivrikas, A., Aljada, M., Burn, P. L. ,Meredith, P., Powell, B. J. (2013).Three-dimensionalcarbazole-based dendrimers: model structures for studying charge transport in organic semiconductor films. *Polym. Chem.,* 4, 916-925.olyl)methyl, N-bis(trimethylsilyl)methyl diazoamides: an entry into the β-carboline ring system. *Chem. Commun.*, 4837-4839. doi: 10.1039/B809890E.

[17] Bayraktar N., Devay, S. D., Taşlipina, M. Y. OmerogluSgumuslu, S., Kavutcu, M., Canbolat, O. (2012). The effects of stobadine on purine metabolism in rats treated with carbon tetrachloride. *Turk. J. Med. Sci.*, 42(5), 894-900. doi:10.3906/sag-1109-37.

[18] Meyer, K. U., Pindur U. (2001). Novel bis (tetrahydropyrrolo[3,4-b]carbazoles) linked with aliphatic chains: synthesis and structural aspects. *J. Chem. Soc.*, 6, 695–700. doi: 10.1039/b009731o.

[19] Bisagmi, E., Ducroq, C., Hung, N. C. (1980). Aromatization des dihydro-3,4-carbazole(2H)ones-1 par le chlorure de pyridinium. *Tetrahedron,* 36(10), 1327-1330. http://dx.doi.org/10.1016/0040-4020(80)85044-7.

[20] Tsang, W. C. P., Zheng, N., Buchwald, S. L. 2005. Combined C-H Functionalization C-N Bond Formation Route to Carbozles. *J. Am. Chem. Soc.,* 127, 14560.doi: 10.1021/ja055353i.

[21] Liu, Z., Larock, R. C. (2007). Tetrahedron. Author manuscript; available in PMC 2008. *Tetrahedron,* 63(2), 347–355. doi: 10.1016/j.tet.2006.10.071.

[22] Song, J. J., Reeves, J. T., Fandrick, D. R., Tan, Z., Yee, N. K., Senanayake, C. H. (2010). Construction of the indole nucleus through C-H functionalization reactions. *ARKIVOC*, 1, 390-449.

[23] Iddon, B., Meth, C. O., Scriven, E. F. V. (2003). Developments in arylnitrene chemistry: Synthesis and mechanism (new synthetic methods). *Angewandte Chemie (Intertional edition in English),* 18(12), 900-917.doi: 10.1002/anie.197909001.

[24] Swenton, J. S., Theodore, J. L., Smyser, G. L. R. (1973). Effect of biphenyl geometry and substituents on the multiplicity and efficiency of the photocyclization reaction of 2-substituted biphenyls, *J. Org. Chem.*, 38(6), 1157-1166. doi:10.1021/jo00946a021.

[25] Larsen, L. K., Moore, R. E., Patterson, G. M. L. (1994). Carbolines from the Blue–Green algae Dichothrixbauerian. *J. Nat. Prod.,* 57(3), 419–421. doi: 10.1021/np50105a018.

[26] Rocca, P., Cochennec, C., Marsis, F., Dit – Dumont, T., Mallet, M., Godard, A., Queguiner, G. (1993).First metallation of aryliodide: Directed o-lithiation of iodopyridines Halogen Dance and application to synthesis. *J. Org. Chem.,* 58(27), 7832-7838.doi: 10.1021/jo00079a031

[27] Rocca, P., Marsis, F., Godard, A., Queguiner, G. (1993). Connection between metallation and cross coupling strategies: A convergent route to azacarbazoles. *Tertrahedron*, 49(1), 49-64.

[28] Okuda, S., Robinson, M. M. (1959). 7-Azaindole. V. Investigation of alternative synthesis of the ring system. *J. Am. Chem. Soc.*, 81(3), 740.doi: 10.1021/ja01512a057.

[29] Moody, C. J. J. (1985). *Chem. Soc., Perkin Trans.*, 1, 2505-2508.

[30] Haider, N., Jbara, R., Kaferbock J. and Traar, U. (2009). Synthesis of tetra- and pentacycliccarbazole-fused imides as potential antitumor agents. *ARKIVOC*, 6, 38-47. doi: http://dx.doi.org/10.3998/ark.5550190.0010.605.

[31] Haider, N., Wanko, R. (1999). 2-Amino-4,10-dimethylpyrrolo[3,4-b]carbazole-1,3(2H,5H)-dione. *Molecules,* 4(10), M105.doi:10.3390/M105.

[32] Agrawal, M., Jain, S., Agarwal, A., Dwivedi, J., Sharma, S., Kishore, D. (2012).Application of novel precursors derived from carbazolo condensed azepinones to the direct single step synthesis of corresponding isoxazole and pyrazole annulated analogues of medicinal importance. *Orient. Pharm. Exp. Med.*, 12,141–150.doi: 10.1007/s13596-012-0060-8.

[33] Kaimanakova, S. I., Kuleshova, E. F., Soloveva, N. P., Granik, V. G. (1982). Reactions of amide and lactam acetals with derivatives of urea and urethane and synthesis of condensed pyrimidines. *Chemistry of Heterocyclic Compounds*, 18(11),1208-1211.doi: 10.1007/BF00506376.

[34] Preston, R. W. G., Tucker, S. H., Cameron, J. M. L. (1942). The Graebe–Ullmann synthesis of carbazole derivatives. Preparation and synthesis of 1-nitrocarbazole. *J. Chem. Soc.*, 500-504.doi: 10.1039/JR9420000500.

[35] Doz, Pielichowski, J., Puszyński, A. (1974). *Eineneue Methodezur Nitrierung des Carbazols und seiner Derivate,* 105(4), 772-774. doi: 10.1007/BF00912975.

[36] Khitam, T. A., Sultani, Al. (2010). *Journal of Al-Nahrain University_Science*, 13(4), 31-38.

[37] [45] Bogdal, D., Pielichowski, J., and Jaskot, K. (2010). New Synthesis Method of N-Alkylation of Carbazole Under Microwave Irradiation in Dry Media. Synthetic Communications: *An International Journal for Rapid Communication of Synthetic Organic Chemistry,* 27 (9), 1553-1560. http://dx.doi.org/10.1080/00397919708006093.

[38] Yaqub, G., Hussian, E. A., Rehman M. A. (2009). Advancements in syntheses of carbazole and its derivatives. *Asian journal of chemistry*, 21(4), 2485-2520.

[39] Pawlu, P., Franczyk, A., Walkowiak, J., Hreczycho, G., Kubicki, M., Marciniec, B. (2011). (E)-9-(2-Iodovinyl)-9 H -carbazole: A New Coupling Reagent for the Synthesis of π -Conjugated Carbazoles. *Organic Letters,* 13, 8. doi: 10.1021/ol200350a.

[40] Tsuchimoto, T., Matsubayashi, H., Kaneko, M., Nagase, Y., Miyamura, T., Shirakawa, E., (2008). Indium-Catalyzed Annulation of 2-Aryl- and 2-Heteroarylindoles with Propargyl Ethers: Concise Synthesis and Photophysical Properties of Diverse Aryl- and Heteroaryl-Annulated[a]carbazoles. *J. Am. Chem. Soc.,* 130, 15823 – 158359. doi:10.1021/ja803954e.

[41] Palmer, B. D., Thompson, A. M., Booth, R. J., Dobrusin, E. M., Kraker, A. J., Lee, H. H., Lunney, E. A., Mitchell, L. H., Ortwine, D. F., Smaill, J. B., Swan, L. M., Denny, W. A. (2006). 4-Phenylpyrrolo[3,4- c]carbazole-1,3(2 H,6 H)-dione Inhibitors of the Checkpoint Kinase Wee1. Structure-Activity Relationships for Chromophore Modification and Phenyl Ring Substitution. *J. Med. Chem.*, 49, 4896-4911. doi:10.1021/jm0512591.

In: Natural Heterocycles
Editors: K. L. Ameta and Saksit Chanthai

ISBN: 978-1-63463-424-3

Chapter 11

EXTRACTION AND BIOLOGICAL EVALUATION OF SOME FLAVONOIDS

Seema Kothari[1], Shiv N. Paliwal[2] and K. L. Ameta[3*]
[1]Department of Chemistry, Pacific College of Engineering, PAHER University, Udaipur, India
[2]Deparment of Chemistry, Career Point Technical Campus, Career Point University, Rajsamand, Rajasthan, India
[3]Department of Chemistry, FASC, Mody Institute of Technology and Science, Lakshmangarh, Rajasthan, India

ABSTRACT

Flavonoids are the most important naturally occurring bioactive compounds present in almost every sort of fruit, vegetable and in the products derived from them. Flavonols may be found mainly in fruits and vegetables, while flavones are abundant in herbs and spices. Flavanones are primarily found in a variety of citrus fruits. Many studies have shown, that intake of fruits and vegetables with high flavonoid content is associated with lowered risk of incidence of some diseases such as cardiovascular or cancer etc. These findings are attributed to experimentally confirmed biological effects of flavonoids such as antioxidant, anti-inflammatory, anti-allergic, anti-cancer, cardio protective, anti-estrogenic, anti-microbial, anti-viral, anti-HIV, anti-biotic, anti-cataract etc.

Keywords: Natural flavonids, extraction, biological activities

1. INTRODUCTION

It is estimated that about 2% of all carbon photosynthesized by plants is converted into flavonoids or closely related compounds [1]. The word flavonoid comes from the Latin flavus, which means yellow; however some flavonoids are red, blue, purple or white [2] in color. Flavonoid thus constitutes one of the largest groups of naturally occurring phenols [3, 4] which is largely responsible for the color of many fruits and flowers. As per the recent study, over 4000 flavonoid compounds have been characterized [5]. Majority of flavonoids occurs in fruits, vegetables and beverages like tea, coffee, beer, wine and fruit drinks [6-9]. The flavonoids have aroused considerable interest recently because of their potential beneficial effects on human health due to their applications such as anti-viral, anti-allergic, anti-platelet, anti-inflammatory, anti-tumor and anti-oxidant agent etc.

2. Chemical Tests of Flavonoids

Following tests are widely used for the identification of flavonoids [10].

(i) Shinoda test: To dry powder or extract, add 5ml of 95% ethanol, few drops of conc. HCl and 0.5g magnesium turnings, pink color observed.
(ii) To small quantity of residue, add lead acetate solution. Yellow colored precipitate is formed.
(iii) Addition of increasing of sodium hydroxide to the residue show yellow coloration, which discoloration after addition of HCl acid.
(iv) Flavonoids give dull green or a red brown color with ferric chloride.

3. Classification of Flavonoids

All flavonoids contain fifteen carbon atoms in their basic nucleus and these are arranged in a C_6-C_3-C_6 configuration, there are, two aromatic rings linked by a three carbon unit which may or may not form a third ring [11]. For convenient the rings are labeled A, B and C and the individual carbon atoms are referred to a numbering system which utilize ordinary numerals for the A and C rings and 'primed' numerals for the B ring (Figure 1).

Flavonoids include a series of subclasses with a common general structure characterized by the presence of two aromatic rings linked by a three-carbon bridge. These subclasses, are classified based on the connection between the two aromatic rings, as well as the oxidation state of the different rings, are: flavones, flavonols, flavanone, isoflavonoid, biflavone, neoflavone, xenthones and anthocyanidine [12-15]. The flavonoids can be divided into nine types all are summarized as in Table 1.1

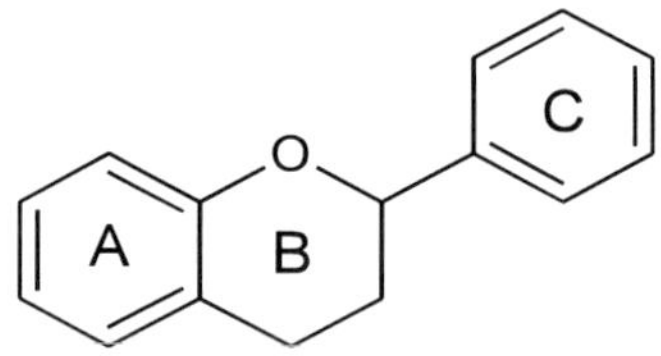

Figure 1. Basic structure of flavonoids.

Table 1.1. Flavonoids and their structure

Types of flavonoid	Structure
1. Flavone	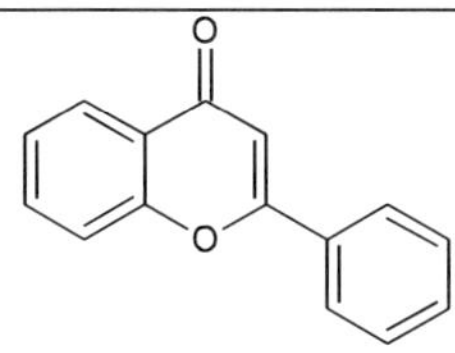
2. Flavonol	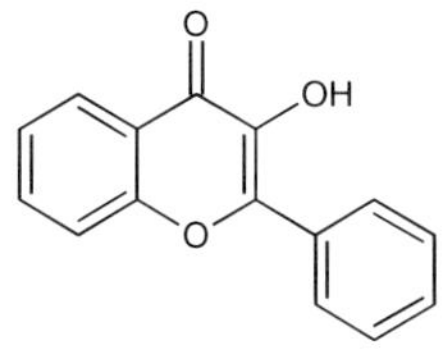
3. Flavanone	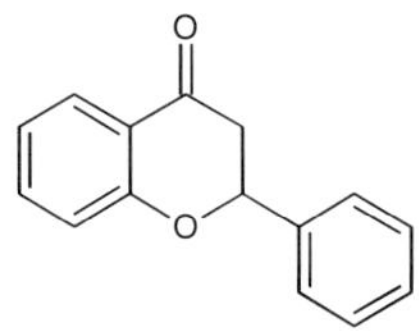
4. Chalcone	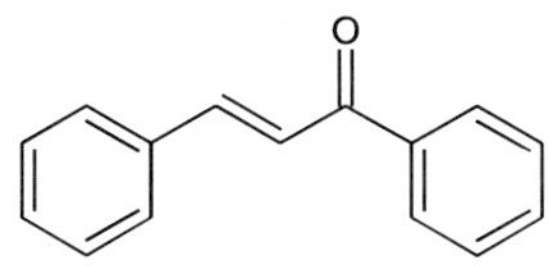
5. Isoflavonoid	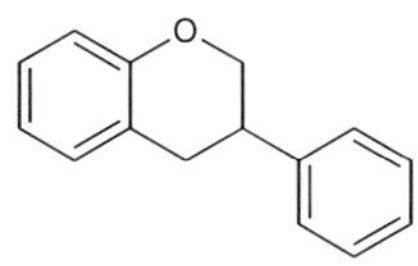
6. Biflavones	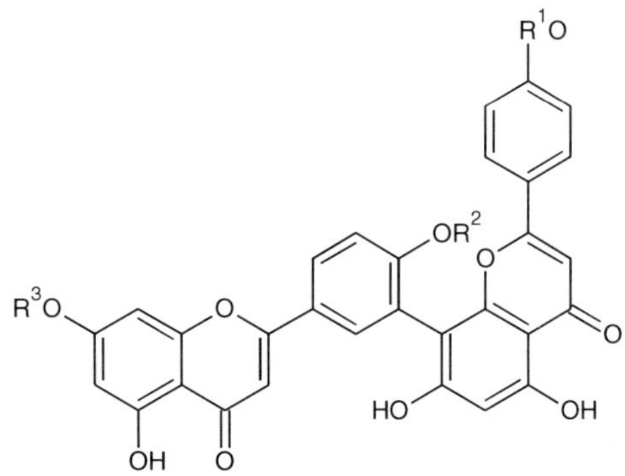
7. Neoflavone	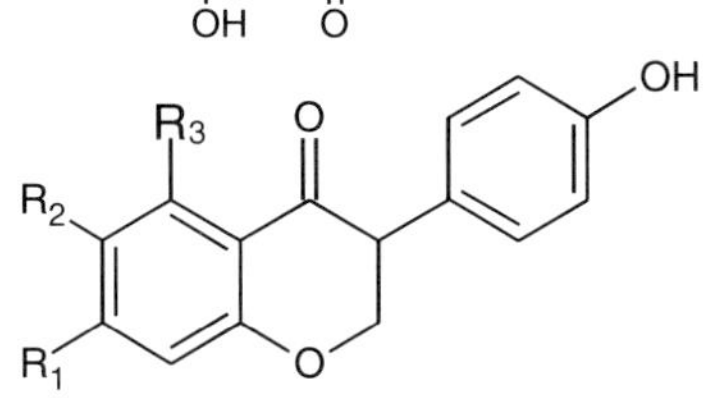
8. Xanthones	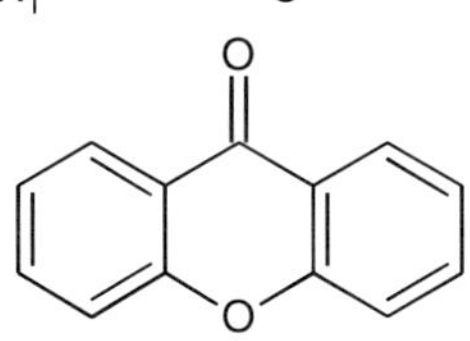

Table 1.1. (Continued)

Types of flavonoid	Structure
9. Anthocyanidin	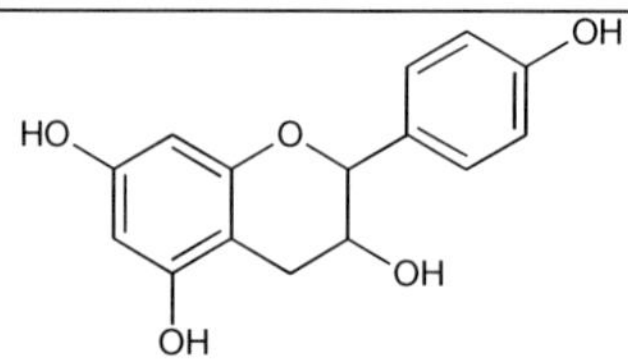

Some examples of naturally occurring flavonoids in plant- derived food stuffs are given in Figure 2.

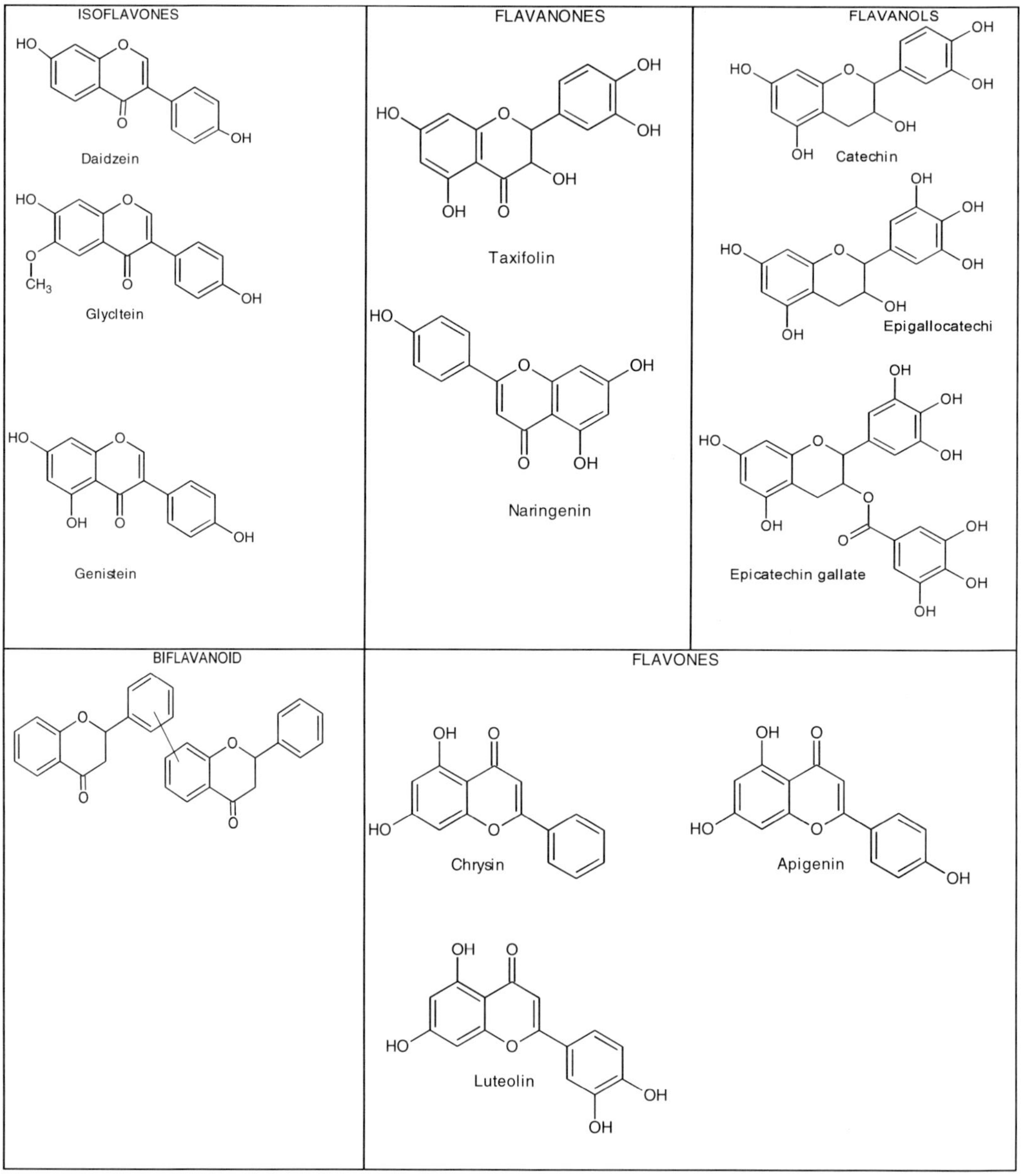

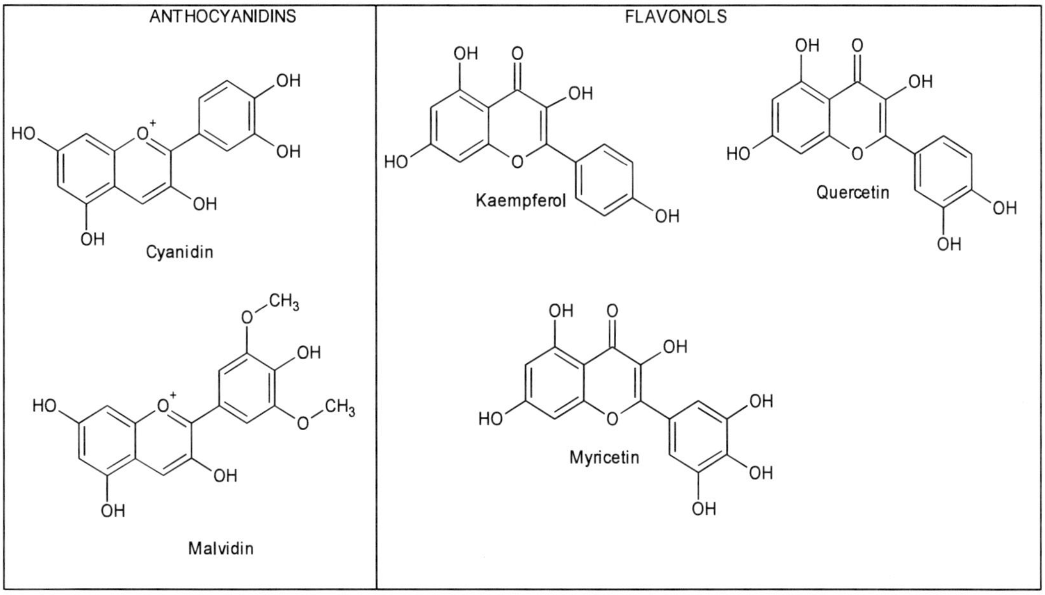

Figure 2. Some examples of natural flavonoids.

The Table 1.2 describe the various flavonoids present in our daily dietary food sources [16, 17].

Table 1.2.

S. No.	**Flavonoid subclass**	**Food source**
1.	Flavonol	Onion, kale, broccoli apples, cherries, berries, tea, red wine
2.	Flavones	Parsley, Thyme
3.	Flavonones	Citrus
4.	Catechins	Apple, tea
5.	Anthocyanidins	Cherries, Grapes
6.	Isoflavones	Soya beans, Legumes

4. The Distribution of Flavonoids in Nature

Flavonoids are characteristic constituents of green plants with the exception of algae and hornworts.

Table 1.3. Distribution of flavonoids in the plants

Division	Common name	Flavonoids occurs
1.Bryophyta	Liver worts	Flavones C-and O-glycosides Flavones O-glycosides* Aurone O-glycosides* Dihydroclalone C-glycosids*
	Mosses	Flavones O-glycosides* Biflavones* Aurone* FlavonesC-and O-glyccosides 3-deoxyanthocynins
2.Tracheophyta	Psilophytes	Biflavones O-g lycosides Flavones O-glycosides** Flavones C-glycosides**
	Lycopods	Flavones o-glycosides Flavones o-glycosides* Biflavones
	Horsetails	Flavones o-glycosides Flavones o-glycosides Peoantho ayanidines Flavones c-glycosides** Flavanones,dihydroflavonds
	Ferns	Flavones o-glycosides Flavones C- and O- glycosides* 3-Deoxyanthocynins Flavonones Chalcones, dihydroxychalcones Biflavones*, Anthocyanins
3.Gymnosperms	Gymnosperms	Flavones, Flavonones Isoflavones*, Biflavonoids Flavonoe C-glycosides Proanthocyanidines Anthocayanidines Flavanols, dihydroflavonols
4.Angiosperm	Angiosperm	Flavones and flavonols,C-and O- glycosides and bisulphates Isoflavones C-and O-glycosides Proanthocyanidins Anthcyanins, Biflavones* Aurones O- glycosides Dihydroflavonols,O- glycosides

* Flavonoid occurs in rare amount.
** Flavonoid occurs in trace amount.

They occur virtually in all the parts of plants including leaves, roots, bark, pollen, nectar, flowers, berries and seeds. In the few recorded cases of flavonoids, found in animals, it is considered that the flavonoids originated from the plants upon which the animals feed rather than being biosynthesized [18]. The distribution of flavonoids in the plant kingdom is presented in table 1.3. Thus, the useful information about the flavonoid type, be encountered in a plant under investigation may often be gained by reference to the literature pertaining to previous flavonoid studies of related plants, e.g., from the same genus or family.

5. Extraction Techniques

5.1. Solubility of Flavonoids

Flavonoid aglycones are basically polyphenols and exhibit similar chemical properties like phenolic compounds. Flavonoids possessing a number of unsubstituted hydroxyl groups and are moderately soluble in polar solvent such as ethanol (EtOH), methanol (MeOH), butanol (BuOH), acetone, dimethylsulphoxide(DMSO), dimethyl formamide (DMF), water etc. The presence of an attached sugar tends to render the flavonoid more water soluble and combinations of the above solvents with water are thus better solvents for glycosides. In contrast, less polar aglycones such as isoflavones, flavanones and highly methoxylated flavones and flavonols tend to be more soluble in solvents such as ether and chloroform.

5.2. Selection, Preparation and Extraction of Plant Material

Fresh vegetation provides the ideal starting material for a flavonoid analysis, although the dried sample which has been carefully stored for many years may still be satisfactory [19, 20]. However, there is a tendency in old plant material for glycosides to be converted to aglycones by fungal action and for sensitive aglycones to be oxidized.

It is recommended that, with fresh plant material after a selected sample has been set aside as a voucher specimen, the remainder be dried rapidly (to prevent enzyme action) in an oven at about 100°C. The dried plant material can then either be stored in a sealed plastic bag for future use or ground to a fine powder for solvent extraction. If the dried plant material does not grind up easily with mortar and pestle, it may be macerated in a wiring blender with or without the solvent to be used for interaction. It is often a wise precaution to extract a sample of undried plant material for a later chromatographic check that the drying process has not altered the flavonoid composition.

After weighing out a portion of the ground plant material, extraction is carried out in two steps, first with MeOH: H_2O (9:1) and secondly with MeOH: H_2O (1:1). At each step, sufficient solvent is added to make liquid slurry and the mixture is left for 6-12 hours. Filtration to separate extract from plant material can be carried out rapidly by using a glass wool or cotton wool plug in the neck of a filter funnel or better by suction using Buchner funnel (recommended paper). The two extracts are then combined and evaporated to about 1/3 the original volume, or until most of the MeOH has been removed. The resultant aqueous extracted can then be cleared of low polarity contaminants such as fats, terpenes, chlorophylls xanthophylls etc. by extraction (in a separating funnel) with hexane or chloroform. This should be done several times and then the extracts are combined. Although unlikely to contain flavonoids, the combined extract should not be discarded until checked by chromatography. The solvent-extracted aqueous layer, containing the bulk of the flavonoids, is then evaporated to dryness under vacuum in a rotary evaporator.

The above procedure is suitable for the extraction of most of flavonoids except anthocyanins or low polarity flavonoids (which sometime occur on the exterior of the plant). For anthocyanins fresh leaves or petals should not be dried but should be crushed with MeOH containing 1% concentrated HCl. Extraction is almost immediate as evidenced by the color in solution, and chromatographic or spectroscopic analyses of the extract should be carried out as soon as after the extraction is possible in order to minimize the risk of glycoside hydrolysis occurring. The low polarity flavonoids occasionally found on the exterior of desert plants, like ferns etc. They are best isolated by simply soaking the fresh plant materials in hexane or ether for a few minutes. The resultant extract will also contain waxes and fat which can be separated later by chromatography [21].

5.3. Paper Chromatogaphy

Paper chromatography (p.c.) is probably the most widely used chromatographic technique available to flavonoid chemists today. Preliminary analysis of a plant extract for the presence of flavonoids is conveniently accomplished by this method and the separation normally carried out by two-dimensional p.c in the descending mode [22].

5.4. Column Chromatogaphy

For the large-scale separation using this technique the isolation of flavonoids can be scaled up to industrial scale. The technique basically involves application of a flavonoid mixture (in solution) to a column of powdered adsorbent (such as cellulose, silica or polyamide) and subsequent sequential elution of individual components with appropriate solvents. The column is simply a glass tube fitted with a top at one end, with dimensions such that the diameter to length ratio is in the range 1:10 to 1:30. The size required for any particular separation can be roughly calculated once the weight of the flavonoid mixture is known. It is generally considered that for separation based on partitioning (i.e. most cellulose and silica chromatography) the sample to column packing ratio should be in the range between 1:50 to 1:500, the latter ratio being more appropriate to complex mixtures and the former to simple mixtures. With polyamide (and to some extent silica) columns, sample loading in the 1:50 range is generally satisfactory.

Column packing should be chosen that are marked specifically for column chromatography as the particle size is important. If it is too small, the flow rate of the eluting solvent may be unacceptably slow and if it is too large then the chromatographic separation of components may be poor. Commercially available packing is usually between 100-300 mesh ranges [21].

5.5. Thin-Layer Chromatogaphy (T.L.C)

For rapid microanalysis of flavonoids the t.l.c is used as a one of the best analytical tools, which requires very small amount of material [23-26]. Thin layer chromatography is particularly used for the following purposes:

- Investigating solvents for possible use in column chromatography;
- Analysis of fractions from column chromatography;
- Identification of flavonoids by chromatography;
- Small scale isolation of pure flavonoids.

For extraction, the solvent is chosen as a function of the type of flavonoid required. Polarity is an important parameter for the isolation of flavonoids. Least polar flavonoids (isoflavones, flavanones, methylated flavones, and flavonols) are removed with chloroform, dichloromethane, diethyl ether, or ethyl acetate, while flavonoid glycosides and more polar aglycones are extracted with alcohol–water mixtures. In general, flavonoids are water soluble compounds and can be extracted with 70% ethanol and remain in alcoholic phase when partitioned with petroleum ether. They usually found as glycosides (attached to sugars). Lipophilic flavonoids of leaf can be extracted by solvents of reasonable polarity followed by solvents such as hexane or petroleum ether (to defat). The glycosides can be extracted using acetone or lower alcohols like ethanol or methanol or hydroalcoholic mixture to ensure maximum removal of flavonoids. After removing the volatile solvent, aqueous remnant is submitted to liquid extraction successively with petroleum ether, diethyl ether and ethyl acetate to get waxy materials, free aglycone and major chunk of glycosides respectively. Flavonoid glycosides are vulnerable to enzymatic degradation during extraction. It is thus required to use dry, lyophilized, or frozen samples. Sometimes boiling water can also be used, while extracting glycosides, to inactivate glycosidase enzyme responsible for enzymatic degradation of glycosides. The mostly cited method for the removal of crude saponins and flavonoids mixture from the plants share few steps, especially if these secondary metabolites are being processed for separation using column chromatography. First, extraction is performed with methanol followed by suspending the dried residue (after concentrating the filtrate) into water. This solution is first made devoid of lipid content (waxes, chlorophyll and fats) by fractionating with n hexane or petroleum ether. This defatted material is finally fractionated in separatory funnel by ethyl acetate and n butanol (liquid –liquid extraction) to produce fractions rich in flavonoids and saponins respectively [27-29].

5.6. High Performance Liquid Chromatography (H.P.L.C)

High performance liquid chromatography is basically a form of column chromatography. This utilizes a column of packing material of small particle size and regular shape. Because of the fineness of the packing, pressures of up to 5000 lb in^{-2} is employed to achieve acceptable flow rates. The technique offers the researcher a method of quantitatively analyzed. Quantification is achieved by automatically monitoring the eluant leaving the column by means of a variable wave length UV spectromoniter and the "chromatograms" is traced out as a series of peaks on a chart [30].

5.7. Recrystallization Techniques

In case of the availibity of sufficient flavonoids (i.e. more than 5mg) a most effective method of purification, (both from other minor flavonoids and from soluble chromatographic material) is recrystallization. The melting point of flavonoids (purified in this manner) is a useful characteristic for comparison of the unknown with previously isolated flavonoids.. Crystalline flavonoids are also the ideal material for sugar analysis since they are free from the soluble polysaccharides which invariably contaminate flavonoids isolated by paper chromatography.

5.8. Conventional Soxhlet Extraction (CSE)

Conventional soxhlet extraction (CSE) is a standard technique. It is still widely used due to its simplicity [31]. Some amounts of dried and ground part of plant (source of flavonoid) were placed in a soxhlet apparatus. Extraction was performed with 150ml of an appropriate solvent for 6h. After extraction a rotary vacuum evaporator (Eyela, A-1000s, Japan) at 40°C was used in order to remove solvent. The solvents are generally used: pure ethanol, methanol, petroleum ether or 70% ethanol etc.

5.9. Supercritical Fluid Extraction (SC-CO_2)

Supercritical fluid extraction with CO_2 is an attractive application for food industry due to its unique properties [32]. The unique characteristic of this system is application of gases above their critical point to extract selective soluble components from a raw material [33]. However CO_2 is not a suitable solvent for the extraction of polar compounds because it behaves as non-polar fluid for certain conditions of temperature and pressure. Therefore, this is the main limiting step in its use for the separation of polar phenolic compounds. But in order to increase the polarity of extraction solvent some food grade modifiers ethanol can be used [34].

Flavonoids are widely distributed in the leaves, seeds, bark and flowers of Spearmint. Mandana Bimakr et al. [35] obtained different bioactive fiavonoid compounds including catechin, epicatechin, rutin, myricetin, luteolin, apigenin and naringenin from spearmint (*Mentha spicata L*) leaves by using conventional soxhlet extraction (CSE) and supercritical carbon dioxide (SC-CO_2) extraction at different extraction schemes and parameter.

Opuntia milpaalta, a member of the Cactaceae family contains many kinds of flavonoids. Weirong CAI at al. [36] observed that flavonoid from *Opuntia milpaalta* with 80% ethanol (v/v), extraction temperature of 90°C and the solvent to raw material ratio of 25:1. After purification with AB-8 resin, the main components of the extract were characterized as isorhamnetin 3-O-(2, 6-dirhamnosyl) glucoside and isorhamnetin 3-O-D-rutinoside. Sameria M. Al-Kalib et al. [37] extracted some flavonoids from *Zingiber officinale* roots by soxhlet extraction method [38, 39]. Citrus processing by products are a potential source of natural flavonoids [40, 41], citrus peels contain a high concentration of phenolic compounds eq. flovanone glycosides, hesperidin, narirutin and neohesperidine etc. Karsheva et al. [42] extracted total polyphenols contents from citrus mandarian peels with ethanol-in-water solution.

Propolis is a resinous natural product produced by bees that is rich in biologically active compounds. Propolis contains flovones, flavonols, flavanoes, dihydroflavonls as well as total phenolic substances [43, 44]. Bartosz Tylkowski at al [45] extracted these compounds with an ethanol water mixture and sequential concentration of the extract by nanofiteration, using two membranes: starmem^{TM} 122 (polyimide) and Duramem^{TM} 200 (modified polyimide). Fenglin Li et al [46] extracted some flavonoids from *lpomoea batatas* leaf (FIBL) by ethanol concentration 60%, ratio of solvent to raw material 1:30, extraction temperature 75°C and extraction time 1.5h, while extraction yield of FIBL was 5.94%.

5.10. Solid-state Fermentation

Extraction/production of bioactive compounds by fermentation is also an interesting alternative that merits attention, since it is able to provide high quality and high activity extracts while precluding any toxicity associated to the organic solvents. In this process, bioactive compounds are obtained as secondary metabolites produced by microorganisms after the microbial growth is completed. Studies on liquid culture show that the production of these compounds starts when growth is limited by the exhaustion of one key nutrient: carbon, nitrogen or phosphate source [47-49]. Fermentation processes may can be divided into two systems: submerged fermentation (SmF), which is based on the microorganisms cultivation in a liquid medium containing nutrients, and solid-state fermentation (SSF), which consists of the microbial growth and product formation on solid particles in the absence (or near absence) of water; however, substrate contains the sufficient moisture to allow the microorganism growth and metabolism [50]. In recent years, SSF has received more interest from researchers since several studies have demonstrated that this process may lead to higher yield and productivities or better product characteristics than SmF.

5.11. Microwave Assisted Extraction (MAE)

Microwave-Assisted Extraction (MAE) technique has been developed for the extraction of antioxidant constituents in plants. Several influential parameters of the MAE procedure such as ethanol concentration, solvent volume, and microwave power and extraction time are to be considered during extraction of plant phyto- constituents. MAE has been shown to be cost-effective when compared to other extraction methods. It surpasses supercritical fluid extraction technique because it is easy to operate at a cheaper cost. The main advantage of MAE over Ultrasonic Assisted Extraction and Soxhlet extraction is due to the fact that, it can be used to extract plant metabolites at a shorter time interval. The use of microwave irradiation is another way of increasing the efficiency of conventional extraction methods [51, 52]. There are a number of factors that one must consider when using microwave assisted extraction method such as the choice of solvent, microwave application time, microwave-power, uniqueness of the matrix; Effect of contact sample surface area, and effect of temperature.

6. Flavonoids in Biological System

The flavonoids have aroused considerable interest because of their potential effects on human health. They have been reported to have antibacterial activity, human immunodeficiency virus (HIV), antiviral activity, antibiotic resistant bacteria, antioxidant effect of flavonoids, K^+ activities (anticancer), anti-inflammatory and antitumor activities etc.

6.1. Antimicrobial Activity

A number of flavones, flavanols, flavanones and isoflavones as well as their methoxy, isoprenyl and acylated derivatives, show antibacterial activity.

Flavones: Rasadah et al. [53] reported that chrysin-(5,7- dihydroxyflavone), in the amount of 5mcg inhibited the growth of the Gram negative bacilli *Escherichia coli* and *Pseudomonas aeruginosa* at a rate comparable to that of streptomycin. Basile et al. [54] reported that apigenin, vitexin, saponaria and lucenin 2-O-glycoside and luteolin-7-O-glycoside acted selectively towards certain Gram- negative bacilli.

Flavonols: Basile et al. [54] reported that among the quercetin mono and diglycosides tested, quercetin-3-O-rhamnoside exhibit the strongest activity against *Pseudomonas maltophilia* and *E. cloacae*, while the other quercetin-O-glycosides exhibited weak activity or no activity against these bacteria.

Acylated flavonols: Waage and Hedin [55] tested a number of acylated kaempferol-3-O-glycosides. The most effective against Gram positive bacteria were compounds having one or two cis-p- coumaroyl gronp.

Flavonones: Liu, H. et al. [56] tested six flavonones for antibacterial activity (which included naringenin, tanifolin and dihydrokaempferide), only naringenin showed activity against E. coli, *S.aureus* and E. faecalis.

Isoflavonones: Malterund et al. [57] showed that the isoflavonones containing prenyl groups had the highest activity against Gram- positive bacteria such as *staphylococcus aureus* and *Bacillus subtiles*. This activity was the greatest when the prenyl groups were located at positions C-6 or C-8 in ring A and C-3' or C-5' in ring B.

Lipophilic flavonoid: Bojase et al. [58] reported that the lipophilic compound 7,8-dihydronyflavone exhibited weak activity against *S.aureus* and no activity against *Proteus vulgaris*. These results may be related to the high lipid content of the cell wall of *P. vulgaris*, which may have trapped the 7, 8- dihydroxyflavone. The cell wall of S. aureus, lacking a lipid layer, could be penetrated. The flavonoids exhibited a stronger effect on DNA synthesis *in P. vulgaris*, while exhibiting a stronger effect on RNA synthesis in S. aureus.

Aglycones of flavonoids: Mori et al. [59] tested the activities of the aglycones of 28 flavones, flavonols, flavanones and isoflavones against *P. vulgaris* and *S. aureus*. The aglycones most active against P. vulgaris were myricetin (MIC=50mcg/ml), morin, quercetagetin and datiscetin.

Structure – Activity Relationship of Flavonoids

Mori et al. [59] show that most of the activities of flavonoid was related to the presence of hydroxyl group 3'4'5' in ring B and at C-3. Zheng et al. [60] showed that methylated favones were active against some Gram- positive bacteria. The most active compound with hydroxyl groups at C-5 and C-7 and with three substitutions in ring B. For example the activity of 5, 7, 4- trihydroxy-3'5' –dimethoxyflavone against Gran- positive *S. aurens* (MIC=500-1000 mcg/ml) was much greater than that of 7, 4'- dihydroxyl-3'5'- dimethoxy flavone. However, against the Gram- negative bacillus *E. coli* former compound was reported to be weakly active and against other Gram- negative bacteria, it was inactive. 5,7,4'-trihydroxy-8-methyl-6-(3-methyl-[2-butenyl])-2S-flavonone isolated from shrub Eysenhardtia texana and flavonoids 7-hydroxy-3',4'-(methylenedioxy)flavan isolated from Termanalia bellerica possess antifungal activity against Candida albicans. 6,7,4'-trihydroxy-3',5'-dimethoxyflavone and 5,5'-dihydroxy-8,2',4'-trimethoxyflavone are effective against Aspergillus flavus [4].

Tapas et al. isolated from peelings of tangerine orange showed fungistatic action towards Deuterophoma tracheiphila while hesperidin stimulates fungal growth slightly. Quercetin has been reported to completely inhibit growth of Staphylococcus aureus [61]. Quercetin, naringenin are reported to be inhibitors of Bacillus subtilis, Candida albicans, Escherichia coli, Staphylococcus nervous, Staphylococcus epidermis, Saccharomyces cerevisiae [62].

6.2. Antiviral Activity and Human Immunodeficiency Virus (HIV)

Some flavonoids show antiviral activity in their glycone form than in their aglycone form. Few flavonoids were found to demonstrate antiviral activity against influenza viruses [63, 64] HSV-1, HSV-2, the rotavirus and even HIV [65-67]. In HIV, their activity was related to a direct effect on the virus or on the enzyme responsible for its replication (HIV-1 reverse transcriptase or HIV -1 integrase). Flavan-3-o1 was more effective than flavones and flavonones in selective inhibition of HIV-1, HIV-2 and similar immunodeficiency virus infections [68, 69].

The antiviral activity of flavonoids was shown in a study by Wang et al. [70]. Some of the viruses reported to be affected by flavonoids are herpes simplex virus, respiratory syncytial virus, para influenza virus, and adenovirus. Quercetin was reported to exhibit both anti infective and antireplicative abilities. There is some evidence that flavonoids in their glycone form seem to be inhibitorier on rotavirus infectivity than the flavonoids in their aglycone form [71].

6.3. Antibiotic-resistant Bacteria

The use of antibiotics is often accompanied by side effect and often the development of resistant strains [72]. The traditional treatment of antibiotic resistant strains of bacteria consists of the administration of vancomycin, a glycopeptide antibiotic used both in treatment of life threatening Gram-positive infections and infections caused by resistant organisms. The flavanoids are non-toxic or have low toxicity [73].

There were some flavonoids active against antibiotic –resistant bacteria. However, these bacteria were inhibited only by those compounds having hydroxyl group in ring B (myricetin >luteolin> quercetin > kaempferol) or by prenylated flavanones. The combination of vancomycin and 3,5,7- trihydroxyflavone (galangin) or 3,5-dihydroxyflavone blend against VRE strains of *E. faecalis* and *E. faecium* showed that the use of these combinations resulted in a decrease in the MIC for vancomycin to a level characteristic of VSE. Flavonoids, in particular flavanone (hesperidin), were found to inhibit the growth of *H. pylori.*

6.4. Antioxidant Effect

As we know that antioxidant compounds protect the cells against the damaging effects of reactive oxygen species, such as singlet oxygen, superoxide, peroxyl radicals, hydroxyl radicals and peroxynitrite. An imbalance between antioxidants and reactive oxygen species result in oxidative stress, leading to cellular damage. Oxidative stress has been linked to cancer, aging, atherosclerosis, inflammation and neurodegenerative diseases. Flavonoid may help provide protection against these diseases by contributing; along with antioxidant vitamins and enzymes, to the total antioxidants defense system of the human body [74, 75].

The general structural requirements for effective radical scavenging and the antioxidant potential of flavonoids are (i) the presence of a catechol group in ring B, capable of readily donating hydrogen (electron) to stabilize a radical species, (ii) the presence of 2,3-unsaturation in conjugation with a 4-oxo-function in the C-ring, which is responsible for electron delocalization, and (iii) the presence of a 3-hydroxyl group in the heterocyclic ring which increases the radical-scavenging activity. The catechol moiety may also contribute to an ability to chelate transition metal ions, such as copper and iron [76, 77]. The scavenging activityof flavonoids has been reported to be in the order: myricetin > quercetin > rhamnetin > morin >iosmetin > naringenin > apigenin > catechin > 5,7-dihydroxy-3',4',5'-trimethoxyflavone > robinin >kaempferol > flavones [78].

Daniel et al. [79] extracted two flovonoid componds, viz- 5,7- dimethylether of leucopelargonidin 3-O- alpha –L-rhamnoside and 5,3'- dimethyl ether of leucocyanidin-3-O-alpha-D-galactosyl cellobioside from *Eicus begalensis* and evaluated for their antioxidants activity in hyperlipidemic rat. N' Gaman kohue et. al [80] showed an excellent antioxidants activity and good stability from osmotic tests on erythrocytes with the total flavonoid extracts from leaves of *Gmelina arborea*. So these flavonoids play an important role in the cell's defense system.

Anokwuru, et al., [81] explained the effect of extracting solvents on flavonoid and antioxidant activities of the bark of *Azadirachta indica*, leaves of *Acalpha wilkesiana* and Solanum scabrum. Al-Katib S.M. at el [37] showed that flavonoid obtained from ginger extract reduced the level of cholesterol, MDA and peroxynitrite and enhancing the level of GSH, vitamin c, albumin and calcium. These flavonoids provided increasing antioxidant defense and suppression free radicals production in the serum. Ginger extract reduced the development of aortic atherosclerotic lesions, along with an impressive reduction in the levels of plasma LDL- cholesterol in atherosclerotic mice.

Olive oil is the principal source of fats in the Mediterranean diet, which has been associated with a lower incidence of coronary heart disease and certain cancers. Phenolic compounds, e.g., hydroxytyrosol and oleuropein, in extra-virgin olive oil are responsible for its peculiar pungent taste and for its high stability. Recent findings demonstrate that olive oil phenolic are powerful antioxidants, both *in vitro* and *in vivo*, and possesses other potent biological activities [82]. Cassia fistula L., a semi-wild Indian Labernum, is widely cultivated in Mauritius as an ornamental tree for its beautiful bunches of yellow flowers and also used in traditional medicine for several indications. The antioxidant activities of reproductive parts were higher than those of the vegetative organs, with the pods having highest total phenolic, proanthocyanidin, and flavonoid contents [83].

Flavonoids found in litchi fruits are comprised of significant amount of flavonoids. The major flavonoids in ripe fruits include flavonols and anthocyanins. The major flavanols are reported to be procyanidin B4, procyanidin B2 and epicatechin, while cyanindin-3-rutinside, cyanidin-3-glucoside, quercetin-3-rutinosde and quercetin-3-glucoside are identified as the important anthocyanins. Litchi flavanols and anthocyanins exhibit good potential antioxidant activity [84].

6.5. K^+ Activities (Anticancer)

In recent years, flavonoids and their synthetic analogues have been intensely investigated in the treatment of ovarian, breast, cervical, pancreatic, and prostate cancer. Flavonoids have important effects on cancer chemoprevention and chemotherapy [85, 86]. In many molecular mechanisms of action for prevention against cancer, flavonoids play a major role by interacting between different types of genes and enzymes. Many mechanisms of action have been identified, including carcinogen inactivation, antiproliferation, cell cycle arrest, induction of apoptosis, inhibition of angiogenesis, antioxidation, and reversal of multidrug resistance or a combination of these mechanisms.

Anti tumor activities of flavonoids:

1) Inhibition of cell growth.
2) Inhibition of protein kinase activities.
3) Induction of apoptosis.
4) Induction of MMP secretion.
5) Induction of tumor cell invasion.
6) Induction of adhesion/ spreading of cell.
7) Anti-angiogenic properties.
8) There are some examples of flavonoids which have been reported as effective anticancer activity [86].

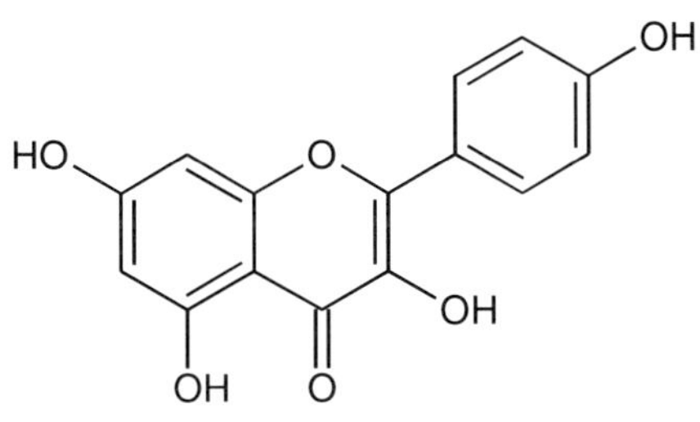

Kaempferol

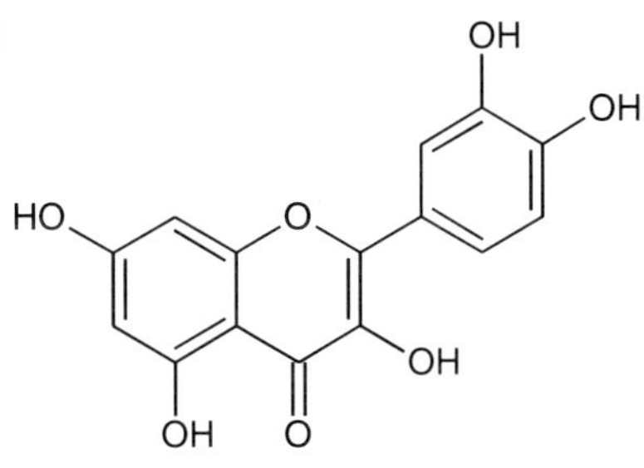

Quercetin

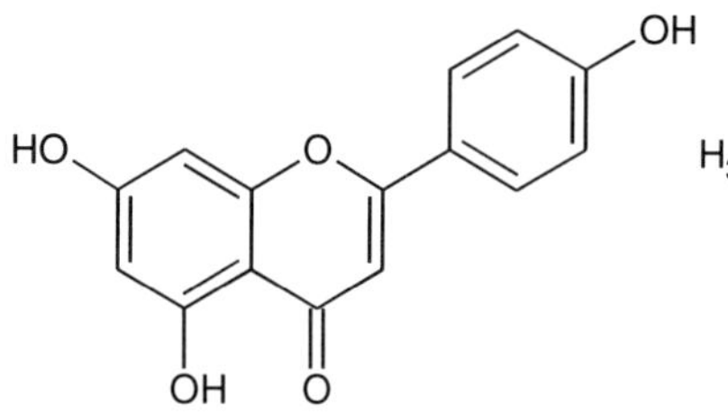

Apigenin

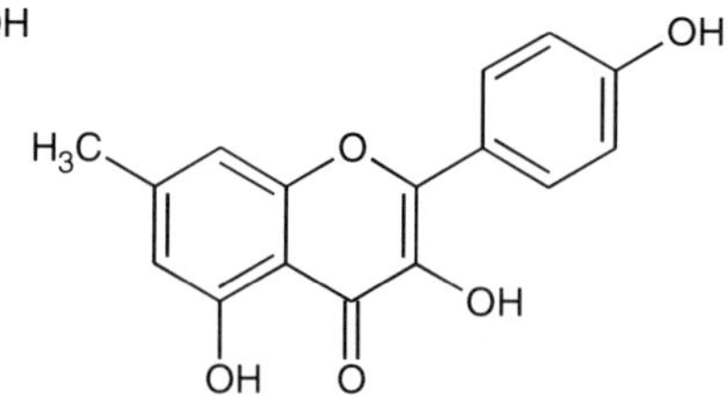

Naringin

Litchi fruit pericarp (LFP) tissues extract displays a dose- and time-dependent inhibitory effect on human breast cancer, which could be attributed, in part, to its inhibition of proliferation and induction of apoptosis in cancer cells through up regulation and down-regulation of multiple genes [83].

6.6. Cardiovascular Effects

The potential beneficial effect of individual phenolics or plant extracts on cardiovascular health includes the different mechanisms implicated on cardiovascular conditions or problems, i.e., hypertension, inflammation, cellular proliferation, trombogenesis, hyperglycaemia and hypercholesterolemia. The naturally occurring flavanols in red wine and purple-grape juice abolished the cyclic flow reductions in coronary blood flow. The biological activity of these compounds could explain the platelet-inhibitory properties of red wine and grape juice independently of the presence of alcohol [87]. Quercetin protects LDL against oxidative modifications effect. 7-monohydroxyethylrutoside and 7', 3', 4'-trihydroxyethylrutoside are reported to be cardioprotective [61].

6.7. Anticataract Action

Several pharmacological actions of natural flavonoids may operate in the prevention of cataract since flavonoids are capable of affecting multiple mechanisms or etiological factors responsible for the development of diabetic cataract. Onion is a flavonoid-rich foodstuff and the major flavonoids contained have been identified as quercetin, quercetin-4'-glucoside and quercetin-3, 4'-diglucoside [88, 89]. In the study of Vadzade et al. [90], the instillation of fresh juice of crude onion into the rat eyes was found to prevent selenite-induced cataract formation by 75%. Flavonoid fractions isolated from natural sources including green and black tea [91, 92], *Ginkgo biloba* [93], grape seeds [94], *Emilia sonchifolia [*95], *Vitex negundo* [96] and broccolib [97] were shown to have anticataract activity. Grape seed extract rich in flavonoids and flavonoid fisetin was found to showed anticataract action [98, 99].

6.8. Anti-Inflammatory Activity

Flavonoids show an anti-inflammatory activity *in vitro* and *in vivo*. Several cellular action mechanisms are proposed to explain their anti-inflammatory activity. The anti-inflammatory activity of flavonoids, establishes not only anti-inflammatory mechanisms, but also gives a platform for developing new class of anti-inflammatory agents [100]. A number of flavonoids are reported to possess anti-inflammatory activity. Hesperidin, a citrus flavonoid possesses significant antiinflammatory and analgesic effects [68]. Recently, apigenin, luteolin and quercetin have been reported to exhibit anti-inflammatory activity [101]. There are some examples of natural flavonoids which show anti-inflammatory activities [102].

Anti-inflammatory flavonoids

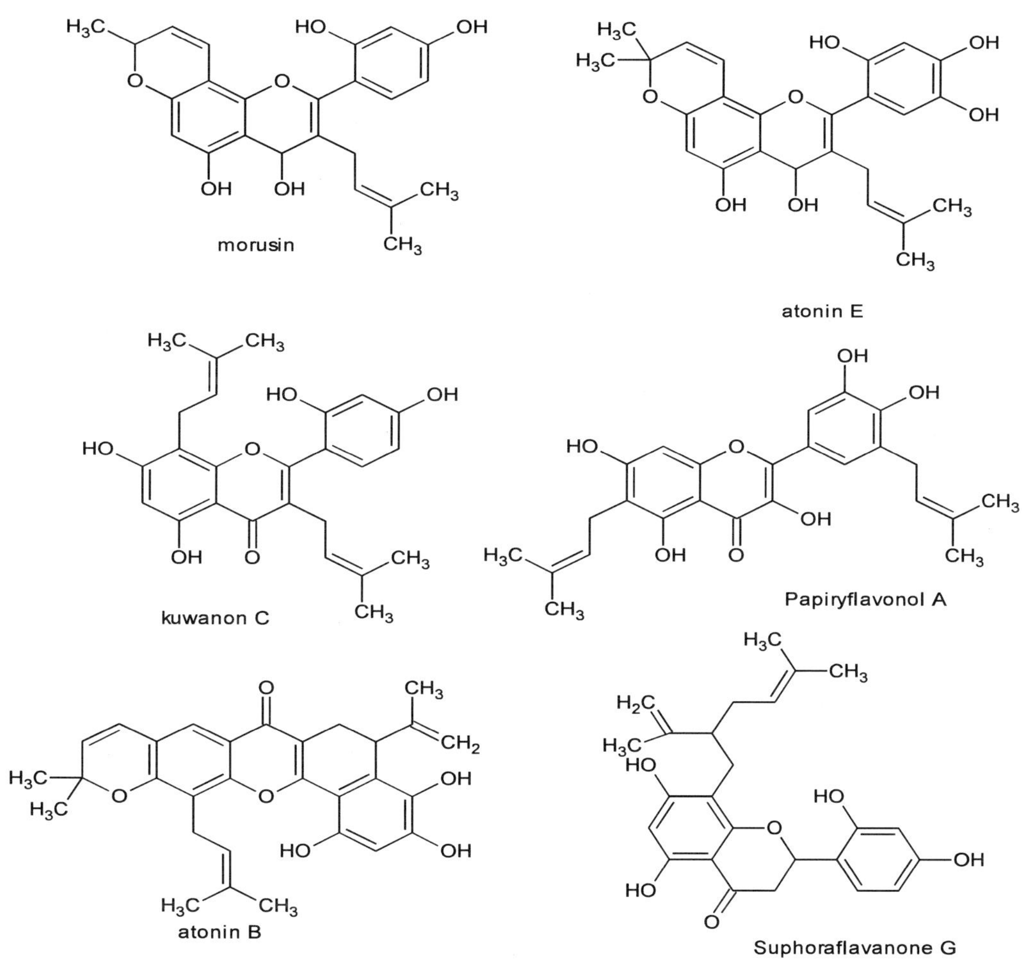

CONCLUSION

Flavonoids have received much attention in the literature over the last 10 year and a variety of potential beneficial effect has been elucidated. However, most of the research involved *in vitro* studies; therefore, it is difficult to draw definite conclusion about the usefulness of flavonoids in the diet. But, currently, the intake of fruit, vegetable and beverages (e.g., tea and moderate amount of red wine) containing flavonoids is recommended, although it is too early to make recommendations on daily flavonoid intake.

REFERENCES

[1] Smith, H. (1972). *In Phytochrome (k. Mitrakos and W. Shropshire, eds).* New York and London: Academic press

[2] Mills, S. and Bone, k. (2000). *Principles and Practice of Phototherapy Modern Herbal Medicine- Churchill Livingstone.* New York: Ivana Vulic.

[3] Geissman, T.A. and Crout, D.H.G. (1996). *Organic Chemistry Of Secondary Plant Metabolism.* San Francisco: Freeman, copper & company.

[4] Cushnie, T.P.T. and Lamb, A.J. (2005). Antimicrobial activity of flavonoids. *International Journal of Antimicrobial Agents,* 26, 343-356.

[5] Murray, M.T. (1996). *Encyclopedia of Nutritional Supplements.* California: Prima Publishing.

[6] Fernandez, S.P., Wasowski, C., Loscalzo, L.M., Granger, R.E., Johnston, G.A.R., Paladini, A.C. and Marder, M. (2006). Central nervous system depressant action of flavonoid glycosides. *European Journal of Pharmacology,* 539, 168-16.

[7] Cook, N.C. and Samman, S. (1996). Flavonoids: Chemistry, metabolism, cardioprotective effects and dietary sources. *Nutritional Biochemistry,* 7, 66-76.

[8] Sahu, S.C. and Gray, G.C. (1996). Pro-oxidant activity of flavonoids: effect on glutathione and glutathione-S-transferase in isolated rat liver nuclei. *Cancer letters,* 104, 193-196.

[9] Prey, J.O., Brown, J., Fleming, J. and Harrison, P.R. (2003). Effect of dietary flavonoids on major signal transduction pathways in human epithelial cells. *Biochemical Pharmacology*, 66, 2075-2088.

[10] Gupta, M.K. (2009). *Chemistry of Natural Product.* 1st ed. Pragati Prakasham.

[11] Robinson, T. (1991). *The Organic Constituents Of Higher Plants – Their Chemistry And Interrelationship.* 6th ed. North Amhers: Cordus Press.

[12] Harborne, J.B. and Baxter, H. (1999). *The Handbook Of Natural Flavonoids.* West Sussex: Wiley.

[13] Harborne, J.B. (1993). *The Flavonoids: Advances In Research Since.* London: Chapman and Hall.

[14] Paul C., Li Y., Mario C., Jia P.H., Kanyanga C., Bart V.P., Luc P., Arnold J.V. and Dirk V.B. (1998). Structure-activity relationship and classification of flavonoids as inhibitors of xanthine oxidase and superoxide scavengers. J. Nat. Prod. 61, 71-76.

[15] Williams, R.J., Spencer, J.P. and Rice-Evans, C. (2004). Flavonoids: Antioxidants or signalling molecules. Free Radic. Biol. Med. 36, 838–849.

[16] Hollman, P.C.H. and Katan, M.B. (1999). Dietary Flavonoids: Intake, Health Effects and Bioavailability. *Food and Chemical Toxicology,* 37, 937-942

[17] Sandbar, H.K., Kumar, B., Prasher, S., Tiwari, P., Salhan, M. and Sharma, P. (2011). A review of phytochemistry and pharmacology of flavonoids. *International pharmaceutica sciencia,* 1(1), 25-41.

[18] Harborne, J.B.(1967). *Comperative Biochemistry of The Flavonoids.* London and New York: Academic Press, 30.

[19] Harborne, J.B. (1967). *Comparative Biochemistry of The Flavonoids.* London and New York: Academic Press, 62.

[20] Niklas, K.J. and Gianassi, D.E. (1978). Angiosperm paleobiochemistry of the succor creek flora (miocene) Oregon. *Am. J. Bot.,* 65, 943-952.

[21] Markham, K.R. (1981). *Techniques of Flavonoid Identification.* London and New York: Academic Press.

[22] Markham, K.R. (1982). *Techniques of Flavonoid Identification.* London: Academic Press.

[23] Kirchner, J.G. (1967). *Thin –layer chromatography,* Inter science publications: New York.

[24] Stahl, E. (1959). TLC-A Laboratory Hand Book. Berlin: Springer Verlag.

[26] Markham, K.R. (1975). *The Flavonoids.* Londan: Chapman and Hall.

[27] Mukherjee, P.K. (2010). Quantity Control of Herbal drug: An Approach to Evaluation of Botanicals. New Delhi: Business Horizones.

[28] Haughton, P.J. and Raman, A.(2011) *Laboratory Handbook for the fractionation of natural Extracts.* New Delhi: Springer.

[29] Gangwal, A., Parmar, S.K., Sheth, N.R., (2010). Triterpenoid, flavonoids and sterols from lagenaria siceraria fruits. *Scholars Research Library,* 2 (1): 307-317.

[30] Kingston, D.G.I. (1979). History of lloydia and the journal of natural products. Journal of Natural Products, 42, 237.

[31] Grigonis, D., Sivik, P.R.V., Sandahl, M. and Eskilsson, C.S. (2005). Comparison of different extraction techniques for isolation of antioxidants from sweet grass (Hierochloe odorata). *J. Supercrit Fluids,* 33,223-233.

[32] Lin, M.C., Tsai, M. J. and Wen, K.C. (1999). supercritical fluid extraction of flavonoids from scutellariae radix. *J. Chromatogram,* 830, 387- 395.

[33] Cavero, S., Garcia – Risco, M., Marin, F., Jaine, L., Santoyo, S., Senorans, F., Reglero, G. and Ibenez, E. (2006). Supercritical fluid extraction of antioxidant compounds from oregano: chemical and functional characterization via lc–ms and *in vitro* assays. *J. Supercrit fluid,* 38, 62-69.

[34] Qingyong Long, C.M.W. (2001). Supercritical fluid extraction in herbal and natural product studies- a practical review. *Talanta,* 53, 771-782.

[35] Bimakr, M., Rahaman, R.A., Taip, F.S., Ganjloo, A., Salleh, L.M., Selamat, J., Hamid, A. and Zaidul, I.S.M. (2011). Comparison of different extraction methods for the extraction of major bioactive flavonoid compounds from spearmint (*Mentha spicata* L.) leaves. *J. Food and Bioproducts Processing,* 89, 67-72.

[36] Weirong, C.A.I., Xiaohong, G.U. and Jian Tang, C. (2010). Extraction, purification, and characterisation of the flavonoids from opuntia milpa alta skin. *J. Food Sci.,* 28, 108-116.

[37] Al – katib, S.M., Al-Khashad, E.M., Kalo M.S. and Hamdoon A.A. (2009). The antioxidant effects of flavonoids and non flavonoid part extracted from ginger (zingiber officinale) roots. *J. Raf, Sci.* 20 (3), 18-31.

[38] Tayler, V.E. (1993). In the Honest Herbal: A Sensible Guide the Use of Herbs and Related Compounds. New York: Pharmacentical Products Press.

[39] Wagnes, H. (1986). *In Plant Flavonoids in Biology and Medicine, Biocem. Pharmacol. and Structure – Activity Relationship.* New York: Alan R. Liss.

[40] Smith, B.B., Li, B. and Hossin, M. M. (2006). Extraction of phenolics from citrus peels in solvent extraction method. *Separation and Purification Technology,* 48, 182 -188.

[41] Manthey, J.A. and Grohmann K. (2001). Phenols in citrus peel byproducts. Concentrations of hydroxycinnamates and polymethoxylated flavones in citrus peel molasses. J. Agric. Food Chem., 49, 3268-3273.

[42] Karsheva, M., Kirova, E. and Alenandrova, S. (2013). Comparison Of Citrus Peels As A Source Of Valuable Components - Polyphenols and Antioxidants .*J Chem Tech and Metallurgy,* 48 (1), 35-41.

[43] Ghisalberti, E.L., Jefferie, P.R., Lanteri, R. and Matisons, J. (1978). Constituents of propolis. *Experientia,* 15, 157-158.

[44] Chena, C. R., Shen, C. T., Wua, J. J., Yangb, H. L., Hsuc, S. L. and Changa, C. M. J. (2009). Precipitation of sub-micron particles of 3,5-diprenyl-4-hydroxycinnamic acid in Brazilian propolis from supercritical carbon dioxide anti-solvent solutions. *J. Supercript Fluids,* 50, 176-182.

[45] Tylrowski, B., Trusheva, B., Bankova, V., Giamberini, M., Peev, G. and Nikolova, A.(2010). Extraction of biologically active compounds from propolis and concentration of extract by nanofiltration. *J. Membrane Science,* 348, 124-130.

[46] Fengli, Li., Qinguang, L., Gao, D. and Peng, Y. (2009). The optimal extraction parameters and anti-diabetic activity of flavonoids from *ipomoea batatas* leaf. *African J. Of Traditional, Complementary And Alternative Medicine,* 6 (2), 195-202.

[47] Nigam, P,S.(2009). Production of Bioactive Secondary Metabolites. Netherlands: Springer.

[48] Nigam, P.S. and Pandey, A. (2009). *Solid-State Fermentation Technology for Bioconversion Of Biomass And Agricultural Residues.* Netherlands: Springer,197–221.

[49] Barrios-Gonzâlez, J., Fernãndez, F.J., Tomasini, A. and Mejía, A. (2005). Secondary metabolites production by solid-state fermentation. *Malays. Jounral of Microbiology,* 1, 1–6.

[50] Pande, A. (2003) Solid state fermentation. *Biochem. Eng. J.,* 13, 81–4.

[51] Starmans, D.A.J. and Nijhuis, H.H. (1996). Extraction of secondary metabolites from plant material: A review. *Trends Food Sci. Technol.,* 7, 191-197.

[52] Afoakwah, A. N., Owusu, J., Adomako, C. and Teye E. (2012). Microwave assisted extraction (MAE) of antioxidant constituents in plant material. *Global J Bio-Sci. & Biotechnology,* 1(2), 132-140.

[53] Rasadah, M.A., Houghton, P.J., Amala, R.and Hoult, J.R.. (1998). Antimicrobial and anti-inflammatory activity of extracts and constitutes of *oroxylum indicum vent. Phytomedica,* 5, 375–381

[54] Basile, A., Girodano, S., lopez–Saez, J. A. and Cobian, C. C., (1999). Antibacterial activity of pure favonoids isolated from mosses. *Phytochemistry,* 52, 1479-1482.

[55] Wage, S.K. and Hedin, P. A. (1985). Quercetin 3-*O*-galactosyl-(1 → 6)-glucoside, a compound from narrowleaf vetch with antibacterial activity. *Phytochemistry,* 24, 243-245.

[56] Liu, H., Orjala, J., Sticher, O. and Rali, T. (1999). Acylated flavonol glycosides from the leaves of stenocleana palustris. *J. Nat. Prod.,* 62, 70-75.

[57] Malterund, K., Bremns, E., Faegri, A., Moe, T. and Sandanger, E. (1985). Flavonoids from the Wood of Salix caprea as Inhibitors of Wood-Destroying Fungi. *J.Nat. Prod.,* 48, 559 -563.

[58] Bojase, G., Majinda, R., Gashe, B. and Wanjada, C. (2002). Antimicrobial flavonoids from bolusanthus speciosus.*Planta Medica,* 68, 615-620.

[59] Mori, A., Nishino, C., Enoki, N. and Tawata, S. (1987). Antibacterial activity and mode of plant flavonoids against proteus vulgaris and staphylococcus aureus. *Phytochemistry,* 26, 223-224.

[60] Zheng, W.F., Tan, R.X. and Lui, Z.L.(1996). Two flavones from *Artemisia giraldii* and their antimicrobial activity. *Plant Medica*, 62, 160-162.

[61] Tapas, A.R., Sakarkar, D.M. and Kakde, R.B.(2008). Flavonoids as nutraceuticals: A Review. *Tropical Journal of Pharmaceutical Research*, 7, 1089-1099.

[62] Taleb-Contini, S.H., Salvador, M.J., Watanabe, E., Ito, I. and Oleveira, D.C.R.D. (2003). Antimicrobial activity of flavonoids and steroids isolated from two Chromolaena species. *Revista Brasileira de Ciencias Farmaceuticas.*39, 403-408.

[63] Almeida, A.P., Miranda, M., Simonia, I., Wigg, M., Lagrota, M. and Costa, S. (1998). Flavonol monoglycosides isolated from the antiviral fractions of *Persea americana (Lauraceae)* leaf infusion. *Phytother Res.,* 12, 562-567.

[64] Arthan, D., Svasti, J., Kippakoop, P., Pittayakhachonwat, D. and Tanticharoen, M. (2002). Antiviral isoflavonoid sulfate and steroidal glycosides from the fruits of *Solanum torvum*. Phytochemistry, 59, 459-463.

[65] Hu,C.Q., Chen, K., Shi, Q., Kilkuskie, R. E., Cheng, Y. C. and Lee, K.H., (1994). Anti-aids agents, 10. Acacetin-7-o-β-d-galactopyranoside, an anti-hiv principle from chrysanthemum morifolium and a structure-activity correlation with some related flavonoids. *J. Nat. Prod.,* 57, 42-51.

[66] Lin, Y.M., Anderson, H., Flavin, M.T., Pai, Y.H., Mata-Greenwood, E., Pengsuparp, T., Pezzuto, J.M., Schinazi, R.F., Hughes, S.H., Chen, F.C. (1997). *In vitro* anti-HIV activity of biflavonoids isolated from Rhus succedanea and Garcinia multiflora. *J. Nat. Prod.* 60(9), 884-8.

[67] Paterson, D.L., Swindells, S,. Mohr, J., Brester, M., Vergic, EN., Squier, C., Wagner, M. M. and Sing, N.(2000). Adherence to protease inhibitor therapy and outcomes in patients with HIV infection. *Ann Intern Med.,* 133, 21-30.

[68] Raj Narayana, K., Sripal Reddy, M., Chaluvadi, M.R. and Krishna, D.R. (2001). Bioflavonoids classification, pharmacological, biochemical effects and therapeutic potential. *Indian Journal of Pharmacology,* 33, 2-16.

[69] Havsteen, B.H. (2002). The biochemistry and medical significance of the flavonoids. *Pharmacology and Therapeutics.* 96, 67-202.

[70] Wang, H.K., Xia, Y., Yang, Z.Y, Natschke, S.L. and Lee, K.H. (1998). Recent advances in the discovery and development of flavonoids and their analogues as antitumor and anti-HIV agents. *Adv Exp Med Biol.,* 439, 191–225.

[71] Bae, E.A., Han, M.J., Lee, M. and Kim, D.H. (2000). *In vitro* inhibitory effect of some flavonoids on rotavirus infectivity. *Biol Pharm Bull.,* 23, 1122–4.

[72] Beali, J.M. (2004). Antibacterial antibiotics. Wilson and Gisvold's Textbook of Organic Medicinal and Pharmaceutical Chemistry. 11th ed. New York, Lippincott Williams & Wilkins, 299-366.

[73] Narayana, K.R., Reddy . M. S. and Chaluvadi, M.R. (2001). *Ind. J. Pharmacol.,* 33, 2-16.

[74] Wegener, T. and Fintelmann, V. (1999). Flavonoids and Bioactivity. *WeinMed Wochem Schr.,* 149, 241-7.

[75] Ares, J.J. and Outt P.E. (1998). Gastroprotective agents for the preventionof NSAID-induced gastropathy. *Curr Pharm Des.,* 4, 7-36.

[76] Bors, W. and Michel, C. (2002). Chemistry of the antioxidant effect of polyphenols. *Ann N Y Acad Sci.,* 957, 57–69.

[77] Amic, D., Davidovic-Amic, D., Beslo, D., Rastija, V., Lucic, B., Trinajstic N. (2007). SAR and QSAR of the antioxidant activity of flavonoids. *Curr Med Chem.,* 14, 827–845.

[78] Robak, J. and Glyglewski, R.J. (1988) Flavonoids are scavengers of superoxide anions. *Biochem Pharmacol.,* 37, 837-41.

[79] Daniel, R.S., Mathew, B.C., Devi, K.S. and Auqusti, K.T. (1998). Antioxidant effect of two flavonoids from the bark of ficus bengalensis linn in hyperlipidemic rats. *Indian J. Exp. Biol.*, 36 (9), 902 -906.

[80] N'Gaman K.C.C., Mamyrbekova-Bero, J.A. and Bekro, Y.A. (2011). Effect of flavonoids of gmelina arborea roxb. (verbenaceae) from côte d'ivoire on the antioxidant activity and osmotic stability of erythrocytes. *J. Appl. Bio. Sci.,* 39, 2626-263.

[81] Anokwaru, C.P., Anyasor, G.N., Ajibaye. O., Fakoya, O. and Okebugwa, P. (2011). Effect of extraction solvents on phenolic, flavonoid and antioxidant activities of three nigerian medicinal plants. *Nature and Science,* 9(7), 53 -61.

[82] Visioli, F., Poli, A. and Gall, C. (2002). Antioxidant and other biological activities of phenols from olives and olive oil. *Med. Res. Rev.,* 22(1), 65-75.

[83] Luximon-Ramma, A., Bahorun, T., Soobrattee, M.A. and Aruoma, O.I.(2002). Antioxidant activities of phenolic, proanthocyanidin and flavonoid components in extracts of Cassia fistula. *J. Agric. Food Chem.*, 28:50(18), 5042-7.

[84] Jiangrong, Li. and Jiang, Y.(2007). Litchi Flavonoids: Isolation, Identification and Biological Activity. *Molecules,*11, 12(4):745-58.

[85] Nijveldt, R.J., Nood, E.V., Hoorn, D.V., Boeleus, P.G., Norren, K.V. and Leeuwen P.A.V. (2001). Flavonoids: a review of probable mechanisms of action and potential applications. *Am. J. Clin. Nutr.,* 74(4), 418-425.

[86] Havsteen, B.H. (2002). The biochemistry and medical significance of the flavonoids. *Pharmacology and Therapeutics,* 96, 67-202.

[87] Demrow, H.S., Slane, P.R. and Folts, J.D. (1995). Administration of wine and grape juice inhibits *in vivo* platelet activity and thrombosis in stenosed canine coronary arteries. *Circulation*, 91, 1182–1188.

[88] Fossen, T., Pedersen, A.T. and Andersen, M. (1998). Flavonoids from red onion (Allium cepa). *Phytochemistry,* 47, 281–285.

[89] Miean, K.H. and Mohamed, S. (2001.) Flavonoid (myricetin, quercetin, kaempferol, luteolin and apigenin) content of edible tropical plants. *J. Agr. Food Chem.*, 49, 3106–1312.

[90] Vadzadeh, A., Ghorbanihaghjo, A., Bonyadi, S., Rashidi, M.R., Mesgari, M., Rashtchizadeh, N.and Argani, H. (2009). Preventive effect of onion juice on selenite-induced experimental cataract. *Indian J. Ophthalmol.*, 57, 185–189.

[91] Thiagarajan, G., Chandani, S., Sundari, S.C., Harinarayana Rao, S., Kulkarni, V.A. and Balasubramanian, D. (2001). Antioxidant properties of green and black tea and their potential ability to retard the progression of eye lens cataract. *Exp Eye Res.*, 73, 393–401.

[92] Gupta, S.K., Selvan, V.K., Agrawal, S.S. and Saxena, R. (2009). Advances in pharmacological strategies for the prevention of cataract development. *Indian J. Ophthalmol.*, 57, 175–183.

[93] Thiagarajan, G., Chandani, S., Harinarayana, Rao. S., Samuni, A.M., Chandrasekaran. K. and Balasubramanian, D. (2002). Molecular and cellular assessment of Ginkgo biloba extract as a possible ophthalmic drug. *Exp Eye Res.*, 75, 421–430.

[94] Durukan, A.H., Evereklioglu, C., Hurmeric, V., Kerimoglu, H., Erdurman, C., Bayraktar, M.Z. and Mumcuoglu, T. (2006). Ingestion of IH636 grape seed proanthocyanidin extracts to prevent selenite-induced oxidative stress in experimental cataract. *J. Cataract Refract Surg.*, 32, 1041–1045.

[95] Lija, Y., Biju, P.G., Reen,i A., Cibin,T.R., Sahasranamam, V. and Abraham, A. (2006). Modulation of selenite cataract by the flavonoid fraction of emilia sonchifolia in experimental animal models. *Phytother Res,* 20, 1091–1095.

[96] Rooban, B.N., Lija, Y., Biju, P.G., Sasikala, V., Sahasranamam, V. and Abrahama, A. (2009). Vitex negundo attenuates calpain activation and cataractogenesis in selenite models. *Exp Eye Res.*, 88, 575–582.

[97] Vibin, M., Siva Priya, S.G. N., Rooban, B., Sasikala, V., Sahasranamam, V. and Abraham, A. (2010). Broccoli regulates protein alterations and cataractogenesis in selenite models. *Curr. Eye Res.*, 35, 99–107.

[98] Jia, Z., Song, Z., Zhao, Y., Wang, X. and Liu, P. (2011). Grape seed proanthocyanidin extract protects human lens epithelial cells from oxidative stress via reducing NF-kappa B and MAPK protein expression. *Mol. Vis.*, 17, 210–217.

[99] Yao, K., Zhang, L., Zhang, Y., Ye, P. and Zhu, N., (2008) The flavonoid fisetin inhibits UV radiation-induced oxidative stress and the activation of NF-kappa B and MAPK signaling in human lens epithelial cells. *Mol. Vis.* 14, 1865–1871.

[100] Kim, H. P., Son, K. H., Chang, H.W. and Kang, S S., (2004) Anti-inflammatory Plant Flavonoids and cellular action mechanism, *J. Pharmacol. Sci.*, 96, 229 – 245.

[101] Raj Narayana, K., Sripal Reddy, M., Chaluvadi, M.R. and Krishna, D.R. (2001). Bioflavonoids classification, pharmacological, biochemical effects and therapeutic potential, *Indian Journal of Pharmacology,* 33, 2-16.

[102] Shahidi F., Yang, Z and Saleemi, Z.O. (1998). Natural flavonoids as stabilizers. *J Food Lipids,* 1, 69-75.

In: Natural Heterocycles
Editors: K. L. Ameta and Saksit Chanthai
ISBN: 978-1-63463-424-3

Chapter 12

A Dibenzofuran: (+)-Usnic Acid: A Heterocyclic Natural Compound Isolated from Lichen Parmelia Perlata

Anil K. Sharma* and Kuldeep K. Sharma
Department of Chemistry, Jagannath University,
Jaipur (Rajasthan), India

Abstract

A dibenzofuran, 2,6-diacetyl-7,9-dihydroxy-8,9b-dimethyl-1,3(2H,9bH)-dibenzo-furandio- -ne, also known as (+) Usnic acid, was identified as heterocyclic natural compound from lichen Parmelia perlata. Structure was elucidated on the basis of different spectroscopic techniques.

Keywords: *Parmelia perlata*, Lichen, phytochemicals, biological activities

1. Introduction

The recent advances in bioassay techniques along with scientific and technological advancements help the researchers to rediscover the benefits of herbs and spices that were so well known to our forefathers. These benefits are directly connected to the increasing awareness of the impact of food on the well-being and health of a person. According to the literature more than 139 drugs of conventional system of medicines owe their origin to herbs and herbal active principles.

Approximately 74% drugs were discovered because of their use in traditional system of medicine. In most of the European and Asian countries especially India, Germany and China, herbal medicines are an integral part of the health care system so the discovery and development of new therapeutic agents is a continuing process.

* E-mail: anilkumarsharma_008@yahoo.com

In this process the authors work with lichen to isolate Phytoconstituents. Lichens are the combination of algae & fungi and they are rich source of different types of natural compounds (Phytoconstituents). Lichens have been used for medical purposes since ancient times. Among the lichen substances, the most widely distributed and extensively investigated is Usnic acid (Heterocyclic Natural Compound) since the activity of Usnic Acid is dependent on its structure (e.g., tautomeric species present, keto and enol form).

The keto-enol tautomerism of Usnic Acid may be monitored by vibrational spectroscopy (both Raman and Inelastic neutron spectroscopy), by focusing on the deformation mode of the C–H oscillator of the chiral center of the molecule, which is exclusive to the keto tautomer [1]. Literature analysis on this topic reveals that Usnic Acid exhibit antibacterial [2], antibiotic [3], enzyme inhibition [4], antioxidant [5], anti-inflammatory [6], antimicrobial [7, 8] activities as well as some other properties such as UV protection, allergenic potential, toxicity, antiviral, antiprotozoal, antiproliferative, and analgesic activities. [9-11]. Ecological effects such as antigrowth, anti-herbivore and anti-insect properties have also been demonstrated [12].

Usnic Acid was isolated first time in 1844 [13] and later on it was isolated by many researchers. Now we isolate this heterocyclic natural compound (Usnic Acid) from lichen *Parmelia perlata*. *Parmelia perlata* belongs to Parmelia genus (Family Parmeliaceae) and commonly known as Charila in India. The beautiful hills of Uttarakhand (In India) are the best source of lichens due to varied temperature differences in day and night.

2. Material and Method

General Experimental Procedure

Melting points was determined in soft glass capillaries in an electrothermal melting point apparatus. Qualitative TLC was conducted on aluminum sheet Kieselgel 60 F254 (E. Merck). Silica gel (E. Merck, 60-120 mesh, 550 gm) used for column (1.5m × 4.0cm) chromatography. The IR spectra were recorded on FTIR SHIMADZU 8400S spectrometer with KBr pellets. The 1H and 13C NMR spectra were recorded in $CDCl_3$ at 300 MHz and 75 MHz on a Brucker NMR instrument, respectively, using TMS as internal standard. FAB mass spectra were recorded on JEOL SX 102 /DA-6000 mass spectrometer using Argon /Xenon as FAB gas.

Plant Material

The plant material, *Parmelia perlata* was collected from the hills of Uttarakhand (India) and the authenticity was confirmed by Incharge of Herbarium, Department of Botany, University of Rajasthan, Jaipur (India).

Extraction and Isolation of the Constituent

The plant material (1.5 kg) was dried under the shade and was powdered until it was very fine and extracted with petroleum ether for 72 hrs on water bath. The extract was filtered hot and solvent was removed under reduced pressure where a semi-solid yellowish mass (7.0 gm) was obtained.

The solvent free extract was chromatographed over silica gel column. For this purpose a column of 1.5 m in height with 4.0 cm in diameter was used and it was charged with 550gm silica gel (60-120 mesh).

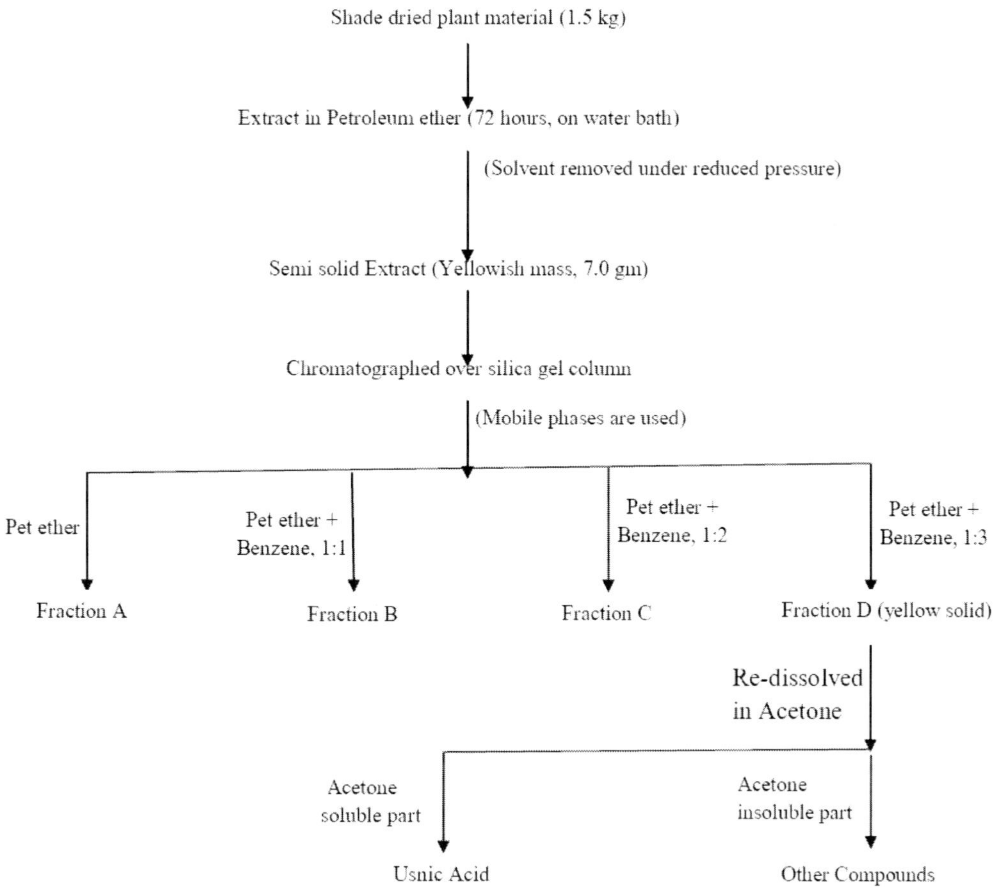

The column was eluted with different solvents in order of increasing polarity where one heterocyclic natural compound was isolated, purified and characterized with other natural compounds.

2,6-diacetyl-7,9-dihydroxy-8,9b-dimethyl-1,3(2H,9bH)-dibenzofurandione, also known as (+) *Usnic acid*:

Compound was isolated as yellow solid when column was eluting with petroleum ether and benzene (1: 3).

This mass was re-dissolved in acetone and acetone soluble part was purified and analysed.

Its R_f value was 0.80 in chloroform and methanol (95% + 5%) system. The m.p. of this crystalline yellow compound was 204°C. IR (KBr, cm^{-1}) 3150 (O–H stretching), 1800 (C=O, str.), 1550 (C=C, str.) etc. 1H NMR (δ ppm, $CDCl_3$) 2.67 (*s*, 3H, -$COCH_3$ at C-2), 5.97(*s*, 1H, C-4, -H), 2.66 (*s*, 3H, -$COCH_3$ at C-6), 13.30 (*s*, 1H, C-7, -OH), 2.11 (*s*, 3H, C-8, -CH_3), 11.01 (*s*, 1H, C-9, -OH), 1.76 (*s*, 3H, C-9b, -CH_3), 18.84 (*s*, 1H, C-3, -OH, enolic form) 13C NMR (δ ppm, $CDCl_3$) 198.02 (C-1), 109.24 (C-2), 200.38 (-$COCH_3$ at C-2), 32.08 (-$COCH_3$ at C-2), 163.90 (C-3), 101.58 (C-4), 179.32 (C-4a), 155.31 (C-5a), 98.37 (C-6), 200.50 (-$COCH_3$ at C-6), 31.22 (-$COCH_3$ at C-6), 163.56 (C-7), 105.33 (C-8), 7.50 (-CH_3 at C-8), 157.38 (C-9), 104.03 (C-9a), 59.13 (C-9b), 27.83 (-CH_3 at C-9b). MS(m/z), 345 (M^++H), 344 (M^+), 261, 260, 234, 233, 232, 217, 215 etc. Molecular formula calculated as $C_{18}H_{16}O_7$.

RESULTS AND DISCUSSION

2,6-diacetyl-7,9-dihydroxy-8,9b-dimethyl-1,3(2H,9bH)-dibenzofurandione, also known as (+) Usnic acid:

The mass spectrum of this compound (Figure 1) showed a molecular ion peak at m/z 344[M^+]. Other prominent ions were observed at m/z : 345 [M^++H], 261, 260, 234, 233, 232, 217, 215 etc. The spectral analysis and molecular weight determination established its molecular formula as $C_{18}H_{16}O_7$. In the IR spectrum (KBr, cm^{-1}) the presence of phenolic (OH) group was observed at 3150 and the absorptions at 1800 confirmed the presence of carbonyl group whereas the absorption at 1550 was assigned for >C=C< stretching.

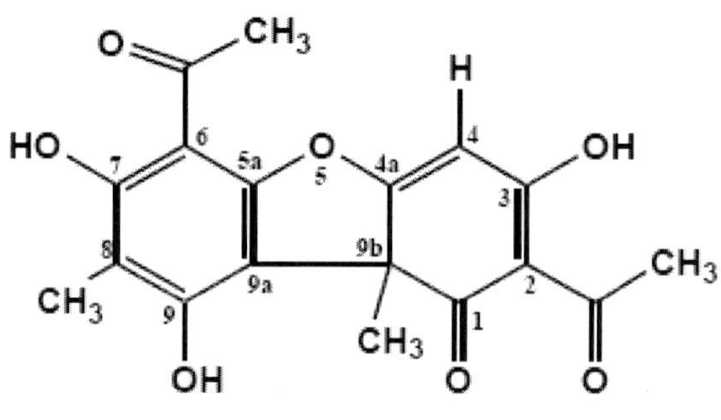

Figure 1. (+)-Usnic Acid.

The 1H NMR spectrum (δppm, $CDCl_3$) showed two sharp singlets at 1.76 (s, 3H) and 2.11 (s, 3H) for two methyl groups and their positions were assigned as C-9b and C-8 respectively as reported [14]. The absorption at 2.66 (s, 3H) and 2.67 (s, 3H) as two sharp singlets confirmed the presence of two acetyl groups at C-6 and C-2 positions respectively. Two sharp singlets were observed at 11.01 (s, 1H) and 13.30 (s, 1H) for phenolic groups at position C-9 and C-7respectively.

Vinylic proton presents at C-4 position showed the absorption at 5.97 as a singlet for one proton. The enolic proton showed intense 1H NMR absorption at 18.84 due to intramolecular hydrogen bonding (Figure 2). In solution form this compound exists in two tautomeric forms i.e. ketonic and enolic (Figure 3). The absorptions at 163.90, 163.56 and 157.38 in 13C NMR spectrum (δppm, $CDCl_3$) also confirmed the presence of three phenolic groups linked at C-3,

C-7 and C-9 positions respectively. Vinylic carbon atom was confirmed by the absorption at 101.58 which was assigned to C-4 carbon atom.

The absorption appearing at 198.02 clearly indicated the presence of carbonyl group at C-1 position. The values of other carbon atoms in this compound were established as 109.24 (C-2), 200.38 (-$COCH_3$ at C-1), 32.08 (-$COCH_3$ at C-1), 179.32 (C-4a), 155.31 (C-5a), 98.37 (C-6), 200.50 (-$COCH_3$ at C-6), 31.22 (-$COCH_3$ at C-6), 105.33 (C-8), 7.50 (-CH_3 at C-8), 104.03 (C-9a), 59.13 (C-9b), and 27.83 (-CH_3 at C-9b). On the basis of above spectral analysis this compound was identified as (+) usnic acid [14-21].

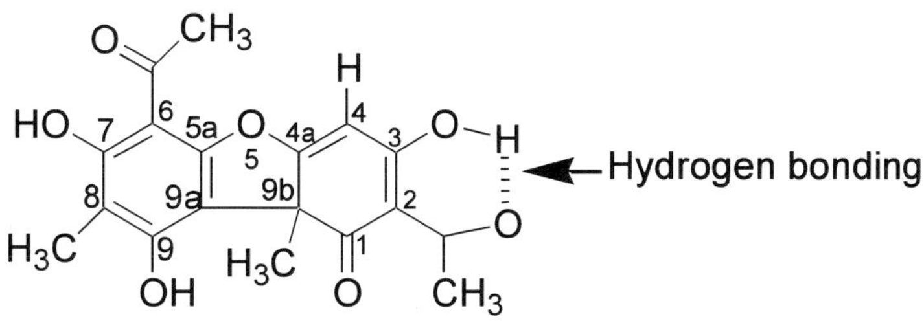

Figure 2. Hydrogen Bonding of (+)-Usnic Acid.

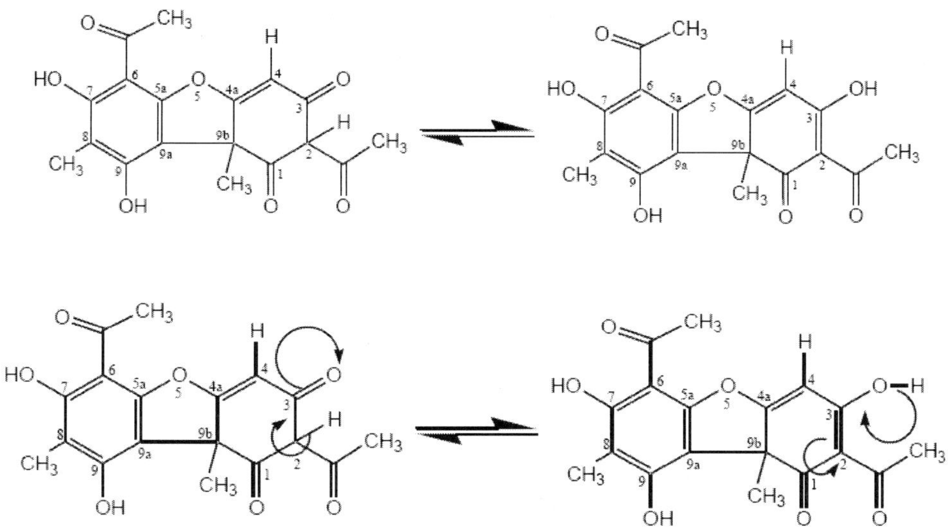

Figure 3. Tautomerism of (+)-Usnic Acid.

Reported Pharmacological Action of Usnic Acid

Antimicrobial Activity [22]

Usnic acid is a very effective antibacterial molecule and it is active against Gram positive bacteria and mycobacterium. In clinical trials, a mouthwash containing 1% (+) usnic acid was given to volunteers at regular intervals. The oral bacterial flora of volunteers was examined and it was revealed that the growth of Streptococcus mutans involved in the etiology of dental caries was selectively suppressed. Usnic acid was also found to be effective against

Mycobacterium aureum. In vitro assays, usnic acid and its salt inhibited the growth of Mycobacterium tuberculosis at very low concentration.

Antiprotozoal Activity [22]

Usnic acid exhibited significant inhibitory effect against the pathogenic protozoan richomonas vagina at comparatively lower concentrations than metronidazole. The compound also showed leishmanicidal properties both in vitro and in vivo studies; intralesional administration produced a reduction in lesion weight as well as in body burden of parasite.

Antiviral Activity [22]

In a cancer chemoprevention assay, (+) usnic acid was found to be significantly effective against tumor-promoter- induced Epistein-Barr virus with an ED50 of 1.0 μg/mL. (+)-usnic acid also inhibited the cytopathic effects of Herpes simplex type 1 and polio type 1 viruses in the infected kidney cells of African green monkey. In a clinical trial, the effect of an intravaginal formulation containing usnic acid and zinc sulfate as an adjuvant therapy to radio surgical treatment and it was evaluated on 100 females infected with genital human papilloma virus. The treatment significantly improved the time of reepithilization one month after the radio surgery.

Antiproliferative Activity [22]

(-)-Usnic acid caused moderate inhibition in the murine P388 leukemia assay, and also exhibited cytotoxic activity against cultured L1210 cells; it was inferred that p-triketone moiety was essential for the optimum activity. On the other hand, (+) usnic acid (50μg/mL) reduced the cell counts of leukemic (K-562) and endometrial carcinoma cell culture. (+) usnic acid exhibited cytotoxic activity against human keratinocyte cell cultures.

Anti-inflammatory Activity [22]

In an acute rat paw edema and a chronic rat cotton pellet assays at 100mg/Kg oral dose level, the anti-inflammatory action of (+)-usnic acid was comparable to ibuprofen at the same dose level.

Analgesic and Antipyretic Activity [22]

In two mice studies, the analgesic and antipyretic effects of usnic acid were evaluated at 100 mg/kg oral dose level and it was observed that usnic acid exhibited a significant analgesic effect as indicated in the acetic acid-induced writhing and tail pressure tests. At oral dose levels up to 300mg/kg, usnic acid also expressed significant antipyretic activity determined through lipopolysachharide-induced hyperthermia.

ACKNOWLEDGMENTS

The authors are thankful to Dr. M. C. Sharma (Associate Professor) and Dr. M. P. Dobhal (Associate Professor), Department of Chemistry, University of Rajasthan, Jaipur, Rajasthan (India) for their necessary support, guidance and encouragement.

REFERENCES

[1] S Maria, A Jorge, A Dora (2010). *Anal. Lett.*, 43, 1713.

[2] T Tay, Ays ¸E Ö Türk, M Yılmazb, H Türk, M Kıvanc (2004). Verlag der Zeitschrift für Naturforschung, Tübingen, 59c, 384.

[3] M Melgarejo, O Sterner, J V Castro, P Mollinedo (2008). *Rev. Boliv. Quim.* 25 (1), 24.

[4] L Arribas, C Vicente (2008). *Curr. Enz. Inhib.* 4, 180.

[5] F Odabasoglu, A Cakir, H Suleyman, A Aslan, Y Bayir, M Halici, C Kazaz (2006). *J. Ethnopharmacol,* 103, 59.

[6] C S Vijayakumar, S Viswanathan, M K Reddy, S Parvathavarthini, A B Kundu, E Sukumar (2000). *Fitoterapia*, 71, 564.

[7] M T Saenz, M D Garcia, J G Rowe (2006). *Fitoterapia,* 77, 156.

[8] M Yılmaz, A Ö Türk, T Tay, M Kıvan (2004). Verlag der Zeitschrift für aturforschung, *Tübingen*, 59C, 249.

[9] C Bézivin, S Tomasi, F L Dévéhat, J Boustie (2003). *Phytomedicine,* 10, 499.

[10] S Karthik, K C Nandini, T R P Kekuda, K S Vinayaka, S Mukunda (2011). nnal. *Biol. Res, 2* (4), 38.

[11] R N Dailey, D L Montoomery, J T Ingram, R Siemion, M Vasquez, M F Raisbeck (2008). *Vet Pathol*, 45, 19.

[12] K Ingolfsdottir (2002). *Phytochemistry*, 61, 729.

[13] W Knop (1844). *Ann. Chem. Pharm.* 49, 103.

[14] C K Wat and G H N Towers (1972). *Phytochemistry,* 11, 3540.

[15] K Takahashi, M Takani and Y Wada (1980). *Chem. Pharm. Bull.,* 28, 1590.

[16] I Francolini, P Norris, A Piozzi, G Donelli and P stoodley (2004). *Antimicro. Agent And Chemoth.,* 48, 4360.

[17] T Sassa and M Igarashi (1990). *Agric. Biol. Chem,.* 54, 2231.

[18] M Takani and K Takahashi (1985). *Chem. Pharm. Bull.,* 33, 2772.

[19] M Yilmaz, O T Aysen, T Tay and M Kivanc (2004). *Z. Naturforsch.*, 59C, 249.

[20] J P Kutney, H Sanchez and T Yee (1977). *Can. J. Chem.,* 55, 1073.

[21] J P Kutney and H Sanchez (1977). *Can. J. Chem.,* 55, 1079.

[22] V H Frankos (2005). NTP nomination for usnic acid and usnea barbata herb. Special Assistant for Science Review FDA-CFSAN-ONPLDS

INDEX

B

C

D

E

F

G

H

I

J

K

L

M

N

O

P

T

U

V

W

X

Y

Z